Monographs on oceanographic methodology 8

In this series:

1. *Determination of photosynthetic pigments in sea-water*
2. *Zooplankton sampling*
3. *A guide to the measurement of marine primary production under some special conditions*
4. *Zooplankton fixation and preservation*
5. *Coral reefs: research methods*
6. *Phytoplankton manual*
7. *Mathematical models in biological oceanography*
8. *The mangrove ecosystem: research methods*

The mangrove ecosystem: research methods

Edited by Samuel C. Snedaker and Jane G. Snedaker

on behalf of the Unesco/SCOR
Working Group 60 on Mangrove Ecology

Unesco

The designations employed and the presentation of material throughout this publication do not imply the expression of any opinion whatsoever on the part of Unesco concerning the legal status of any country, territory, city or area or of its authorities, or concerning the delimitation of its frontiers or boundaries.

Published in 1984 by the United Nations Educational, Scientific and Cultural Organization
7 place de Fontenoy, 75700 Paris
Typeset and printed by Richard Clay (The Chaucer Press) Ltd, Bungay, United Kingdom

ISBN 92-3-102181-8

Preface

At the request of Unesco, the Scientific Committee on Oceanic Research (SCOR) formed Working Group 60 in 1978 to address two major topics of interest to mangrove scientists throughout the world. The first was to develop a concise summary of the state of knowledge concerning mangroves and, based on that, to prepare a handbook on methodological procedures for the study of mangroves. The working group was composed of scientists of international reputation in the field, and represented a variety of disciplines. The founding members of the working group were: François Blasco (France), Valentine J. Chapman (New Zealand), Hansa Chansang (Thailand), Antonio Lot-Helgueras (Mexico), Federico Pannier (Venezuela), Samuel C. Snedaker, Chairman (United States) and Bruce G. Thom (Australia). The first meeting, held in 1978 in San José, Costa Rica, with all members in attendance, was deemed highly successful for, among other reasons, there was a unity of opinion concerning the needs of mangrove researchers. The working group heartily endorsed the original terms of reference and undertook to establish other terms of reference unanimously considered to be of equal importance. It was also recognized that to be successful in its mission, the working group would have to include a zoologist with experience with mangrove fauna. Dr A. Sasekumar (Malaysia) was proposed and accepted by the working group. The meeting concluded with the assignment of research and writing tasks which were willingly assumed by each member.

Over the next two years, working group members began the preparation of their assigned manuscripts and frequently communicated with one another on topics of mutual interest. As would have been expected, V. J. (Val) Chapman was the first to submit a completed manuscript. In 1980, when the other members had completed the basic work, a second meeting was organized. By that time, A. Sasekumar had accepted membership, but regrettably, Antonio Lot-Helgueras had to decline further participation owing to pressing professional obligations in Mexico. The meeting was held on Loloata Island in the Gulf of Papua, Papua New Guinea, with all current members in attendance.

The progress of each individual member was reviewed and discussed, and plans were made for completion of the terms of reference. As a replacement for Antonio Lot-Helgueras, Gilberto Cintrón (Puerto Rico) was proposed and accepted by the working group. During the meeting, it was decided that it would be preferable to combine both sets of terms within a single published volume. As a result of that decision, this volume contains a summary of knowledge in a variety of disciplines, as well as a set of recommended procedures to be used in the study of mangroves and the mangrove environment.

Following the Papua New Guinea meeting, manuscripts were prepared, submitted, reviewed and returned to the authors preparatory to completion of the final manuscript.

Working group members, as well as the scientific community at large, were shocked to learn of the death of Professor Chapman in late 1980. It was decided through correspondence that Professor Chapman could not be replaced and that the working group chairman would undertake to complete his original manuscript. It was further recommended that this volume be dedicated to the advances he made in the field of halophytes in general, and mangroves in particular.

The work that followed suffered from a variety of problems (frequently involving the international postal system), but in spite of these problems this volume was completed, and submitted for publication in 1983.

Unesco is highly appreciative of the efforts of the scientists who prepared the present volume and wishes to express its particular thanks to Professor Snedaker, the scientific editor, Jane G. Snedaker, general editor, and Catherine E. Sikkema, illustrator, for their dedication to the project.

The scientific opinions expressed in this work are those of the authors and are not necessarily those of Unesco. Equipment and materials have been cited as examples of those most currently used by the authors, and their inclusion does not imply that they should be considered as preferable to others available at that time or since developed.

Contents

Foreword

Land's beginning, ocean's end; oceans begin where the land doth end; where does it begin and where does it end?

Mangroves and mangrove forests along most tropical shorelines have attracted man's curiosity from the earliest times and the literature now exceeds 7,000 titles. In spite of this long history and abundant recorded knowledge, it was not until the 1970s that man began to understand and appreciate the role of this unique vegetation. Now in the 1980s we see mangroves being protected or managed in many areas of the world, for the fisheries they support, the forest products they yield and the stability they contribute to the coastal zone. Yet, at the same time, mangroves and mangrove forests are being destroyed for reasons which are frequently illogical.

Because of the rapidly developing interest in mangroves, scientists throughout the world in universities, government agencies and philanthropic institutions are making an increasing commitment to broaden our knowledge of this biological community. Unesco and the Scientific Committee on Oceanic Research (SCOR) have made a similar commitment to support the scientific community by assisting it in the pursuit of this goal. One of the key steps is the creation of this volume, the eighth in the series of Monographs on Oceanographic Methodology which has been prepared to provide a general understanding of the mangrove ecosystem and to review and recommend basic research procedures in the study of the ecosystem. The volume represents a current concensus of worldwide scientific opinion on the relevant directions for new research, and on the methods and techniques which might be helpful in acquiring new knowledge. Furthermore, it is hoped that the worldwide availability of this volume will encourage scientific communication and co-operation.

The identification of profitable areas for mangrove research and the recommendation of useful methods are the opinions of the contributing authors of Unesco/SCOR Working Group 60. The omission of pertinent research topics and methods does not imply a lack of relevancy or value. In fact, diversity is probably no less important in the research directions and methodology of the scientific community than it is to the maintenance and perpetuation of nature.

SAMUEL C. SNEDAKER,
Chairman, Unesco/SCOR Working Group 60

List of contributors

Kevin G. Boto,
Australian Institute of Marine Science,
PMB No. 3,
Townsville, M.S.O. 4810,
Queensland, Australia.

François Blasco,
Université Paul Sabatier,
Institut de la Carte Internationale du Tapis Végétal,
39 allée Jules-Guesde,
31400 Toulouse, France.

Iver M. Brook,
Division of Biology and Living Resources,
Rosenstiel School of Marine and Atmospheric Science,
University of Miami,
4000 Rickenbacker Causeway,
Miami, Florida 33149,
United States.

Melvin S. Brown,
Division of Marine Affairs,
Rosenstiel School of Marine and Atmospheric Science,
University of Miami,
4000 Rickenbacker Causeway,
Miami, Florida 33149,
United States.

Valentine J. Chapman †
Department of Botany,
University of Auckland,
Auckland, New Zealand.

Gilberto Cintrón,
Department of Natural Resources,
Government of Puerto Rico,
P.O. Box 5887,
Puerta de Tierra,
Puerto Rico 00906,
United States.

Keith E. Cooksey,
Division of Biology and Living Resources,
Rosenstiel School of Marine and Atmospheric Science,
University of Miami,
4000 Rickenbacker Causeway,
Miami, Florida 33149,
United States.

† *In memoriam.*

Jack W. Fell,
Division of Biology and Living Resources,
Rosenstiel School of Marine and Atmospheric Science,
University of Miami,
4000 Rickenbacker Causeway,
Miami, Florida 33149,
United States.

Federico Pannier,
Universidad Central de Venezuela,
Facultad de Ciencias,
Escuela de Biología,
Departmento de Botanica,
Apartado 80.390,
Caracas, Venezuela.

A. Sasekumar,
Department of Zoology,
University of Malaya,
Kuala Lumpur 22–11,
Malaysia.

Yara Schaeffer Novelli,
Instituto Oceanographico,
Universidad de São Paulo,
Cidade Universitaria,
Butanta, CEP–05508,
São Paulo, Brazil.

Catherine E. Sikkema
(scientific illustrator),
Kendall Graphics,
Miami, Florida,
United States.

Jane G. Snedaker
(general editor),
Tropic House International,
7549, S.W. 58th Avenue,
South Miami, Florida 33143,
United States.

Samuel C. Snedaker
(scientific editor),
Division of Marine Affairs,
Rosenstiel School of Marine and Atmospheric Science,
University of Miami,
4000 Rickenbacker Causeway,
Miami, Florida 33149,
United States.

Bruce G. Thom,
Department of Geography,
University of New South Wales,
Duntroon, ACT,
Australia.

Introduction

Samuel C. Snedaker
Division of Marine Affairs
Rosenstiel School of Marine and Atmospheric Science
University of Miami
4600 Rickenbacker Causeway
Miami, Florida, United States

Over the past decade the number of research studies focusing on mangroves and the mangrove environment has significantly increased as a result of public and scientific interest in their role in nature and their value to mankind. Although the total number of published reports exceeds some 7,000 titles (cf. Rollet, 1982), many major gaps in our scientific understanding persist. Confounding the problem are the difficulties of conducting research in the mangrove environment and the absence of recommended research protocols which would facilitate comparative evaluations, and general syntheses on specific topics.

In Part I, 'Characterization of the mangrove environment', Bruce Thom outlines important features of the major coastal landforms that frequently provide optimum habitats for mangrove forest development. This overview demonstrates the fact that the spatial patterns of mangrove development are closely associated with landform types and that local mangrove formations tend to be controlled by the constantly changing coastal geomorphology. Field research that does not take these fundamental processes into account can lead to erroneous conclusions concerning species zonation, successional status and the causes of differences in regional patterns. François Blasco takes a more global view in which he characterizes the differences in mangrove species distribution as a function of broad latitudinal gradients in climate. In addition to suggesting that the primary reasons why mangroves dominate certain climatic regimes, as opposed to others, he defines practical techniques for characterizing regional climatic patterns as they relate to mangrove distribution and development. In this context, François Blasco discusses the evolution of mangroves and the importance of palynology in the study of earlier floras, and the dispersal and distribution of the major species. The main emphasis of the chapter is that mangrove assemblages and their distribution are not fixed features, but, instead, undergo constant change and evolution in both space and time.

Many research efforts focus on 'Community structure and description' for the purposes of preparing general inventories of 'what's there' and in what number (or mass), describing how the components are structured, and making general assessments of the abiotic conditions. Valentine Chapman outlines the traditional approach to characterization which might be considered to represent the minimum requirements of any general survey. François Blasco draws attention to the fact that the taxonomy of the mangrove species is very poorly developed; most researchers can identify the genera, but species determinations in many areas remain questionable. The basic protocol for quantitatively determining the physiognomic structure of mangrove communities is presented by Gilberto Cintron and Yara Schaeffer Novelli. Their approach places emphasis on relatively simple techniques that yield much useful data and information that have significant value in both basic ecological studies and the management of forested areas. Widespread use of their recommended techniques and interpretative approaches would solve one of the major problems in comparing data from different parts of the world. Kevin Boto draws attention to the fact that mangrove soils, or sediments, are unique to saline intertidal areas and that an understanding of the biogeochemical processes is a prerequisite for understanding the structure and function of mangroves which appear to be so well adapted to these scientific conditions. Federico Pannier provides a checklist of analytical methods that are commonly employed in the study of mineral nutrient stocks and cycling processes. Frequently, mangrove researchers place greater emphasis on the dominant plant form—the mangroves—than they do on the animal forms that are so dependent upon, and prevalent within, the mangrove habitat. A. Sasekumar describes the methods recommended for studies of the mangrove fauna and Iver Brook reviews the methods and techniques that have been used for determining the nearshore animal biomass. Both authors emphasize that quantification is essential in developing a scientific understanding of the faunal associates and their relationship to the mangrove ecosystem. Keith Cooksey draws attention to the role of diatoms, which among other lower life-forms are frequently ignored in mangrove ecosystem studies. Many of the smaller plants and animals, including the microflora and microfauna, have biochemical roles in the ecosystem that are disproportionate to their size and mass. Studies of the processes that they control may be one of the more important areas of investigation that remains to be developed.

Mangrove ecosystems are reputed to be highly productive in spite of the relatively harsh environment in which they occur, and frequently dominate. Their productive basis is, of course, a function of their photosynthetic ability. As one of the main international researchers in this area, Federico Pannier describes the major parts of the protocol for the study of photosynthesis. He also outlines an equivalent protocol for the study of water relations in mangroves. Both areas of investigation are relatively sophisticated and thus are infrequently attempted. However, research on mangrove photosynthesis and water relations must be considered a fundamental and basic foundation for the interpretation of overall ecology and biology of this diverse taxonomic group. The high productivity of mangroves is also expressed in the production of leaf litter which, following decomposition and nutrient enrichment, forms a major food source for most estuarine animals. It is this one aspect of the functioning of mangroves that has led to the widespread interest of their valuable role in the ecosystem. Melvin Brown outlines the basic approach for determining the rates of production of leaf litter, and Jack Fell et al. review the protocols for evaluating the processes of decomposition and nutrient enrichment.

Although a variety of topics and methods are covered in this monograph, they are, collectively, highly selective and non-exclusive. Active and serious researchers will identify additional topics worthy of investigation and will have recommendations for their study. The value of this monograph, therefore, extends beyond the limitation in topics and treatments to the extent that it sets a direction for further scientific work on the mangrove ecosystem.

Part I

Characterization of the mangrove environment

1

Coastal landforms and geomorphic processes

Bruce G. Thom
Department of Geography
University of New South Wales
Duntroon, ACT, Australia

Introduction

A basic task confronting the mangrove ecologist is to explain the development of mangrove communities through time. This problem has been approached in several different ways depending on the background of the researcher, the conceptual models prevailing at a given time, and the existence and quality of information on environmental conditions in any given area of study. An ecologist with a background in physiology or taxonomy would be less concerned with the role of habitat dynamics compared with one having a strong feel for geology, geomorphology or soils. There is also the likelihood that fashionable concepts would influence modes of thinking; the dominance of climax theory in plant ecology during the early part of this century is an example of this point. Knowledge of habitat change in an area involving changing patterns of sedimentation would clearly strengthen the hand of an ecologist seeking to study community dynamics.

In order to explain the evolution of mangrove communities through time, it is necessary to develop an understanding of past as well as present conditions. Techniques of the geomorphologist, stratigrapher, pedologist, palaeontologist and palynologist have been used in this regard. Patterns of species distribution and variations in physiognomy at the present time have been evaluated in relation to physical environmental conditions. Functioning aspects of communities, including studies of reproductive strategies and mechanisms to combat physiological stress, further provide a basis for research into problems of mangrove persistence and change. Therefore, any attempt to comprehensively 'model' a mangrove ecosystem to explain community change, involving prediction of what may happen in the future, is fraught with complexities and uncertainties. The problem is multi-dimensional in space as well as time, besides being multi-disciplinary in scope (Lugo and Snedaker, 1974). Yet this has not prevented ecologists from undertaking field studies of mangroves with the objective of explaining apparent 'zonation', 'stepped sequences', 'mosaic patterns' or discontinuities. Conclusions developed in such studies are often severely limited by an inadequate understanding or the dynamics of environmental conditions and the relation of those conditions to vegetation dynamics. Snedaker (1982) has recently published an extensive review of the literature on mangrove species zonation. His paper should be read in conjunction with this chapter in order to provide a broader biological perspective.

In this chapter an attempt is made to outline various approaches used to explain mangrove distribution. This will include a discussion of the successional approach and the application of gradient analysis. The writer's own interest in 'physiographic ecology' (Zimmermann and Thom, 1982) will then be used to provide the basis for an evaluation of the role of habitat change in producing community patterns in different environmental settings. These settings will be first discussed broadly. It will be argued that a 'global' perspective only is necessary to determine general patterns and trends in mangrove ecosystem behaviour. In order to have a better explanation, it is necessary to examine specific patterns of mangrove species distribution and structural types in relation to dominating geomorphic processes and products, that is to determine just how mangrove communities are patterned in response to varying substrate and energy conditions (Stoddard, 1980).

Historical perspective

SUCCESSIONAL MODEL

For many years the conceptual basis of mangrove ecology was derived from the work of Clements. Many mangrove ecologists have been content to utilize climax theory by relating observed zonation patterns to successional processes. This approach can be termed the 'classical successional view of mangrove dynamics'. As Snedaker (1982, p. 112) has recently pointed out:

> The interpretation of mangrove species zonation as being a spatial expression of plant succession seems to have its origin in the Western Hemisphere where much of the early work focused on the land-building role of mangroves.

The successional model emphasizes the role of the biotic rather than the geomorphic processes. Habitats change because plants induce sediment accumulation, both organic (autochthonous) and inorganic (allochthonous or terrigenous). Plant communities develop from one stage to another as a particular type 'prepares the ground' for another type. Finally, a 'climax' community is attained. Many authors see this final stage as being a non-mangrove, often tropical rain forest (Watson, 1928, p. 134; Davis, 1940). Where the model is applied it is usually assumed that the shoreline is migrating seaward (prograding) forming what sedimentologists term a regressive stratigraphic sequence. Progradation is encouraged by deposition around mangrove roots. This is an old concept succinctly expressed in a well-quoted book by Warming (1909, p. 236): 'In as much as Rhizophoraceae form the outposts of the mangrove they entangle mud between their roots and thus add to the land.'

More recently Davis, working in Florida, and Chapman who has studied mangroves in many parts of the world, have made extensive use of the successional concept. Davis (1940) argued that complex succession develops between various mangrove communities leading to a tropical-forest climax. However, this interpretation was reduced to being relevant to only local areas by the work of Scholl (1964), who showed the effect on the mangroves of southern Florida of a transgressing sea inducing extensive shoreline erosion (see also Spackman et al., 1966).

Chapman (1970, 1976) has been responsible for generalizing many ecological analyses of mangrove zonation into successional models. Although he has stressed the role of particular environmental factors which influence the establishment and growth of particular species, he tried to simplify patterns into successional sequences. This has even been done where classical succession has been suggested as an inappropriate model (Egler, 1950; Thom, 1967). In recognizing that difficulties may arise with the classical model, Chapman (1976, pp. 39–40) in reviewing the work of Egler admitted:

> In making these suggestions one is aware that zonation pattern is not the sole criterion in determining any plant succession. Much information can be obtained by excavation, pollen analysis and study of old maps.

It appears that mangrove succession may occur where a steady input of mud facilitates shoreline progradation. A number of cases exist where sequences have been studied historically, geologically and ecologically and there is evidence for successional change. The west coast of Malaysia is a good example. Near the mouths of rivers, extensive mud-flats are accreting seawards upon which a succession of different mangrove types colonize, regenerate, become replaced and finally develop into tropical rainforest (Watson, 1928; Carter, 1959; Coleman et al., 1970; Diemont and van Wijngaarden, 1975). In Borneo, the ecological and palynological work of Anderson and Muller (1973) clearly demonstrates succession. Thom (1967) demonstrated that at particular sites around lagoon shores in Tabasco, Mexico, an abundance of mud creates habitats for saltmarsh and mangrove succession. In general, such studies highlight the active role of mud accumulation. The classical model involving a land-building role for mangroves is given secondary consideration.

From a geomorphic point of view, succession is potentially demonstrable where sedimentary regression has occurred. In terms of the array of habitats colonized by mangroves it represents one type of habitat change where either continuous or episodic mud deposition is the dominant process. Ecologists must be very cautious in applying the succession model to a broad spectrum of environments where mangroves exist. The accreting mud-flat is only one of many. As noted by Sauer (1961, p. 222), 'without independent evidence of shoreline change, it is gratuitous to infer forward motion of the shore or vegetation merely from banded vegetation structure'. Snedaker (1982, p. 113) takes an even stronger stance in concluding 'that insufficient data and information are available for mangroves which would allow for a sound appraisal of their successional status'.

GRADIENT ANALYSIS

An alternative, albeit sometimes complementary approach to explaining mangrove patterns involves the study of environmental factors. The basis for this approach lies in the analysis of species change along environmental gradients; in the words of Whittaker (1953), there is no absolute climax for any area, and climax composition has meaning only relative to position along environmental gradients and to other factors.

In mangrove research it has been shown by several authors that environmental factors influencing pattern in mangrove vegetation act in a complex fashion (Clarke and Hannon, 1967, 1969, 1970, 1971); Buckley (1982, p. 105) states:

> The distributions of individual mangrove species may be controlled by the precise patterns of variation of factors such as water level, salinity, pH, sediment flux, oxygen potential, form and abundance of a range of anions and cations, hydrodynamic stress and crown exposure, together with interspecific competition and successional factors.

Watson (1928) made one of the first attempts to utilize the gradient concept in mangrove research. He divided the western Malayan mangrove forest into five types which he correlated with tidal inundation classes. 'It should be noted that the guiding principle in the allotment of a species to any particular inundation class is its *ability to regenerate itself*, and not merely to exist, under the conditions obtaining within that class.' (Watson, 1928, p. 130) This type of work has been extended to include an appreciation of functional properties of different species and assemblages of species along environmental gradients (Lugo and Snedaker, 1974). More recently, Bunt and Williams (1980) have applied quantitative methods in a phytosociological study of tidal forests in north Queensland (cf. also Buckley, 1982). They show that mangroves do not contain consistent non-overlapping species groups occupying separate topographic sections of the tidal range. However, individual species are associated with particular sections of the tidal range (cf. also Bunt and Williams, 1981).

In general, the gradient concept is conventionally used in mangrove ecology to relate degrees of surface inundation to gradients in environmental factors which influence plant physiology. An association is sought between certain factors and the spatial limits or boundaries of particular species or species groups. Authors such as Rabinowitz (1978) seek a physiological explanation which involves a 'mode of action' of physical factors. Salinity is often used as a key factor which links the physical environment through the physiology of the mangroves to patterns of spatial organization. According to Snedaker (1982, p. 119):

It can be stated that each mangrove species probably has an optimum salinity, which at higher concentrations induces a proportionately higher respiration making it relatively less competitive, and a lower concentration is forced to compete with species better adapted to maximize photosynthesis.

The gradient is undoubtedly of great value in the study of the ecophysiology of species. However, it does not reveal how the community evolved to its present state, nor does it provide much information on future directions of change.

Physiographic ecology

DEFINITION

The prime purpose of this chapter is to highlight an approach to mangrove ecology which focuses on changes in habitats and plants occupying those habitats. It is similar in content and scope to 'geomorphic botany' (Goodlett, 1969), or 'physiographic plant geography' (Zimmermann and Thom, 1982, p. 47):

Physiographic plant geography regards physical and botanical landscapes as mosaics of landforms, and result of processes acting on materials with different properties and of assemblages of species (plant communities) which segregate on the various landforms with their different growth conditions.

As pointed out by these authors, the physiographic approach takes for granted that plant establishment and growth is an exceedingly complex phenomenon (Grubb, 1977). However, this complexity is not seen as being particularly relevant to explanations of distributions on landform (physiographic) scales. According to Zimmerman and Thom (1982, p. 48):

Distribution on these scales is often the result of narrower sets of relationships largely the coincidence of phenological events with specific physiographic processes. . . . The physiographic approach rests on the assumption that clues to distributional controls are obtained from the study of presence/absence of species within a particular climate, and from the identification of specific environmental factors common to all those habitats. These factors relate primarily to the phenology of reproduction.

In the study of plants distributed at the scale of slopes, flood plains, terraces, river embankments, swales of beach-ridge plains, etc., the ecologist is concerned with two basic problems. The first is identifying the presence/absence of species at particular stages of growth as determined primarily by geomorphic processes and conditions. This involves mapping for the definition and analysis of plant–habitat relationships. The second problem concerns identification of direction of change. Essentially this assumes that landforms are subject to continuous change over time. Past, present and future trends must be evaluated in order to explain courses of change in plant distribution.

Three physiographic 'states' constitute the framework in which directional change can be discerned. Dominance of a sediment influx or accretion, leading to lateral and/or vertical accumulation of a landform, forms one state. At the other extreme, erosional conditions may lead to habitat destruction by removing land surfaces or lowering their elevation. The third state represents a quasi-equilibrium or steady-state condition in which there is a balance of erosion and deposition within any given area. In the discussion which follows, it will be pointed out that any one of the three states can characterize a region as a whole, for instance, a river delta or barrier-lagoon complex. Alternatively, individual landforms can be considered at a more local scale in these dynamic terms. For example, a tidal flat can be undergoing erosion by wave processes. Using both local and regional scales, it is possible to utilize the physiographic approach in a flexible way to interpret habitat dynamics and to explain changing mangrove distributions.

APPLICATIONS

In plant ecology there have been a number of studies which have utilized the physiographic approach. Uncertainty with the use of successional theory led Whittaker (1953) to comment that community fluctuations must be seen in the context of 'long-range, directional trends' imposed by physiographic and climatic changes. Hack and Goodlett (1960) provide an excellent case study of spatial and functional relationships between physiographic factors and forest vegetation in the central Appalachians. Specific relationships between geomorphic processes and the distribution of valley-floor species have been examined in a semi-arid basin in Arizona (Zimmermann, 1969). Examples of the physiographic approach in colder climates include the evolution of muskeg in northern Minnesota (Heinselman, 1963, 1970). These studies cast doubt on the validity of conventional views on succession in waterlogged areas.

In mangrove research, a number of studies have demonstrated that the evolution of vegetation patterns is closely related to the dynamics of shoreline development and sedimentation. A recent example is the development of an hypothesis to explain the periodic stabilization of shifting mud-flats on the Guyanan coast of South America (Wells and Coleman, 1981). The present author's own work in Mexico, north-west Australia, Gulf of Papua, the Great Barrier Reef and southern New South Wales is based on the interpretation of the ecology of mangroves as the response of the plants to habitat change induced primarily by geomorphic processes (Thom, 1967, 1975, 1982; Thom et al., 1975; Thom and Wright, 1982). Stoddard (1980), Woodroffe et al. (1980) and Woodroffe (1982) have shown that in reefal areas mangrove communities are 'opportunistically patterned in response to varying substrata and energy conditions' (Stoddart, 1980, p. 269). In these studies, it is possible to see why the generalized zonal succession schemes of Chapman (1976) can be replaced by models dependent on type and degree of geomorphic activity influencing plant distribution and physiognomy (Thom, 1982).

Environmental settings for mangroves

On a global scale, mangrove ecology studies may appear to be a chaotic mess of special cases. However, an attempt has been made to classify regional settings on the basis of dominant geomorphic processes and landform assemblages (Thom, 1982). Repetition of processes and landforms in different regions where mangroves occur provides a conceptual basis for the analysis of mangrove species and community patterns.

The three essential components of an environmental setting are defined as: (a) 'background' geophysical; (b) geomorphic; and (c) biologic (Thom, 1982). The first component includes various factors such as relative sea-level history, climatic conditions and tidal properties of a region. The geomorphic component is subdivided into three spatial scales, the broadest reflecting the general character of sedimentation (i.e. carbonate, terrigenous, etc.), the second reflecting dominance of particular processes (e.g. wave or river, tidal or river, etc.) at the regional level, and the third level relates to 'microtopography' or the topographic expression and composition of particular landforms. It is this last level which provides the mappable unit responded to by plants at various stages of growth. The final component of an environmental setting is biologic, which expresses ecologic factors arising from the 'pool' of species occurring in a particular region. As Thom (1982, p. 7) points out:

> Interspecific competition plays an important role in determining the species diversity and distribution patterns in a given area. The influence of physical factors of the environment, especially those controlling landform conditions, on the relative competitive abilities of mangrove species probably accounts for the distribution of mangroves in typical zoned patterns.

The first two components of an environmental setting, geophysical and geomorphic, can be combined to produce an array of physical settings in which mangroves grow. Five settings repeated in nearby localities throughout the world are recognized on shorelines where terrigenous sediment inputs are dominant. On a global scale carbonate sediments do not represent large areas of mangrove growth; nevertheless several settings are identified. There are many variations of each setting depending on the relative importance of deposition and erosional processes.

TERRIGENOUS SETTINGS

Thom (1982) depicted five terrigenous settings and possible habitats of mangrove growth (Fig. 1.1). These types are a modification of the delta classification of Wright et al. (1974) (cf. also Coleman and Wright, 1975), and the bedrock-embayment classification of Roy et al. (1980).

Setting 1 is characteristic of allochthonous coasts of low tidal range. In such cases river discharge of freshwater and sediment leads to the rapid deposition of terrigenous sands, silts and clays to form deltas. These deltas are building seawards over flat offshore slopes composed of fine grained 'pro-delta' sediments. Such slopes help dampen wave energy and any tendency for longshore drift. The delta geometry consists of multiple branching distributaries forming elongate, finger-like protrusions. The result is a highly crenulate coastline with shallow bays and lagoons between and adjacent to distributaries. The active distributary region is predominantly an area of high freshwater discharge so that salt-tolerant plants are not common. However, there may be abandoned distributary regions within the deltaic plain into which penetrate saline waters either seasonally or more frequently. The area marginal to these distributaries is also relevant to this setting as longshore drift of muds and wave reworking of sands and shells will influence plant establishment and generation (e.g. on chenier plains). Thus parts of the deltaic plain may contain an array of habitats which mangroves colonize or are maintained. Such deltaic plains are often subject to rapid rates of subsidence and switches in centres of freshwater discharge and deposition, so that this environmental setting is characterized by a high degree of morphologic diversity and rapid habitat change. One example is the Mississippi delta and the adjacent chenier plain (Fisk, 1960). The Atrato delta in Colombia is another example (Vann, 1959).

Setting 2 is also associated with allochthonous coasts, only here the dominant physical process is the high tidal range with strong bidirectional tidal currents. These currents are responsible for the dispersion of sediments brought to the coast by rivers, and in the offshore zone they form elongate sand bodies (Wright et al., 1973). Wave power is often quite low because of frictional attenuation over broad intertidal shoals. Typically, the main river channels are funnel-shaped and are fed by numerous tidal creeks. These creeks are separated by extensive tidal-flat surfaces. Where relative sea-level has been stable for 5,000 years or more, these channels appear to be fixed in position and the surfaces of the tidal flat accrete vertically to the high-water spring-tide level (Thom et al., 1975). These surfaces may be subjected to sheet and lateral erosion (Semeniuk, 1980, 1981). They also prograde seawards along the promontories which separate the tidal channels. An example in a dry climate is the Ord River delta in north-western Australia (Thom et al., 1975); a wet climate example is the Klang delta of western Malaysia (Coleman et al., 1970), another is the Kikori deltaic complex in the Gulf of Papua (Floyd, 1977).

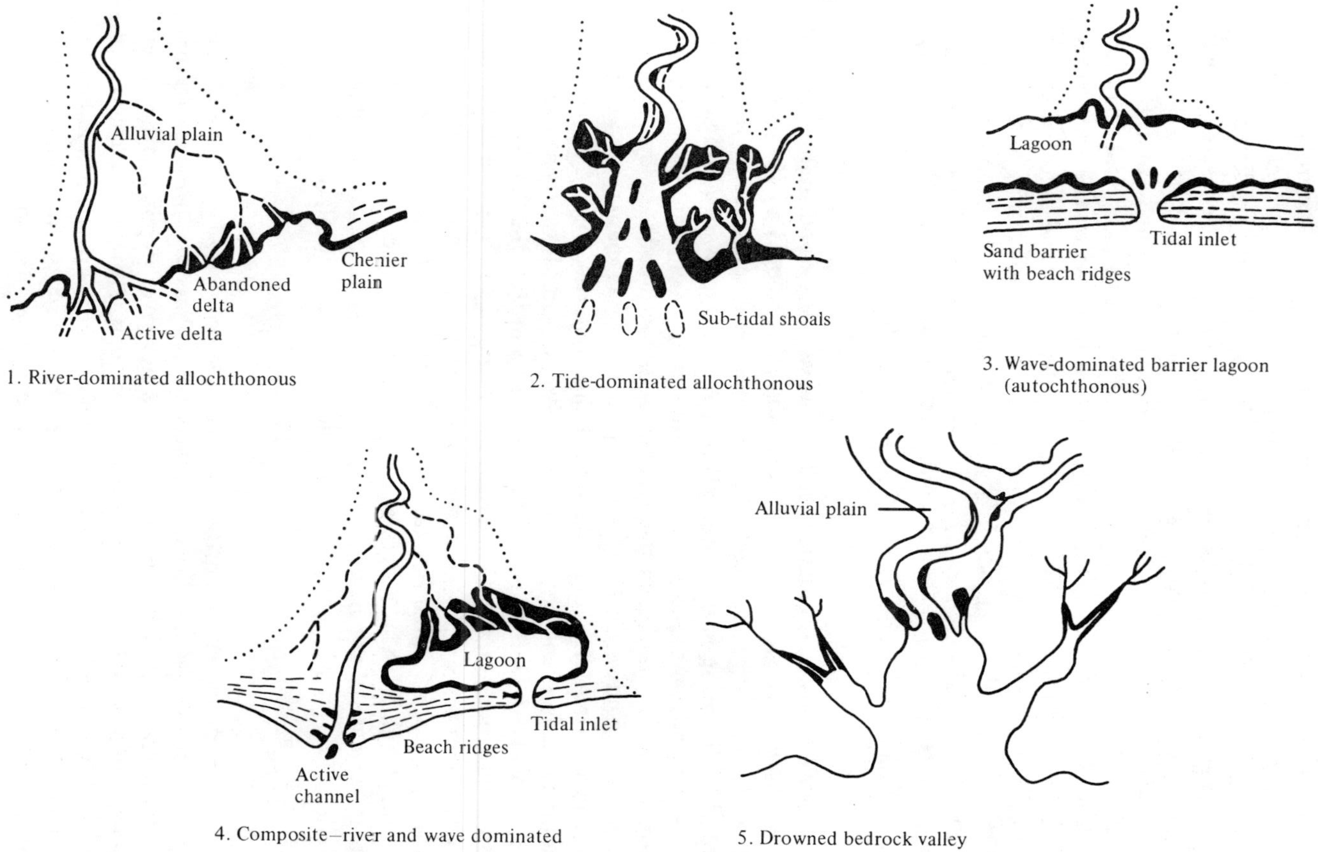

Figure 1.1
Generalized environmental settings for mangrove colonization and development (shaded). The five settings occur on coasts dominated by terrigenous deposition and reworking of sand, silt and clay sediments. After Thom (1982)

Setting 3 is characterized by a higher wave energy at the shoreline and relatively low amounts of river discharge. The slope of the inner continental shelf in such a case would be steeper due to operation at different sea-levels of the high wave energy which reworks sediments delivered to the coast by rivers. This type of coast is referred to as autochthonous (Swift, 1976). Offshore barrier islands, barrier spits or bay barriers are typical of this setting. Barrier islands enclose broad elongate lagoons, whereas bay barriers enclose drowned river valleys. Small digitate deltas prograde into these water bodies without significant opposition from marine forces. The degree of tidal modification of landforms of this setting can be quite variable. Salt-tolerant plants occur around the margins of the lagoon in a variety of habitats. An example of this setting is the linear barrier coastline of El Salvador or the bay barrier coastline of New South Wales. Mangrove distribution in the latter case is dependent in part on the efficiency of tidal inlets (Roy et al., 1980).

Setting 4 represents a combination of high wave energy and high river discharge. Sand debouched by the river is rapidly redistributed by waves alongshore to form extensive sand sheets. Much of the sand deposited on the inner continental shelf during lower sea-levels is reworked landward during marine transgressions and subsequent sea-level 'stillstands'. The result is a coastal plain dominated by sand beach ridges, narrow discontinuous lagoons with an alluvial plain to landwards. Salt-tolerant plants such as mangroves are concentrated in distribution along abandoned distributaries, and in areas near river mouths and adjacent lagoons. The Grijalva delta in Mexico is an excellent example of this type. Here regional subsidence accentuates the development of environmental gradients, distributary diversion and lagoon expansion (Thom, 1967). The Purari delta of Papua New Guinea is a similar setting, but appears to lack the geomorphic (and habitat) diversity involving lagoon development (Thom and Wright, 1982). Where the tidal range is greater and the climate drier, as in the case of the Burdekin delta of Queensland, there is a spread of saline habitats to interdistributary areas which are periodically inundated by high spring tides.

Setting 5 can be described as a drowned river-valley complex. The depositional setting is defined by a bedrock valley system which has been drowned (transgressed) by a rising sea-level. Neither marine nor river deposition has been sufficient to infill what is an open estuarine system. However, the heads of valleys may contain relatively small river deltas which are little modified by waves. At the mouth of the drowned valley bordering the open sea, a tidal delta may occur composed of 'marine' sand reworked landward during a marine transgression (Roy et al., 1980). Broken Bay in New South Wales is a good example; mangroves flourish in fine sediments at the heads of drowned tributary valleys, and in lagoons behind bay barriers near the mouth of the estuary.

CARBONATE SETTINGS

Environments of deposition dominated by the accumulation of carbonate, due to either wave/tide reworking or growth of a skeletal framework, constitutes another group of settings for mangrove growth. Several studies have been undertaken of such environments where they form mangrove habitats. These include work in Florida (e.g. Vaughan 1909; Davis, 1940; Scholl, 1964; Smith, 1968; Spackman et al., 1966; and Wanless and Meeder, in press), in the Caribbean (Chapman, 1944; Woodroffe, 1980, 1982; Woodroffe et al., 1980), and on the Great Barrier Reef (Spender, 1930; Steers, 1937; Macnae, 1966; Thom, 1975; Stoddart, 1980). Three distinct settings are distinguished in this chapter (to be added to those listed above).

Setting 6. On low-energy coasts in the tropics, carbonate platforms may be slowly accreting due to the accumulation of lime muds (marl) and peat. The shoreline may be quite indented with 'promontories' of mangroves protruding into shallow water. Coral-reef or sand barriers to seaward may serve to dampen wave energy, but in this setting the mangrove fringe directly abuts an extensive shallow water area (e.g. Florida Bay, or Laguna dos Terminos in Mexico). Direction of shoreline change in such a setting can be quite variable, in places exhibiting a mangrove 'cliff' (Egler, 1950), or elsewhere a stepped age sequence of plants representing colonization of the mud-flats. Long-term subsidence of the platform may help promote mangrove peat accumulation in this setting (Wanless and Meeder, in press). However, there have been problems in the interpretation of mangrove zonation in this setting.

Setting 7. Mangroves commonly occur in carbonate environments behind a mobile but protective sand or shingle barrier. The barrier can be of various types. On the Great Barrier Reef it may consist of cemented or non-cemented coral detritus ('ramparts') which may contain a patchy cover of mangroves (Thom, 1975; Stoddart, 1980). Shallow peats develop in the lee of these 'ramparts' beneath a dense mangrove cover dominated by *Rhizophora*. Wanless and Meeder (in press) demonstrate how mangrove peats in south Florida have accumulated during a sea-level rise behind regressive, transgressive or oscillating sand barriers (e.g. Cape Sable). In such cases, the sand mainly consists of carbonate particles. Peat development is quite meagre where the platform surface is at present sea-level and is not subsiding, or sea-level is not rising (e.g. northern Great Barrier Reef).

Setting 8. In low-energy embayments that sometimes lacks protective barriers, Pleistocene carbonate surfaces have been transgressed by rising Holocene sea-level. Mangroves may develop on such surfaces leading to the burial of a Pleistocene subaerial 'soil' crust. A cover of mangrove peat may result as documented on Grand Cayman Island (Woodroffe, 1981, 1982). The 'drowning' of an old reef complex can lead to an irregular patchwork of mangroves often reflecting topographic irregularities in the underlying substrate.

Conclusion

Any attempt to explain the development of mangrove communities through time requires an evaluation of changes in magnitude and frequency of coastal geomorphic processes. These processes create variable environmental conditions on which mangroves colonize, reproduce and grow. The geomorphologist views the habitat of plants in coastal regions as that which represents a set of depositional environments subject to constant change. These changes involve directional movement of vegetated surfaces depending on the net balance of erosional and depositional forces.

Three basic approaches are recognized to the problem of persistence and change in mangrove ecosystems: classical succession theory, the gradient concept and that which may be called physiographic ecology. The last-mentioned focuses on change in habitats and community properties through time by seeking to understand the development of landforms or physiography. Species distribution and physiognomy are explained by associating plants directly with diverse and dynamic landform and substrate conditions.

Changes in mangrove communities can be measured by documenting the geomorphic or physiographic history of depositional regions. Within specific physiographic settings the changes are dictated by the relative dominance of specific geomorphic processes, such as wave erosion, river-channel switching, mud-flat accumulation, etc. It is possible to determine how changes in landforms at various spatial scales will induce varying physiological responses in different species of mangrove, which in turn leads to change in community structure. Stresses on plants created by these changing conditions lead to changes in the competitive abilities of different species to survive in physiographically defined habitats. Three basic directions of change can occur within any region, the dominance of any particular direction determining the dynamics of the mangrove community structure in that region. Under conditions of high sediment influx accretion will occur and habitat change is induced by shoreline progradation and vertical accumulation. Stepped plant sequences are the result. Erosion by waves, tidal currents or river channels leads to destruction of communities and local deposition and regrowth. Thirdly, steady-state conditions may prevail perhaps involving cyclic short-term instability, or long-term self-maintenance and persistence of a specific community. There now exists a number of case-studies which highlight the importance of understanding geomorphic processes and products in the interpretation of temporal variation in mangrove communities (Thom, 1967; Thom et al., 1975; Semeniuk, 1980, 1981; Augustinus, 1980).

The physiographic (or geomorphic) approach highlights the multidimensional character of mangrove ecosystems. Given sufficient field data, it is possible to predict direction and magnitude of changes in communities through time. This requires that adequate geomorphological research be undertaken as part of any effort to unravel the dynamics of vegetation in mangrove areas. Ecologists working in such environments, in this writer's view, cannot afford to overlook the ever-changing character of landforms upon which grow the mangroves they are studying. These landforms, either directly or indirectly, influence so many of the other environmental factors which are discussed in this monograph.

References

ANDERSON, J. A. R.; MULLER, J. 1973. Palynological Study of a Holocene peat and a Miocene Coal Deposit from NW Borneo. *Pollen et Spores*, Vol. 15, pp. 291–351.

AUGUSTINUS, P. G. F. 1980. Actual Development of the Chenier Coast of Surinam (South America). *Sedimentary Geology*, Vol. 26, pp. 91–113.

BUCKLEY, R. C. 1982. Patterns in North Queensland Mangrove Vegetation. *Australian J. of Ecology*, Vol. 7, pp. 103–106.

BUNT, J. S.; WILLIAMS, W. T. 1980. Studies in the Analysis of Data from Australian Tidal Forests ('Mangroves'). I: Vegetational Sequences and their Graphic Representation. *Australian J. of Botany*, Vol. 5, pp. 385–90.

——. 1981. Vegetational Relationships in the Mangroves of Tropical Australia. *Marine Ecol.* (Progress Series), Vol. 4, pp. 349–59.

CARTER, J. 1959. Mangrove Succession and Coastal Change in Southwest Malaya. *Transactions of the Institute of British Geographers*, Vol. 26, pp. 79–88.

CHAPMAN, V. J. 1944. 1939 Cambridge University Expedition to Jamaica. *J. Linnean Soc.* (Botany), Vol. 52, pp. 407–534.

——. 1970. Mangrove Phytosociology. *Tropical Ecology*, Vol. 11, pp. 1–10.

——. 1976. *Mangrove Vegetation*. Vaduz. J. Cramer. 447 pp.

CLARKE, L. D.; Hannon, N. J. 1967. The Mangrove Swamp and Salt Marsh Communities of the Sydney District. I: Vegetation, Soils and Climate. *J. of Ecology*, Vol. 55, pp. 753–71.

——. 1969. The Mangrove Swamp and Salt Marsh Communities of the Sydney District. II: The Holocoenotic Complex with Particular Reference to Physiography. *J. of Ecology*, Vol. 57, pp. 213–34.

——. 1970. The Mangrove Swamp and Salt Marsh Communities of the Sydney District. III: Plant Growth in Relation to Salinity and Waterlogging. *J. of Ecology*, Vol. 58, pp. 351–69.

——. 1971. The Mangrove Swamp and Salt Marsh Communities of the Sydney District. IV: The Significance of Species Interaction. *J. of Ecology*, Vol. 59, No. 535–53.

COLEMAN, J. M.; GAGLIANO, S. M.; SMITH, W. G. 1970. Sedimentation in a Malaysian High Tide Tropical Delta. In J. P. Morgan (ed.), *Deltaic Sedimentation Modern and Ancient*, Vol. 15, pp. 185–97 (Special publication), Society of Economic Paleontologists and Mineralogists.

COLEMAN, J. M.; WRIGHT, L. D. 1975. Modern River Deltas: Variability of Processes and Sand Bodies. In M. L. Broussard (ed.), *Delta Models for Exploration*, pp. 99–149. Houston, Tex., Houston Geological Society.

DAVIS, J. H. 1940. *The Ecology and Geologic Role of Mangroves in Florida. Papers of the Tortugas Laboratory*, No. 32, pp. 303–412. (Carnegie Institute Publication, 517.)

DIEMONT, W. H.; VAN WIJNGAARDEN, W. 1975. Sedimentation Patterns, Soils, Mangrove Vegetation and Land Use in the Tidal Areas of West Malaysia. In: G. Walsh, S. Snedaker and H. Teas (eds.), *Proceedings of the International Symposium on Biology and Management of Mangroves*, Vol. II, pp. 513–28.

EGLER, F. 1950. Southeast Saline Everglades Vegetation, Florida and its Management. *Vegetatio*, Vol. 3, pp. 213–65.

FISK, H. N. 1960. Recent Mississippi River Sedimentation and Peat Accumulation. *Compte Rendu, 4th International Congress on Carboniferous Stratigraphy and Geology, Heerlen, Netherlands, 1958*, pp. 187–99.

FLOYD, A. G. 1977. *Ecology of the Tidal Forests in the Kikori–Romilly Sound Area, Gulf of Papua*. Lae (Papua New Guinea), Division of Botany, Office of the Forests, Department of Primary Industry. 59 pp. (Biology Report, 4.)

GOODLETT, J. C. 1969. Vegetation and the Equilibrium Concept of the Landscape. In K. N. H. Greenidge (ed.), *Essays in Plant Geography and Ecology*, pp. 33–44. Halifax, N.S., Nova Scotia Museum.

GRUBB, P. J. 1977. The Maintenance of Species-richness in Plant Communities: The Importance of the Regeneration Niche. *Biological Review*, Vol. 52, pp. 107–45.

HACK, J. T.; GOODLETT, J. C. 1960. *Geomorphology and Forest Ecology of a Mountain Region in the Central Appalachians*, pp. 1–66. U.S. Geological Survey. (Professional Paper, 347.)

HEINSELMAN, M. L. 1963. Forest Sites, Bog processes and Peatland Types in the Glacial Lake Agassiz Region, Minnesota. *Ecological Monographs*, Vol. 33, pp. 327–73.

——. 1970. Landscape Evolution, Peatland Types, and the Environment in the Lake Agassiz Region, Minnesota. *Ecological Monographs*, Vol. 40, pp. 235–61.

LUGO, A. E.; SNEDAKER, S. 1974. The Ecology of Mangroves. *Annual Review of Ecological Systems*, Vol. 5, pp. 39–64.

MACNAE, W. 1966. Mangroves in Eastern and Southern Australia. *Australian J. of Botany*, Vol. 14, pp. 67–104.

RABINOWITZ, D. 1978. Early Growth of Mangrove Seedlings in Panama, and an Hypothesis Concerning the Relation of Dispersal and Zonation. *J. of Biogeography*, Vol. 5, pp. 113–33.

ROY, P. S.; THOM, B. G.; WRIGHT, L. D. 1980. Holocene Sequences on an Embayed High-energy Coast: An Evolutionary Model. *Sedimentary Geology*, Vol. 26, pp. 1–19.

SAUER, J. 1961. *Coastal Plain Geography of Mauritius*. Baton Rouge, Louisiana State University Press. (Coastal Studies series, 5.)

SCHOLL, D. W. 1964. Recent Sedimentary Record in Mangrove Swamps and Rise in Sea-Level over the Southwest Coast of Florida, Part I. *Marine Geology*, Vol. 1, pp. 344–66.

SEMENIUK, V. 1980. Mangrove Zonation along an Eroding Coastline in King Sound, Northwestern Australia. *J. of Ecology*, Vol. 68, pp. 789–812.

——. 1981. Long-term Erosion of the Tidal Flats, King Sound, Northwestern Australia. *Marine Geology*, Vol. 43, pp. 21–48.

SMITH, W. G. 1968. Sedimentary Environments and Environmental Change in the Peat-forming Area of South Florida. Pennsylvania State University, 254 pp. (Ph.D. thesis.)

SNEDAKER, S. C. 1982. Mangrove Species Zonation: Why? In: D. N. Sen and K. S. Rajpurohit (eds.), *Tasks for Vegetation Science*, pp. 111–25. The Hague, W. Junk.

SPACKMAN, W.; DOLSEN, C. P.; RIEGEL, W. 1966. Phytogenic Organic Sediments and Sedimentary Environments in the Everglades Mangrove Complex. Part I: Evidence of a Transgressing Sea and its Effects on Environments of the Shark River Area of Southwestern Florida. *Palaeontographica*, Vol. 117, pp. 118–52.

SPENDER, M. A. 1930. Island Reefs of the Queensland Coast. *Geographical J.*, Vol. 76, pp. 194–214, 273–97.

STEERS, J. A. 1937. The Coral Islands and Associated Features of the Great Barrier Reef. *Geographical J.*, Vol. 89, pp. 1–28, 119–39.

STODDART, D. R. 1980. Mangroves as Successional Stages, Inner Reefs of the Northern Great Barrier Reef. *J. of Biogeography*, Vol. 7, pp. 269–84.

SWIFT, D. J. P. 1976. Continental Shelf Sedimentation. In: D. J. Stanley and D. J. P. Swift (eds.), *Marine Sediment Transport and Environmental Management*, pp. 311–50. New York, John Wiley.

THOM, B. G. 1967. Mangrove Ecology and Deltaic Geomorphology: Tabasco, Mexico. *J. of Ecology*, Vol. 55, pp. 301–43.

——. 1975. Mangrove Ecology from a Geomorphic Viewpoint. In G. Walsh, S. Snedaker and H. Teas (eds.), *Proc. International Symposium on Biology and Management of Mangroves, Honolulu, Hawaii, 1974*. Vol. 2, pp. 469–81. Gainesville, Fla., University of Florida.

——. 1982. Mangrove Ecology: A Geomorphological Perspective. In B. Clough (ed.), *Mangrove Ecosystems in Australia, Structure, Function and Management*, pp. 3–17. Canberra, A.N.U. Press.

THOM, B. G.; WRIGHT, L. D. 1982. Geomorphology of the Purari Delta. Purari River (Wabo) Hydroelectric Scheme. *Environmental Studies*, Vol. 17, pp. 1–24.

THOM, B. G.; WRIGHT, L. D.; COLEMAN, J. M. 1975. Mangrove Ecology and Deltaic Estuarine Geomorphology: Cambridge Gulf–Ord River, Western Australia. *J. of Ecology*, Vol. 63, pp. 203–32.

VANN, J. 1959. Landform-Vegetation Relationships in the Atrato Delta. *Annals Assoc. Am. Geographers*, Vol. 49, pp. 345–60.

VAUGHAN, T. W. 1909. The Geologic Work of Mangroves in Southern Florida, *Smithsonian Misc. Collection*, Vol. 52, pp. 461–64.

WANLESS, H. R.; MEEDER, J. F. *Mangrove Sedimentation in Geologic Perspective*. (In press.)

WARMING, E. 1909. *Oecology of Plants*. Oxford, Oxford University Press. 422 pp.
WATSON, J. D. 1928. Mangrove Forests of the Nalay Peninsula. *Malaya Forest Records*, Vol. 6, pp. 1–275.
WELLS, J. T.; COLEMAN, J. M. 1981. Periodic Mudflat Progradation, Northeastern Coast of South America; A Hypothesis. *J. of Sedimentary Petrology*, Vol. 51, pp. 1069–75.
WHITTAKER, R. H. 1953. A Consideration of Climax Theory: The Climax as a Population and Pattern. *Ecological Monographs*, Vol. 23, pp. 41–78.
WOODROFFE, C. D. 1982. Geomorphology and Development of Mangrove Swamps, Grand Cayman Island, West Indies. *Bull. of Marine Science* (University of Miami), Vol. 32 pp. 381–98.
WOODROFFE, C. D.; STODDART, D. R.; GIGLIOLI, M. E. C. 1980. Pleistocene Patch Reefs and Holocene Swamp Morphology, Grand Cayman Island, West Indies. *J. of Biogeography*, Vol. 7, pp. 103–13.
WRIGHT, L. D.; COLEMAN, J. M.; THOM, B. G. 1973. Processes of Channel Development in a High-Tide-Range Environment: Cambridge Gulf-Ord River Delta. *J. of Geology*, Vol. 81, pp. 15–41.
WRIGHT, L. D.; COLEMAN, J. M.; ERICKSON, M. W. 1974. *Analysis of Major River Systems and Their Deltas: Morphologic and Process Comparisons*. Baton Rouge, Coastal Studies Institute Louisiana State University Press. 114 pp. (Tech. Report No. 156.)
ZIMMERMANN, R. C. 1969. *Plant Ecology of an Arid Basin. the Tres Alamo–Redington Area South-eastern Arizona*, U.S. Geological Survey. 51 pp. (Professional Paper, 485-D.)
ZIMMERMANN, R. C.; THOM, B. G. 1982. Physiographic Plant Geography. *Progress in Physical Geography*, Vol. 6, pp. 45–59.

2

Climatic factors and the biology of mangrove plants

François Blasco
Institut de la Carte Internationale du Tapis Végétal
39 allée Jules-Guesde
31400 Toulouse, France

Introduction

The ecological amplitude of mangrove species is fascinating and often disconcerting. Until recently, it has been difficult or impossible to determine with certainty the controlling climatic, edaphic and hydric factors, primarily because these factors are usually interdependent. It is, however, a banal finding that mangroves develop best in tropical estuaries which receive heavy rainfall evenly distributed throughout the year, whereas aridity is a limiting factor in many regions of the world.

Regarding zonation and serial succession of vegetation in a mangrove ecosystem, it is generally agreed that they are related to complex local factors among which hydrology and climate are dominant, particularly because the competitive ability of each species is related to its climatic requirements. This has been expressed by Walter (1971, p. 161): 'Under different climatic conditions and different floristic compositions, different zonations would be expected.'

The basic climatic factor governing the geographical distribution of species is probably air temperature. Many biogeographers state that the number of mangrove tree species declines from about thirty in equatorial Indo-Malesian countries to only one north of the African Red Sea Coast at 27°40′ N. (*Avicennia marina*), whereas *A. marina* var. *resinifera* appears to be the most cold-resistant species occurring near Auckland at about 37° S. This species can be considered as a vicariant of *A. germinans* which is the most tolerant of low temperatures in North America.

An advanced knowledge of climatic conditions is also essential to soil scientists, not only because low precipitation and high evaporation rates cause hypersaline conditions, but also because the main mineralogic constituents and pedogenetic processes are related to the prevailing climatic factors. For example, Moorman and Pons (1975) note that in coastal sediments of

> humid and sub-humid tropical areas, the dominant constituents are kaolinite, iron oxides and quartz . . . In the few sub-arid areas where mangroves occur (such as the Senegal and Indus rivers) the sediments are less kaolinitic and contain less iron oxides; illite and minerals of the smectite group dominate the clay fraction.

It also appears that climatic factors have a direct biological implication on mangrove plants, but it is however a complex factor difficult to assess. Research on the epidermis of mangrove plants (e.g. stomatal frequency, indices and behaviour) has to be conducted within a climatic context because, according to Sidhu (1975, p. 569),

> the variations of stomatal and epidermal patterns of fully expanded leaves are functions of genetic differences among species as well as variations in environmental factors such as temperature, humidity, light intensity.

These few examples clearly show that an accurate knowledge of bioclimatic conditions, for each mangrove ecosystem, represents the fundamental ecological information required for an understanding of most problems related to the biology of mangrove plants.

Present knowledge on temperature and rainfall

The present knowledge concerning major relationships between the distribution of mangrove forests and climatic regions has been broadly summarized by Pannier and Pannier (1977) and by several contributors to the book *Wet Coastal Ecosystems* (Chapman, 1977), notably West for the Americas (pp. 193–213), Saenger et al. for Australasia (pp. 293–345) and Walter under the title 'Climate' (pp. 61–7). The expressed results are not new; they concern the biogeographical distribution of mangrove species, which is mainly related to temperature conditions, whereas rainfall conditions often make it difficult to determine the various types of zonation within the intertidal zone.

According to Walter (1977), mangrove ecosystems are mainly found in three climatic divisions of the earth: (a) the equatorial zone, between approximately 10° N. and 5–10° S.; (b) the tropical summer-rainfall zone, north and south of the equatorial zone, to approximately 25–30° N. and S., partly in the subtropical dry zone of the deserts, still further poleward; and (c) partially in warm temperate climates that do not have really cold winters, and only on the eastern border of the continents.

At this point, it may be noted that the conclusions given above are not wrong but they are extremely general and lack accuracy. They are based on data provided by some 8,000 climate diagrams collected in the *Klimadiagramm-Weltatlas* (Walter and Leith, 1960–67). Since temperature and rainfall are essential bioclimatic factors for mangroves as well as for terrestrial plants, it was necessary to combine them in a single diagram. This research was carried out by two teams: Bagnouls and Gaussen (1953) and Walter and Leith (1960–67).

They have suggested that a month is 'dry' when its precipitation (*P*, expressed in mm) is equal to, or less than twice, the mean temperature (*T*, in degrees centigrade). A dry month is a month in which *P* is less than 2*T*. This definition can be applied all over the world. In order to show the seasonality of each wet and dry period, they have suggested the use of very simple ombrothermic diagrams or climatic diagrams (Fig. 2.1). On the same graph, the months in ordinates are quoted in abscissa, with the scale of precipitation (in millimetres) on the right side and, on the left side, the scale of temperatures (in degrees centigrade), which is double the precipitation scale. When the precipitation falls below the temperature curve, the period is considered dry (dotted on the graph). When the precipitation is above the temperature curve, the period is considered to be humid. In addition, Walter proposes that precipitation above 100 mm be printed in the scale of 1:10 and marked in black. These rainfall-temperature diagrams have been used by several scientists for mangrove study sites in India (Blasco, 1975), in Florida, Puerto Rico, Mexico and Costa Rica (Pool et al., 1977) and also by Walter (1977) and Pannier and Pannier (1977) for South American mangroves.

Tall, dense and floristically diverse mangroves are almost exclusively found either in the 'equatorial zone' (Malaysia, Indonesia, Colombia, etc.) or in the 'tropical summer-rainfall zone' (most coastal areas of India, Burma, Thailand, Indochina, etc.). Thickets of low, scattered or sporadic mangrove species prevail in the 'subtropical dry zone' (north-western Indian coast, Pakistan, African Red Sea coast, etc.) and in 'warm temperate' climates (Australia, New Zealand); their floras are extremely simplified.

As already stated, *A. germinans* marks the poleward limit of mangrove formations in the Americas (see Figs. 2.2, 2.3 and 2.4). Moreover, Pannier and Pannier (1977) point out the substitution of *A. germinans* (=*nitida*) by *A. schaueriana* in the subtropical zone of South America; this observation is open to further taxonomic investigation.

In Australia, Saenger et al. (1977, p. 297) point out that *A. marina* var. australasica is the only species occurring south of Merimbula Estuary (36°50′ S.) on the east coast and south of Carnarvon (25° S.) on the west coast.

The localities and area of distribution of each mangrove species can be considered to be reasonably well documented; they are given in modern floras such as *Flora Malesiana* (van Steenis, 1955–58), whenever the group has been botanically revised (cf. Breteler, 1977; Tomlinson, 1978). Finally, according to the synthesis of Saenger et al. (1977), all mangrove species disappear when the average temperature of the coldest months (January or July) drops to 16 °C or lower. Moreover, mangrove species are extremely sensitive to frost and with the exception of some *Avicennia* (*A. marina* var. *resinifera* and *A. germinans*) which tolerate occasional light frosts (up to −2 °C or −4 °C), they do not survive frequent or lengthy periods of frost.

Tulear

Longitude: 43°41′ E.
Latitude: 23°20′ S.
Altitude: 8 m

Number of years of observation: 56

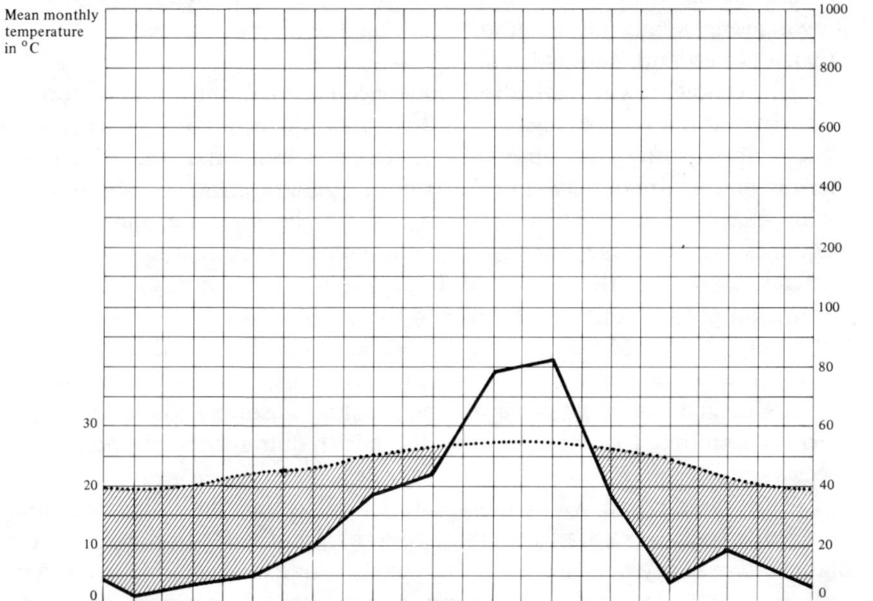

Hemisphere North	J	F	M	A	M	Ju	Jy	A	S	O	N	D	Year
Hemisphere South	Jy	A	S	O	N	D	J	F	M	A	M	Ju	
Rainfall (*R*)	3	6	9	19	35	44	77	82	37	7	18	11	348
Number of rainy days	1	1	1	2	2	3	4	6	2	2	2	1	27
Mean temperature	19.5	20.5	22	23	25	26.5	27.5	27.5	26.5	25	22	20	24
Mean of maxima	26.9	27.3	28.7	29.4	30.3	31.1	32.4	32.7	31.9	30.6	29.0	27.2	29.8
Mean of minima	13.2	14.0	15.8	17.8	20.5	21.4	22.7	22.4	21.5	19.1	16.3	14.3	18.3
Absolute maximum	35	35	35	36.5	39.8	40	40	42	40	39	37	35	37.9
Absolute minimum	6.5	6.0	7.5	10	12	12	13	12	13	10	9	8	10.7
Air humidity (%)	65	64	62	66	64	64	71	71	68	68	66	62	66
Etp (Penman)	193	174	161	126.7	94.4	71	72.4	89.8	118.4	165.5	180	191.2	1,637.7
R/*Etp*	0.02	0.03	0.06	0.15	0.37	0.62	1.06	0.91	0.31	0.04	0.10	0.06	0.21[1]
Winds: Direction	S., *SW.* or SE.					SW., S. or SE.						S. or *SW.*	
Winds: Number of days	55					65						25	
Winds: Speeds	Frequently between 20 and 50 km/h, reaching exceptionally 75 km/h												
Cyclonic storms	0	0	0	0	0	0	0	0	0	0	0	0	0

1. This is the limit between arid and semi-arid regions according to Unesco's classification 'World Distribution of Arid Regions'.

Figure 2.1
Example of a climatic identity card.

The typical responses of mangrove communities to decreasing temperature and increasing thermic amplitude (daily and annual) are reductions in species richness, forest height and the maximum size of trees. This has been clearly shown along a latitudinal gradient along the Florida coast of the Gulf of Mexico (Lugo and Zucca, 1977).

Up to now, few investigations have been carried out on the autecology of mangrove species with regard to their thermic optimum or requirements. According to some field observations it seems that small variations in some thermic factors are decisive in restricting the geographical distribution of some Asian species. These thermic factors include the absolute minimum and its frequency, the average minimum of the coldest month, the daily amplitude (difference between the hottest and coldest hour of the day) and the annual thermic amplitude (difference between the temperature of the coldest and hottest months), etc. Most probably, there exists a thermic threshold for each species during the seedling stage. When this threshold is reached, establishment is made difficult or impossible, or the seedling is simply killed. Experimental work is required to determine the thermic requirements of each mangrove species.

Whereas air and water temperatures determine the latitudinal limits of mangrove species, rainfall generally governs the distribution and zonation of plants along many non-mountainous coasts. For example, in tropical Asia, the wettest coasts are located in Malaysia, Sumatra, Kalimantan (Borneo); the driest are in the Gulf of Kutch (north-western coast of India) and the Pakistan coast. Plant zonation in both cases is entirely different. Under humid climatic conditions, the less salt-resistant species (*Heritiera fomes, Sonneratia caseolaris, Nypa fruticans* and several *Pandanus*) are found in the landward zone, in contact or partly mixed with luxuriant freshwater swamp-forest elements. In dry regions, affected every year by long periods of severe drought, most salt-tolerant species (*Salvadora persica*, *Avicennia marina*, *Salicornia brachiata*, *Suaeda monoica*, *Atriplex stocksii*, etc.) occupy the landward zone in contact with extensive areas of 'blanks' in which the salt concentration is so high that practically all plants are excluded.

The explanation for these contrasting patterns is rather simple. Under humid climatic types, soils are almost continuously leached by heavy rains. Under arid conditions such as in Gujarat, India, where the influence of freshwater is almost nil, the evaporation of water is extremely rapid between two high tides; this leads to an unvarying high concentration of salt in the landward zone. Naturally, in the inner zones which are flooded only twice a year during the equinoctial tides, hypersalinity is so pronounced in the topsoil that even halophytes disappear; these are 'blanks' without any vegetation. These examples demonstrate the importance of rainfall as a determining ecological factor in mangrove areas. It is necessary to evaluate this climatic factor for every mangrove site study.

Evaluation of various degrees of aridity

Ombrothermic diagrams can help to quickly determine the length of the dry season, but they are not designed to determine how dry a station or a season may be. That is why agriculturists usually prefer to compute the potential evapotranspiration (*Etp*), particularly for the dry season. The water deficit, which expresses aridity, results from an excess water loss, compared with the water received. Hence, it is possible to classify climatic regions according to the degrees of aridity, i.e. the values of the ratio P/Etp, in which P is the mean annual rainfall and *Etp* the mean annual potential evapotranspiration. The major procedural problem is to select one among the numerous available *Etp* formulae (Thornthwaite, 1948; Turc, 1954; Penman, 1956; Papadakis, 1965; etc.). For several reasons, both scientific and practical, we have selected Penman's procedure. Thornthwaite, who introduced potential evapotranspiration on climatology, has elaborated a formula that is complete, exceedingly empirical and has given rise to much controversy. Papadakis' formula is simpler:

$$E = 0.5625(e_{ma} - e_d)$$

where E = monthly potential evapotranspiration in cm

e_{ma} = saturation vapour pressure corresponding to the average daily maximum in millibars

e_d = average vapour pressure of the month in millibars.

We have further simplified this formula:

$$E = 0.5625(e_{ma} - e_{mi_{-2}})$$

where $e_{mi_{-2}}$ = saturation vapour pressure corresponding to the average daily minimum minus 2 centigrades in millibars; 2 is the 'normal' difference between average daily minimum and dew-point.

With this formula the only data that are necessary are the average maximum and average minimum temperatures; these data are usually available in all stations (Papadakis, 1966).

Although this formula leads to a satisfactory classification of Indian climatic zones, it is difficult to accept without criticism. For example, the climate of Bombay (six dry months) is similar to that of Singapore or Palembang (no dry months). Penman's formula seems to be a more reliable alternative.

Penman's formula is best suited for estimating water loss from a free water surface or for a crop constantly supplied with water. Thus, for mangrove study sites it appears that this formula is certainly one of the best adapted since many analogies can be found between a wet soil surface of an intertidal mangrove area and a free water surface. Moreover, it includes extremely important climatic parameters prevailing in dry zones such as wind and its drying power, solar radiation and air humidity. Finally, it has been selected by Unesco (1979, p. 10) which argues soundly that 'the great advantage of Penman's formula is that it has been used in numerous biological and physical studies of climate, the results of which have been widely diffused. In addition, it is today considered more satisfactory than the formula used in the Meigs map' (Thornthwaite's index). It should also be stated that the idea of considering evapotranspiration as an essential ecological factor for mangrove studies is not new. Davis (1940) had already mentioned it, and Jennings and Bird (1967) did the same for aridity.

Penman's formula is based on the following considerations: Evaporation from a free water surface is a function of surface saturation deficit and wind:

$$E_0 = f(u)(e_s - e_d) \tag{1}$$

where E_0 is evaporation
$f(u)$ is a function of wind speed (u)
e_s is saturation vapour pressure at surface temperature (T_s)
e_d is mean vapour pressure of the atmosphere.

If surface temperature (T_s) were equal to air temperature (T_a), evaporation would be a function of air saturation deficit ($e_a - e_d$).

This hypothetical evaporation is termed by Penman as the drying power of the air (E_a):

$$E_a = f(u)(e_a - e_d) \tag{2}$$

where E_a is the drying power of the air
e_a is saturation vapour pressure at air temperature (T_a)

By dividing (1) by (2) we obtain:

$$E_0/E_a = (e_s - e_d)/(e_a - e_d) \tag{3}$$

This equation shows that the ratio of evaporation to the drying power of the air is equal to that of two saturation deficits, one computed on the basis of surface temperature and the other on the basis of air temperature.

Since the difference $T_s - T_a$ is seldom considerable, equation (3) provides a sound basis to compute evaporation (Papadakis 1965, pp. 7–8).

As far as we know this formula has never been applied to mangrove ecological studies. However, it promises to yield interesting correlations between the values of *Etp* and the biogeographical distribution of mangrove ecosystems or mangrove species, which may be required for the exact mapping of the main mangrove areas of the world. When we compare the world distribution given by West (1956), Kiener (1973), Chapman (1977) with our distribution map, we find important disparities which restrain reliable correlations with available bioclimate maps. It is suggested that the question of mapping the exact distribution of mangroves be re-examined relative to the available vegetation maps for each country, island or continent.[1]

Based on Penman's formula, the *Map of the World Distribution of Arid*

1. A new map, based on satellite imageries (Landsat 1 and Landsat 2), covering all the countries of South America has been published by Unesco (1981). Catalogues of available vegetation maps for each country have been published by Kuchler (1968–70).

Regions (Unesco, 1979) can be compared with that of the world distribution of mangrove ecosystems. The interesting features shown on the map are four degrees of aridity distinguished on the basis of the ratio P/Etp, temperature regimes and drought period, where, P = mean annual rainfall, Etp = potential evapotranspiration, calculated according to Penman's formula, and tm = temperature of the coldest month.

It can be concluded that (see Figs. 2.2, 2.3, 2.4):

1. At least 90 per cent of the world area in mangroves is found in warm humid regions, where $P/Etp > 0.75$, winters are warm or mild ($tm > 10$ °C), summers are also warm with an average of the warmest month exceeding 20 °C.
2. Under sub-humid climates ($0.50 < P/Etp < 0.75$) mangroves are occasionally found, mainly in Kenya, Tanzania, India, Australia, Venezuela, Mexico.
3. Under semi-arid conditions ($0.20 < P/Etp < 0.50$), mangroves are exceptional, mainly found in the Indus delta (Pakistan) and in Gujarat (India), Australia (Northern Territory and Western Australia), Ecuador.
4. Under arid climates ($0.03 < P/Etp < 20$), mangroves are practically unknown, with the exception of those of the Red Sea (mainly Ethiopian and Egyptian coasts), the Persian Gulf and the Gulf of California. It is noteworthy that these latter regions receive winter rains which are more effective than the summer rains which evaporate rapidly.

These provisional conclusions illustrate the methodology. The general results given above can be considerably broadened and improved if all available data are thoroughly studied and if *Etp* is applied to each mangrove site. This method has to lead to a classification of mangrove ecosystems according to climatic parameters of their environment. The task is already relatively simplified since many of the mean values of *Etp* have been calculated and are available, monthly and yearly, for a great number of coastal tropical stations.[1]

Naturally, this is not a blameless approach for two main reasons. First, even if Penman's formula has a worldwide reputation, it is rather complex and nevertheless not perfect. It probably overestimates *Etp* under humid climates (Papadakis, 1965) whereas it probably underestimates *Etp* in dry zones. However, it has been proved that it operates satisfactorily on a regional scale. On a local scale, some other factors should be taken into account. Secondly, *Etp* values used until now, were based on the means of the main climatic factors. They conceal temporal variability of essential factors (rainfall, temperature, saturation deficits, etc.).

Climatic variability

All available climatic classifications are based on mean values of climatic data, according to averages of at least twenty years. In tropical countries, it is often

1. They can be obtained from the General Secretary, World Meteorological Organization, CP1-CH1211, Geneva 20, Switzerland.

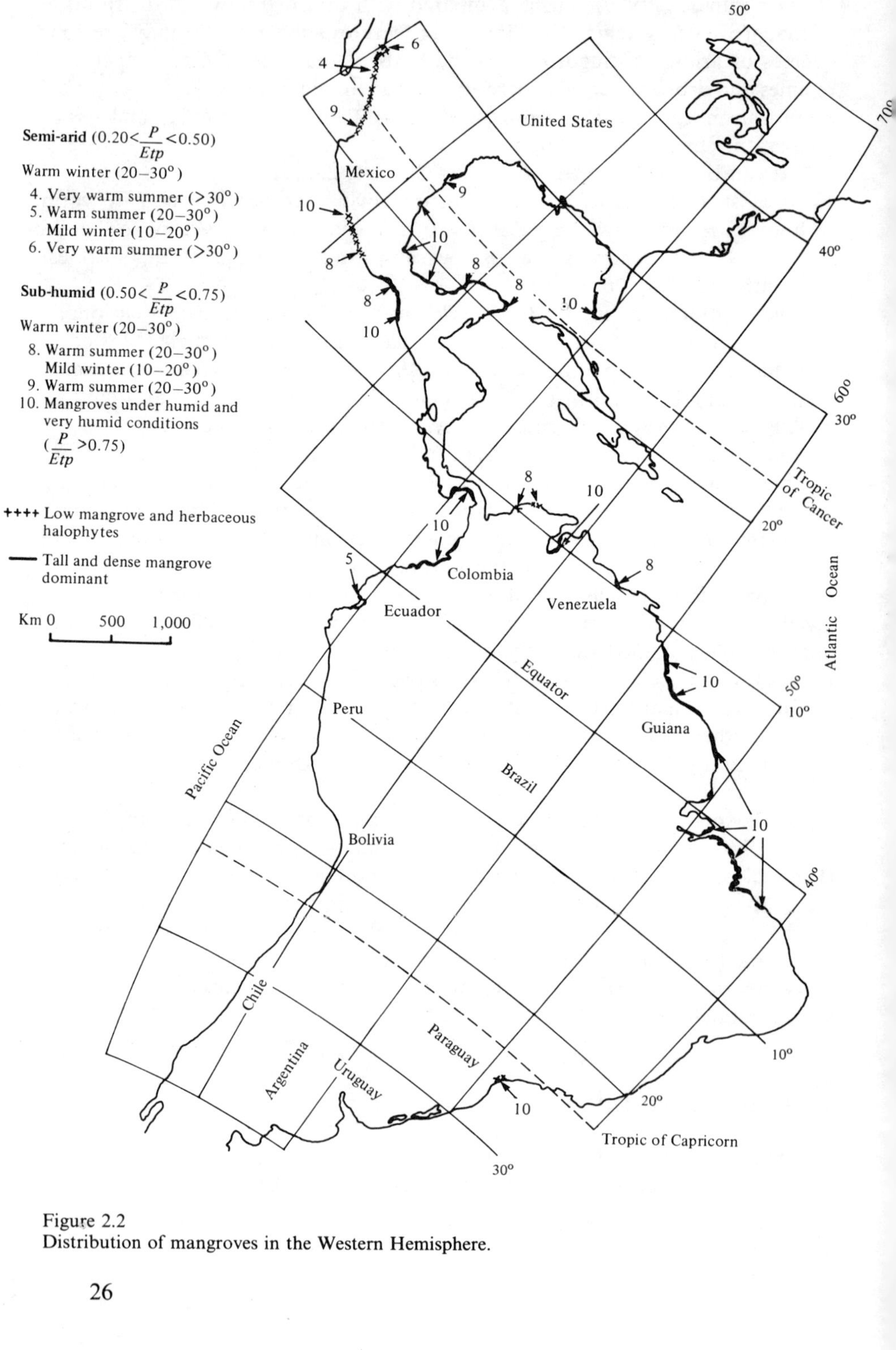

Figure 2.2
Distribution of mangroves in the Western Hemisphere.

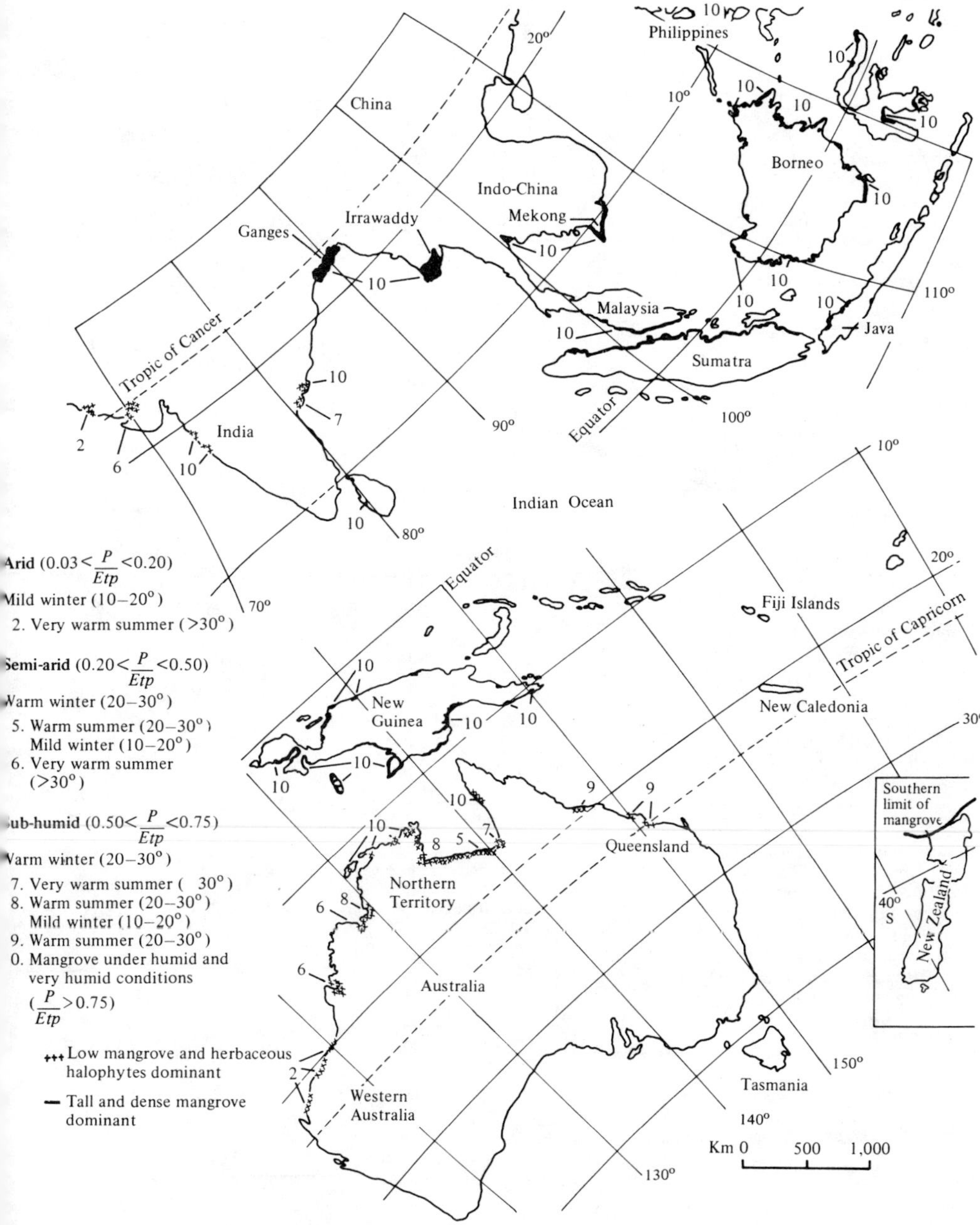

Figure 2.3
Distribution of mangroves in Asia and Oceania.

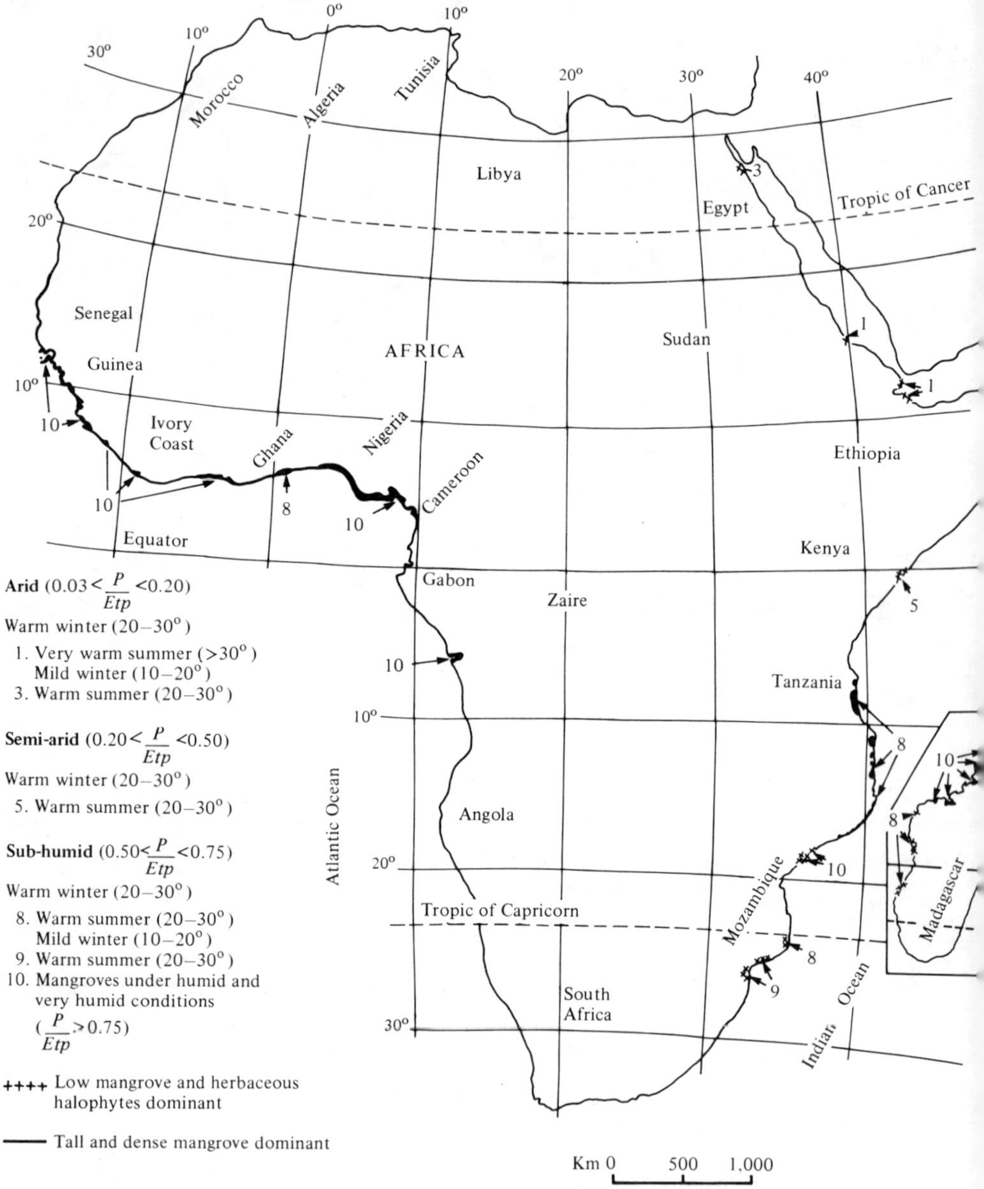

Figure 2.4
Distribution of mangroves in Africa.

necessary to refer also to actual yearly values, at least for rainfall and aridity, so that the exact meaning of averages, as well as the degree of stability of climatic factors, can be easily discerned.

Variability in precipitation is a general phenomenon in tropical countries. On south-eastern Indian coasts, for instance, in the delta of the Cauvery (Cuddalore), the annual rainfall may vary from 3,400 mm in 1895 to only 690 mm in 1968, with the average for the period 1892–1970 being 1,380 mm. It is essential also to consider the variability in the length of the dry season. Again, in the case of the Cauvery delta, the average length is five consecutive dry months, though the actual duration often reaches seven months. In addition, it is usual at tropical latitudes to find a great disparity in the number of rainy days from one year to another. At Cuddalore, it usually varies from 50 to 100 days. The above example brings out the variability of common climatic factors in a mangrove area. These variations have repercussions on the biology of halophytes and on soil properties. As a general rule, most unstable climatic types are found far from the equator, near the subtropical zones, where mangroves, when present, are not luxuriant. In contrast, mangroves develop best in regions experiencing rather regular climates with abundant rainfall evenly distributed throughout the year; this is the case for Malaysia, Indonesia and Papua New Guinea, where mangroves reach their maximum in size, density and specific diversity.

Thus it is necessary to analyse, in addition to the average temperature/rainfall diagrams and *Etp*, the annual variations of rainfall and length of the dry season (actual number of dry months) and to express the results in a synoptic table as illustrated in Figure 2.5.

In some cases, it will be useful to include in the same table data provided by inland stations located in the drainage area system. As a matter of fact, it could happen that even if an estuary is rainless for a part of the year, the normal salinity in the estuary is maintained due to the rain and runoff from the drainage system. This is the case for the mangroves of southern Ecuador which are luxuriant according to Eggers (1892) and West (1956), even though they are located on an arid coast at the foothills of the Andes; these neighbouring slopes, however, are humid. The situation is somewhat similar for the mangroves of Tuléar (Madagascar).

No relevant literature on the effects of climatic variability on mangrove stands is available. We can, however, hypothesize that a natural cause of the mangrove change and disappearance in the Sahelian and Arabian sea regions could be the effect of climatic instabilities. This question was raised by Marius (1979) who has been working for years on the mangroves of Senegal and Gambia (at about 13° N.) where the total mangrove areal extent is about 7,000 km^3.

The average mean of annual rainfall in this area is rather high (1,400 to 1,500 mm), mainly concentrated in July, August and September. According to meteorological records, 1971 and 1972 were exceptional years with annual rainfalls of only 904 and 655 mm. In addition, the pluviometric data for 1977 also show a deficit due to a rainfall of only 855 mm.

STUDY OF CLIMATIC VARIABILITY

Annual variations of rainfall and length of the dry season in the delta of the Cauvery (south-eastern India)

Station	*R* (mean annual rainfall, mm)	Variability of *R* (mm)	Standard deviation	*r* (maximum monthly recorded, mm)	Mean length of the dry season	Maximum length of the dry season	Usual number of rainy days per year	Maximum daily rainfall
Cuddalore (1892–1970) sed. level	1,381	3,428 in 1895 693 in 1968		1,874 in Nov. 1895 and 1,010 in May 1943	Dry months	7 dm	65–90 (56)	200–300 mm

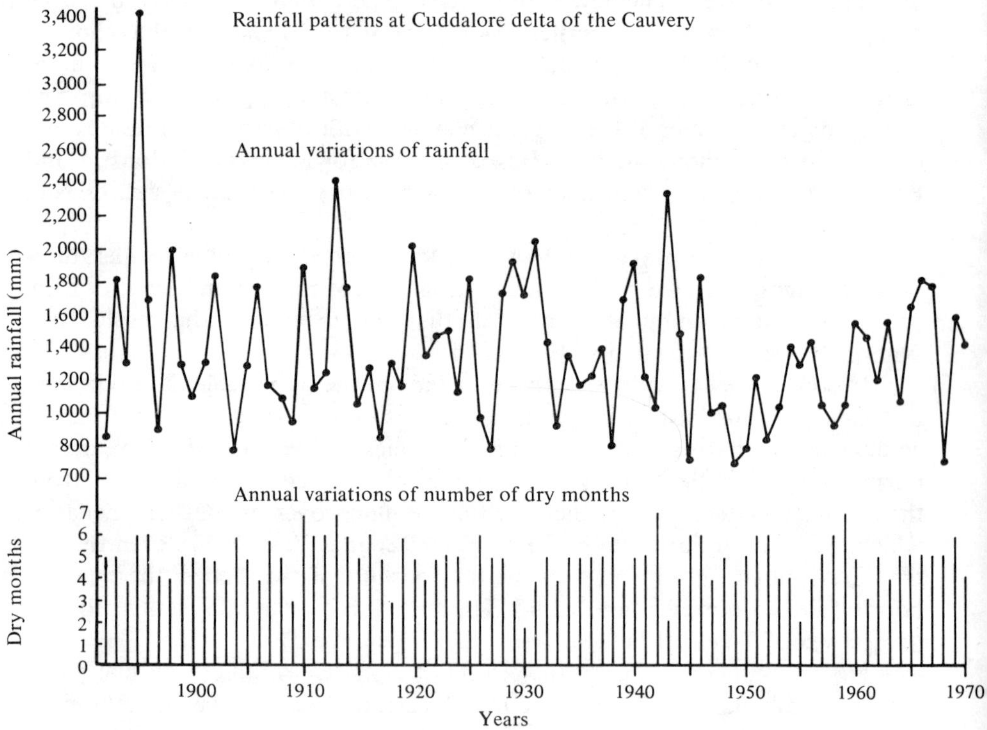

Figure 2.5
Graphic illustration of climatic variability.

These three almost consecutive and pronounced dry periods have drastically affected the mangrove communities. In some areas (such as the Bignona River), the species zonation has been either disturbed or entirely changed. *Rhizophora* species (*R. racemosa* and *R. mangle*) have suffered the most from dryness and have simply disappeared from the area and have been replaced by herbaceous patches of *Sesuvium portulacastrum* and *Paspalum vaginatum*. The *Avicennia* community has also been greatly affected and has widely been invaded by *S. portulacastrum*. However, though only a few bushes of *Avicennia* have survived, it was noted in 1978 that *Avicennia* seedlings were recolonizing the areas previously dominated by *Avicennia*. The case of *Rhizophora* is different as they are unable to recover from the effects of dryness within a short period of time.

These changes in vegetational patterns and species composition have been related to simultaneous edaphic modifications. Changes in soil properties are both morphological and chemical (increase of salinity, decrease of pH, etc.). To some extent, it appears that the natural response of mangrove forests to exceptionally dry years is comparable to the artificial effects of periodic diversion of freshwater inputs.

It is noteworthy that climatic parameters, including their variability, affect mangroves in a variety of ways such as physiological impacts of persistent dry winds, increase in water and soil salinity, chemical changes in groundwater, etc. None of these mechanisms is fully understood at present.

Study of winds and microclimatology

No systematic studies of wind have been carried out in relation to mangrove biology. However, information concerning prevailing direction and speed are usually obtainable from data which has been collected from local meteorological stations. To some extent, biological properties of winds (mainly the drying power of the air) is integrated into Penman's formula. However, the mechanical effect of wind is far from being negligible as has been emphasized for almost every important mangrove ecosystem of the world, for several reasons.

1. Sand-carrying winds rapidly modify the coastal morphology and deeply affect the evolution of adjacent mangrove ecosystems, causing trees to be progressively buried; this is particularly evident along the southwestern coast of Madagascar.
2. The prevailing winds impart an almost constant direction to longshore currents, transporting and depositing a considerable amount of clay, sand and silt (for example, the eastern Indian coasts, Colombia, etc.) (It is to be noted that winds also determine pollen dissemination patterns.)
3. The impact of tropical cyclones and hurricanes on the eastern coastlines of tropical landmasses is extremely important.

Data concerning coastal winds must include at least the following components:

(a) prevailing monthly direction and mean monthly velocity; and (b) inventory of the number, dates and strength of cyclonic storms, which have impacted the mangrove area during the previous decades. This inventory can throw some light on the explanation of the present status of some mangrove stands, not only because some of them are bluntly destroyed by storm surges but also because the dispersal of propagules either inland or far away from the parent trees is often explained by the effect of these surges. For in the New World, as Chapman (1975, p. 213) has indicated,

only Borgesen (1909) reported destruction of mangrove forest by hurricanes in the Danish West Indies; Davis (1940) recorded the destruction of Florida mangroves in the hurricanes of 1926, 1929 and 1935. Chapman (1944) suggested that hurricanes could account for relict patches of dead mangrove in Jamaica and, more recently, Egler (1950) stressed the importance of this factor.

Cyclonic storms are particularly numerous and dangerous to the mangroves and people around the Bay of Bengal mainly in India and Bangladesh (Blasco, 1975). For the mangroves of the Cauvery in south India, Venkatesan (1966, p. 30) has given the following account:

During the cyclone of November 1952, which was accompanied by a tidal wave, the whole of Muthupet and Chattram Forests were overrun by the sea and remained submerged to a depth of about 2 m for a fortnight. As a result of this, *Avicennia* was killed over large areas. The young growth below 3 years, a few large trees in the older crops and the growth on narrow belts along the creeks have, however, survived this onslaught.

Cyclonic winds and associated tidal surges are common phenomena in New Caledonia (Baltzer, 1975), northern Australia (Chapman, 1977, p. 297), Reunion and Madagascar islands, etc.

The intensity of cyclones, in relation to their impacts on the mangroves, can be roughly classified as minor (defoliation of some mangrove components), severe (partial destruction of the ecosystem), and destructive (almost all mangrove trees are killed). So far, only minimal research in this field has been undertaken.

Microclimatic studies

It is obvious that most macroclimatic studies are based on data provided by meteorological stations and that the published observations of humidity, temperature, wind, etc., may differ from those prevailing in the canopies of the various vegetation types. However, local and microclimatic measurements are still extremely rare in swampy areas for several reasons, among which is the logistic difficulty of the task.

In addition, it is undeniable that microclimatic conditions are narrowly related to physiological problems which are usually studied in climatically controlled chambers.

Nevertheless, for research workers who are interested in simulation, vertical profiles of air temperature and humidity through mangrove stands, or for those who desire to study the diffuse solar radiation at several levels of mangrove canopies, it is suggested that reference be made to the work conducted at the Department of Biology, San Diego State University, California (Miller, 1973, 1975).

Conclusions

It is true that among abiotic ecological factors, soil salinity and soil structure, and hydrological conditions, are the main agents controlling the distribution of mangroves. However, it is uncontrovertible that temperature sets the latitudinal distribution of halophytes and that regional rainfall and evapotranspiration are essential factors because of their direct biological implications as well as their decisive influence on soils and pedogenesis.

It is suggested that each mangrove ecosystem must be characterized by its climatic identity card which would integrate all fundamental climatic factors that have been discussed in the text.

In order to provide a graphic expression of the average seasonality of each wet and dry period, it is recommended that climatic diagrams, such as those of Bagnouls and Gaussen (1953) or Walter and Leith (1960–67), be used. However, these diagrams, which express the length of dry seasons, do not give information concerning the degree of aridity. That is why each mangrove climatic identity card must give the monthly values of the potential evapotranspiration computed with Penman's formula as has been done by Unesco (1979). It would be possible to classify the climatic or natural (almost untouched by man) mangroves of the world, or of each continent or country according to the degree of aridity of their site. Naturally, other important climatic data which determine some soil properties, distribution of species, adaptative characters, etc., should also be given on the card.

Finally, the degree of stability of the main climatic factors is expressed by a synoptic table devoted to the variability in precipitation, length of the dry season, number of rainy days and potential evapotranspiration.

With regard to water temperatures, the available data are extremely fragmentary. Presumably, cold marine currents may play an inhibitory role, but to what extent? There is a strong feeling among West African 'mangalogists' that the mangroves of the Kouilou estuary (Congo) are poorly developed because of the close proximity of the cold Benguela current (Moguedet, 1980) which is responsible also for the aridity of West African coastal climates in Angola and southward.

References

BAGNOULS, F.; GUASSEN, H. 1953. Saison sèche et indice xérothermique. *Documents pour les cartes des productions végétales* (Toulouse), Vol. 3, pp. 1–47.

BALTZER, F. 1975. Solution of Silica and Formation of Quartz and Smectite in Mangrove Swamps and Adjacent Hypersaline Marsh Environments. In: G. Walsh, S. Snedaker and H. Teas (eds.), *Proceedings of International Symposium on the Biology and Management of Mangroves*, pp. 482–98. Gainesville, University of Florida.

BLASCO, F. 1975. Mangroves of India. *Inst. Fr. Pondichery, Trav. Sect. Sci. Tech.*, Vol. 14, p. 180.

BRETELER, F. J. 1977. America's Pacific Species of *Rhizophora*. *Acta Bot. Neerl.*, Vol. 26, pp. 225–30.

CHAPMAN. V. J. 1975. *Mangrove Vegetation*. Leutershausen, Strauss & Cramer. 425 pp.

——. (ed.). 1977. *Ecosystems of the World; Wet Coastal Ecosystems*. New York, Elsevier. 428 pp.

DAVIS, J. H. 1940. The Ecology and Geologic Role of Mangroves in Florida. *Pap. Tortugas Lab.*, Vol. 32, pp. 303–412. (Carnegie Institution Publication, No. 517.)

EGGERS, BARON H. VON. 1892. Die Manglares in Ecuador. *Botanisches Centralblatt*, Vol. LII, pp. 49–52.

JENNINGS, J. N.; BIRD, C. F. 1967. Regional Geomorphological Characteristics of some Australian Estuaries. In: G. H. Lauff (ed.), *Estuaries*. Washington, D.C., AAAS.

KIENER, A. 1973. Les mangroves du globe. Aspects écologiques, biocénotiques et physiologiques particuliers; mise en valeur. *Bull. Mus. Hist. nat.*, No. 164, pp. 319–31.

KUCHLER, A. W. 1968–70. *International Bibliography of Vegetation Maps*. Vol. 3: *USSR, Asia, Australia*. 389 pp. Vol. 4: *Africa, South America and World Maps*, 561 pp. University of Kansas Libraries.

LUGO, A. E.; ZUCCA, C. P. 1977. The Impact of Low Temperature Stress on Mangrove Structure and Growth. *Trop. Ecol.*, Vol. 18, No. 2, pp. 149–61.

MARIUS, C. 1979. Effets de la sécheresse sur l'evolution pytogéographique et pédalogique de la mangrove en basse Casamance (Senegal). *Bull. I.F.A.N.*, Vol. 41, No. 4, pp. 669–91. (Series A.)

MILLER, P. C. 1973. A Model of Temperatures, Transpiration Rates and Photosynthesis of Sunlit and Shaded Leaves in Vegetation Canopies. *Plant Response to Climatic Factors*, pp. 427–34 (Proceedings of Symposium 21–26 September 1970, Uppsala, Sweden). Paris, Unesco.

——. 1975. Simulation of Water Relations and Net Photosynthesis in Mangroves in South Florida. *Proceedings of International Symposium on the Biology and Management of Mangroves*, Vol. 2, pp. 615–31. Gainesville, University of Florida.

MOGUEDET, G. 1980. *Le milieu de mangrove au Congo*, pp. 3–20. Talence, CEGET. (Travaux et documents de géographie tropicale, No. 39.)

MOORMAN, F. R.; PONS, L. J. 1975. Characteristics of Mangrove Soils in Relation to their Agricultural Land Use and Potential, *Proceedings of International Symposium on the Biology and Management of Mangroves*. Vol. 2, pp. 529–47. Gainesville, University of Florida.

PANNIER, F.; PANNIER, R. F. 1977. Interpretación fisioecológica de la distribución de manglares en las costas del continente Suramericano. *Interciencia*, Vol. 2, No. 3, pp. 153–62.

PAPADAKIS, J. (ed.). 1965. [Potential Evapotranspiration], Buenos Aires. 54 pp.

—— (ed.). 1966. [Climates of the World and their Agricultural Potential], Buenos Aires. 170 pp.

PENMAN, H. L. 1956. Evaporation: An Introductory Survey. *Neth. J. Agr. Sci.*, Vol. 4, pp. 9–29.

POOL, D. J.; SNEDAKER, S. C.; LUGO, A. E. 1977. Structure of Mangrove Forests in Florida, Puerto Rico, Mexico and Costa Rica. *Biotropica*, Vol. 9, No. 3, pp. 195–212.

SAENGER, P.; SPECHT, M. M.; SPECHT, R. L.; CHAPMAN, V. J. 1977. Mangal and Coastal Saltmarsh Communities in Australasia. In: V. J. Chapman (ed.), *Ecosystems of the World. Wet Coastal Ecosystems*, pp. 293–345. New York, Elsevier. 428 pp.

SIDHU, S. S. 1975. Structure of Epidermis and Stomatal Apparatus of some Mangrove Species. In: G. Walsh, S. Snedaker and H. Teas (eds.), *Proceedings of International Symposium on the Biology and Management of Mangroves*, pp. 569–78. Gainesville, University of Florida.

THORNTHWAITE, C. W. 1948. An Approach towards a Rational Classification of Climate. *Geogr. Rev.*, Vol. 38, pp. 55–94.

TOMLINSON, P. B. 1978. *Rhizophora* in Australasia. Some Clarification of Taxonomy and Distribution. *Journal of the Arnold Arboretum*, Vol. 59, pp. 156–69.

TURC, L. 1954. Le bilan de l'eau des sols. *Annales agronomiques* (Paris), Vol. 5, pp. 491–596; Vol. 6, pp. 5–131.
UNESCO. 1979. *Map of the World Distribution of Arid Regions* (1:25,000,000). Paris, Unesco. 54 pp. (MAB Technical Notes, 7.)
——. 1981. *Vegetation Map of South America* (1:15,000,000; 2 sheets with explanatory book). 183 pp. (*Natural Resources Research*, XVII.)
VAN STEENIS, C. G. G. J. 1955–58. *Flora Malesiana*. Noordhoff Ltd.
VENKATESAN, K. R. 1966. The Mangroves of Madras State. *Indian Forester*, Vol. 92, No. 1, pp. 27–34.
WALTER, H. 1971. *Ecology of Tropical and Subtropical Vegetation*. Edinburgh, Oliver & Boyd. 539 pp.
——. 1977. Climate. In: V. J. Chapman (ed.), *Ecosystems of the World. Wet Coastal Ecosystems*, pp. 61–7. New York, Elsevier.
WALTER, H.; LIETH, H. 1960–67. *Klimadiagramm–Weltatlas*. Jena, Veb. Gustav Fisher Verlag.
WEST, R. C. 1956. Mangrove Swamps of Pacific Coast of Colombia. *Ann. Ass. Amer. Geogr.*, Vol. 46, pp. 98–121.
——. 1977. Tidal Salt-marsh and Mangal Formations of Middle and South America. In: V. J. Chapman (ed.), *Ecosystems of the World. Wet Coastal Ecosystems*, pp. 193–211. New York, Elsevier.

Bibliography

ANON. Acid Sulphate Soils. In: H. Dost (ed.), *Proceedings of the International Symposium of the International Institute for Land Reclamation and Improvement, August 1972, Wageningen*. 2 vols.
JOSHI, G. V. *Studies in Photosynthesis under Saline Conditions*. Kolhapur (India), Shivaji University 1976. 195 pp.
KASSAS, M. On the Ecology of the Red Sea Coastal Land. *J. Ecol.*, Vol. 45, 1957, pp. 187–203.
KASSAS, M.; ZAHRAN, M. A. Studies on the Ecology of the Red Sea Coastal Land. II: The District from El Galada El Quibliya to Hurghada. *Bull. Soc. Géogr. Égypte*, Vol. 37, 1965, pp. 155–93.
PENMAN, H. L. *Vegetation and Hydrology*. Farnham Royal, Commonwealth Agricultural Bureau No. 53. 1963.
THOM, B. G.; WRIGHT, L. D.; COLEMAN, J. M. Mangrove Ecology and Deltaic-Estuarine Geomorphology: Cambridge Gulf–Ord River, Western Australia. *J. Ecol.*, Vol. 63, 1975, pp. 203–32.
WALSH, G. E. Mangroves: A Review. In: R. J. Reimold and W. H. Queen (eds.), *Ecology of Halophytes*, pp. 51–174. New York, Academic Press. 1974. 605 pp.
WEISS, H.; KIENER, A. Observations rélatives à la nature chimique des eaux de la région de Tulear. *Tethys Suppl.*, Vol. 1, 1971, pp. 215–36.

3

Mangrove evolution and palynology

François Blasco
Institut de la Carte Internationale du Tapis Végétal
39 allée Jules-Guesde
31400 Toulouse, France

Introduction

Several attempts have been made to deduce from palynological studies either the presence of mangroves in the past and their evolution or simply the location of ancient shorelines in a geomorphological context. Until recently, it has been agreed that palynology has been one of the most sensitive tools for an evaluation of any floristic change in a mangrove area such as: (a) sudden or gradual vanishing of one or several species in a given locality; (b) reduction of number of individuals; and (c) species substitution and/or colonization by alien species.

The aim of palynological research is to relate pollen, spores and other microfossils found in soil samples to the source vegetation. The interpretation of reconstructed flora assemblages and vegetational development usually leads to palaeo-environmental conclusions, as it is generally agreed that changes in vegetation, whether climatic, edaphic or biotic, reflect local ecological changes.

In any event, the reconstruction of vegetation types on the basis of pollen analysis and pollen diagrams is a highly specialized task that can be carried out only by qualified palynologists who work in experienced laboratories, and have a collection of recent acetolysed pollen slides prepared from duly identified herbarium material. A reference collection is absolutely necessary for the identification of microfossils. As far as tropical coastal ecosystems are concerned, five main departments of palynology devote part of their activities to mangrove studies:

Rijksherbarium, Schelpenkade 6, Leiden, The Netherlands.

Centre D'Études de Géographie Tropicale (CEGET), Domaine Universitaire de Bordeaux, 33405 Talence, France.

Université des Sciences et Techniques du Languedoc, Laboratoire de Palynologie, 2 place Eugène Bataillon, 34060 Montpellier Cedex, France.

Bose Institute, Division of Palynology and Environmental Biology, 93/1, Acharya Prafulla Chandra Road, Calcutta 700 009, India.

French Institute, P.B. 33, Pondichery 605 001, India.

In these laboratories, sediment samples are generally prepared using the chemical techniques described by Faegri and Iversen or with variations and adaptations appropriate to each laboratory or specific type of the sediments.

Pollen analysis, however, has its own defaults and insufficiencies.

It is a time-consuming technique, requiring skillful and highly specialized scientists and technicians.

The identification (diagnosis) of leaving (and *a fortiori* fossil) pollen grains often lacks in precision (generic level only) or remains dubious.

Considerable intraspecific pollen variability exists in certain groups (e.g. *Rhizophora racemosa*).

Investigations on aeropalynology and waterborne pollen are very scarce; very little is known regarding production, dispersal and fossilization capacities of the various species.

Many problems remain to be solved regarding the contamination of samples; the age of the sample by radiocarbon dating is not always the same as that of the pollen grains, particularly when the sediments have been reworked.

The exact relation between the fossil-pollen-spore composition and its corresponding vegetation cover is often hypothetical.

These methodological problems have been discussed by several palynologists and biogeographers: Fagerling (1952), Muller (1959), Faegri (1966), Groot and Groot (1966), among others.

Nevertheless, because of the rather well-defined ecological conditions and the simplified and characteristic floristic composition, mangrove vegetation is one of the most suitable ecosystems for which palynology can help in tracing history and evolution. In this case, the investigator can interpret the results on the basis of qualitative rather than quantitative modifications. Naturally, one can imagine the simplest case in which the pollen of mangroves species (*Rhizophora*, *Avicennia*, *Sonneratia*) completely disappear at one level, being replaced by that of other species. In this case, the interpretation is simple; it reflects the disappearance of mangroves and its parallel replacement by another vegetation type.

Many reports, dealing with the pollen analysis of samples collected at various depths from all over the tropical world, have been published. Very few, however, have been devoted to the vegetational history of estuaries and coastal zones. We should consider as pioneer works the analytical investigations of: Muller (1959) in the Orinoco delta, Venezuela; Aoutin (1967) in the Ogoue delta, Gabon; Assemien (1969) in Bogue, Senegal; Bartlett and Barghoorn (1973); Blasco and Caratini (1973); and Tissot (1979) in Cauvery, India, with the classical masterpiece being Muller's study.

Pollen grains and pollen spectra

The majority of pollen grains produced by mangrove trees and shrubs are of stratigraphic interest. This is particularly the case for *Rhizophora*, *Sonneratia* and *Avicennia*. In some cases like *Excoecaria agallocha* the male plants seem to produce a very small quantity of pollen grains that are rare in sediments. The inability to find *E. agallocha* pollen grains in sediments thus does not prove that the species was absent. Moreover, at least forty species of *Excoecaria* L. are presently known from the tropics (Asia and Africa); most of them are terrestrial glycophytes. Therefore, the identification of their pollen beyond the generic level is meaningless. A similar example is that of *Xylocarpus* Koenig (syn. *Carapa* Aubl.), belonging to the Meliaceae; distinguishing between their pollen grains with those from members of other families, such as the Sapotaceae, is hazardous, if not impossible.

In any case, the description and identification key must be extremely complete and accurate to permit a reliable identification of fossil pollen grains. The present knowledge on the pollen morphology of mangrove species is summarized below:

Acanthaceae: *Acanthus ilicifolius* L. (Blasco and Caratini, 1973; Tissot, 1979)
Avicenniaceae: *Avicennia* (Blasco and Caratini, 1973; Rao and Tian 1974; Sowunmi, 1974; Tissot, 1979).
Combretaceae: *Lumnitzera* (Huang, 1972; Blasco and Caratini, 1973).
Euphorbiaceae: *Excoecaria agallocha* L. (Tissot, 1979).
Meliaceae: *Xylocarpus* (Pennington and Styles, 1975).
Myrsinaceae: *Aegiceras corniculatum* (L.) Blanco (Dunbar and Erdtman, 1969; Blasco and Caratini, 1973).
Palmae: *Nypa fruticans* Wurmb. (Muller, 1961; Thanikaimoni, 1970; Thanikaimoni, et al., 1974); *Phoenix paludosa* Roxb. (Thanikaimoni, 1970).
Plumbaginaceae: *Aegialitis* (Friedrich 1956; Mallik and Chaudhuri, 1968).
Rhizophoraceae: *Rhizophora mucronata*, *R. stylosa*, *R. apiculata*, *R. lamarckii*, *R. mangle*, *R. racemosa* (Muller and Caratini, 1977); *Bruguiera cylindrica*, *B. gymnorrhiza*, *B. parviflora*, *B. sexangula*, *Ceriops decandra*, *Kandelia candel* (Tissot, 1979).
Rubiaceae: *Scyphiphora hydrophyllacea* (Mitra, 1969; Leopold, 1969).
Salvadoraceae: *Salvadora* (Bonnefille, 1971; Maheswari Devi, 1972).
Sonneratiaceae: *Sonneratia* (Muller, 1969).
Sterculiaceae: *Heritiera fomes* (Mallik and Chaudhuri, 1968; Sharma 1969; Tissot, 1979).

Practically all pollen of the mangrove species have been studied, either with the light microscope (LM)—magnification 1,000, resolution 0.1 μm or 1,000 Å, or with the scanning electron microscope (SEM)—magnification about 40,000 or even more, resolution 60–100 Å), or with the transmission electron microscope (TEM)—magnification 200,000–300,000, practical resolution 5–10 Å.

Pollen grains of *Rhizophora* and *Sonneratia*

In spite of extremely accurate descriptions, pollen grains of a given genera of mangroves cannot be identified to the specific level. The best examples are provided by *Rhizophora* and *Sonneratia* members. The pollen morphology of seven species of *Rhizophora* (*mucronata, stylosa, apiculata, lamarckii, mangle, racemosa* and *harrisonii* which is the presumed hybrid between *R. racemosa* and *R. mangle*) have been intensively investigated by Muller and Caratini (1977) from pollinic material taken from herbarium sheets identified by Ding Hou, specialist in this taxonomic group. Their conclusion is that it is 'extremely difficult to find characters which unequivocally permit the recognition of species of *Rhizophora* on their pollen' (p. 36). The members of this genus are noted as being prolific pollen producers.

Nevertheless, the presence of Rhizophoraceae pollen in a sediment implies a warm climate and coastal alluvial deposition because the greatest fraction of pollen grains produced by *Rhizophora* members are dispersed by tidal currents, deposited and fossilized in a coastal area; few of them are carried airborne (anemochorous) or by bees (entomophilous). Presently, the main difficulty is that it is not known the extent to which these pollen grains are separable from those of *Carallia*, a related genera (ten species known in Madagascar, the Indo-Malaysian region and northern Australia) which are terrestrial Rhizophoraceae, sometimes with aerial roots and commonly encountered in evergreen forests, far from coastal areas. In addition, pollen grains of *Bruguiera*, *Ceriops* and *Rhizophora* are very much alike and their separation is a difficult task, for the time being. Records of fossil Rhizophoraceous pollen types are abundant. In Venezuela, Nigeria, Malaysia and Borneo, the earliest occurrence seems to be in the Oligocene.

The main characteristics of *Rhizophora* pollen types, as they can be observed with a light microscope, have been reported by Muller and Caratini (1977, p. 379):

Tricolporate, 15–30 μm, equatorially elongated endoaperture.

Sculpture always fine, elements 1 μm.

Columellae somewhat longer and more clearly visible on poles than on equatorial zone.

Endoapertures bordered over their entire length by costae, from which feature the generic name Zonocostites [Germeraad et al., 1968] for fossil dispersed pollen grains of this type has been derived; in optical section, the asymmetrical rounded shape of the costae is also characteristic.

With regard to *Sonneratia*, which is represented by five living species (*apetala, caseolaris, ovata, alba, griffithii*), intensive studies have been carried out by Muller (1969) on their pollen morphology. This genus plays an essential role in the mangrove ecosystems of tropical Asia. Moreover, *S. alba* is recorded from East Africa, northern Madagascar and northern Australia. It appears that *Sonneratia* pollen grains have an extremely complex morphology which is mainly due to the presence of a poorly understood source of intraspecific pollen variability, mostly in *S. alba, S. caseolaris* and, to a lesser extent, in *S. ovata*. Muller is of the opinion that these differences are based on the biogeography of plants and probably also differences among the genotypes. This indicates the need for research on the variability of pollen morphology on individual trees according to the position of branches, flowering season, temperature and humidity conditions, etc.

Meanwhile, a rather simple morphological description has been given by Muller (1969, p. 225):

Basically the pollen of *Sonneratia* is single, subprolate and has three equatorial porate apertures. The apertures are closed by an ektexinous membrane. The wall is radially differentiated into four layers. Tangential differentiation of structure and sculpture offers the main diagnostic characters by which the pollen of the five species can be recognized. When fresh, the pollen is dry or slightly sticky and has a pale yellowish-white colour. It is released at dusk when the flowers open and produced in copious amounts. According to Faegri and van der Pijl (1967)[1] the flowers are bat-pollinated.

Muller's key for the five living species, based on the main differences in the pollen characters, is the following:

I. Tectum continuous over entire grain . . . . *S. apetala*
II. Tectum broken up into separate verrucae on porate fields:
   1. Polar caps without distinct structure, meridional ridges absent . . . *S. caseolaris*
   2. Polar caps with distinct intra-aerolate or columellate structure, meridional ridges generally present.
      A. Polar tectum irregularly rugulate-verrucate, columellate structure indistinct . . . . *S. ovata*
      B. Polar tectum psilate, columellate structure distinct on equatorial belt:
         (a) intra-aerolate structure distinct and columellate reduced on polar caps, meridional ridges irregularly rugulate or grooved, large porate fields . . . . *S. alba*
         (b) columellate distinct and intra-aerolate structure reduced on polar caps, meridional ridges pilate, small porate fields *S. griffithii*

Some members of the Lythraceae have closely related pollen types but the primary source of confusion is associated with the young undeveloped pollens of *Sonneratia caseolaris* and those of *Duabanga* species which are terrestrial Sonneratiaceae in the Indo-Malaysian region.

With reference to the largest mangrove genus, *Avicennia*, a few species have been palynologically investigated (Caratini, 1974) and no attempt has been made to characterize the pollen of each species. It is a difficult task from the palynological point of view, as well as the taxonomic (Moldenke, 1960).

Conocarpus and *Laguncularia* (Combretaceae) produce atypical pollen grains, difficult to separate from those of the Boraginaceae (e.g. *Heliotropium*) or from the terrestrial Combretacae like *Terminalia*, commonly associated with

1. Reference cited in Muller (1969).

beach ridge vegetation. In some cases, however, the distinction seems to be possible (Riegel, 1965).

Examples of pollen diagrams and interpretations

Palynological studies of core sediments from mangrove areas have been conducted in the Americas, Africa and Asia for the purpose of investigating vegetation changes during the recent past. The following comments may be considered as indicative of our present knowledge in the various parts of the world.

THE AMERICAS

Bartlett and Barghoorn (1973) have given a very didactic explanation of the practical methods presently used for sample preparation-preservation, and on pollen, diagram construction and interpretation. They state that, 'Pollen diagrams have been constructed in the traditional manner, with the percentage of each sporomorph plotted against the depth of the sample from which it was extracted.' Among the significant results obtained in the Isthmus of Panama is the change in the pollen diagram, about 5,000 years ago, from domination by *Rhizophora* pollen grains to a clear dominance of freshwater plant pollen. *Rhizophora* pollen is not present in samples taken from above 8–9 metres below current sea-level. This means that there was a natural substitution of the mangroves along the banks of Gatun Lake, near Escobal, by freshwater swamp vegetation (location of the core: 9°11′ N., 79°57′ W.).

In his masterly work on transport and deposition of pollen grains in the Orinoco delta, Muller (1959) reached several fundamental conclusions, among which are the following, summarized by Bartlett and Barghoorn (1973, p. 284):

(1) Pollen assemblages in analyzed sediments accurately reflect the ecological frequency of *Rhizophora*, and ecological boundaries similarly are generally fairly accurately reflected in the pollen assemblages of surface sediments.

(2) The percentage of pollen of *Rhizophora* plus *Avicennia* in a mangrove swamp may be between 45 and 95%. Narrow fringes of mangrove have lower percentages, such as 30% (Van der Hammen, 1963).[1] Nearshore sediments in front of mangrove swamps may contain from about 30 to 50% *Rhizophora* pollen.

(3) Sediments in swamp forests immediately behind the mangrove may contain 45–10% or less *Rhizophora* pollen, while those farther inland may have 10% or less (Van der Hammen, 1963).

(4) Large, heavy grains such as those of *Symphonia* (Guttiferae) tend to settle close to the coast line, and are rarely found in offshore sediments in any appreciable percentage (Van der Hammen, 1963). The lighter and smaller grains, on the other hand, are transported more easily, and may settle far from shore.

1. Reference cited in Bartlett and Barghoorn (1973).

WEST AFRICA

Assemien (1969) has used fossil pollen grains of *Rhizophora* in order to determine the local climatic and sea-level changes at Bogue, on the lowest part of the Senegal River. At present, Bogue is located at about 250 km inland on the northern river bank (Mauritania). Local climatic conditions are arid according to Unesco's classification. First of all, Assemien (1969) compared the fossil pollen grains of *Rhizophora* with those of the three West African living species: *R. racemosa*, *R. mangle* and *R. harrisonii*; his main conclusion is that these fossil mangrove pollen grains are excellent indicators of sea-level change and that some 5500 B.P., during a marine transgression of recent Quaternary (Nouakchottian), the coastline was located in the vicinity of the Bogue City.

Assemien (1971) also investigated the evolution of the vegetation near Abidjan (Agneby; Ivory Coast) which is partially summarized in Figure 3.1. It is clear from this pollen sequence that a mangrove (*Rhizophora racemosa* and *Avicennia nitida*) and a swamp forest (dominance of *Mitragyna* spp. and *Symphonia globulifera*) coexisted in the same area based on the assemblages in the oldest sediments investigated (core taken from 500 cm dated 4900 B.P.). Very noticeable changes appear at a depth of 275 cm with a pronounced accumulation of mangrove elements which disappear at the level of 150 cm. The total predominance of swamp forest elements is established from 200 cm upwards.

ASIA

For Asia there are several studies conducted in India that are of interest. In the Cauvery delta region, an attempt was made to correlate the present vegetation types with the representation of their pollen grains in recent sediments (Caratini et al., 1973). It appears that the pollen distribution in the recent sediments do not accurately reflect the actual floristic composition. This deviation is chiefly due to an over-representation of the pollen grains of Rhizophoraceae and Sonneratiaceae members whereas the pollen grains of *Avicennia officinalis* L. and *A. marina* (Forsk.) Vierh., found in the sediments, are far below that expected. Both of the latter species are common or dominant, whereas the percentage of their pollen encountered does not exceed 2 per cent; we do not know yet the main cause of this rarity (e.g. poor pollen production, rapid decay in some circumstances, etc.). Almost the same result was obtained with *Excoecaria agallocha*.[1] Some other pollen grains are simply not represented. Apparently abnormal nonrepresentation has been established for species like *Aegiceras corniculatum*, *Clerodendrum inerme* and *Derris trifoliata* which are common in the area but their pollen grains, though easily identifiable, were not found in the sediments. An apparently normal representation

1. Muller (1959) and Wiijmstra (1971) found very high percentages of *Rhizophora* pollen grain in their surface samples from the Orinoco and Guiana coastal basin near the mangrove fringe but they do not comment on a possible over-representation.

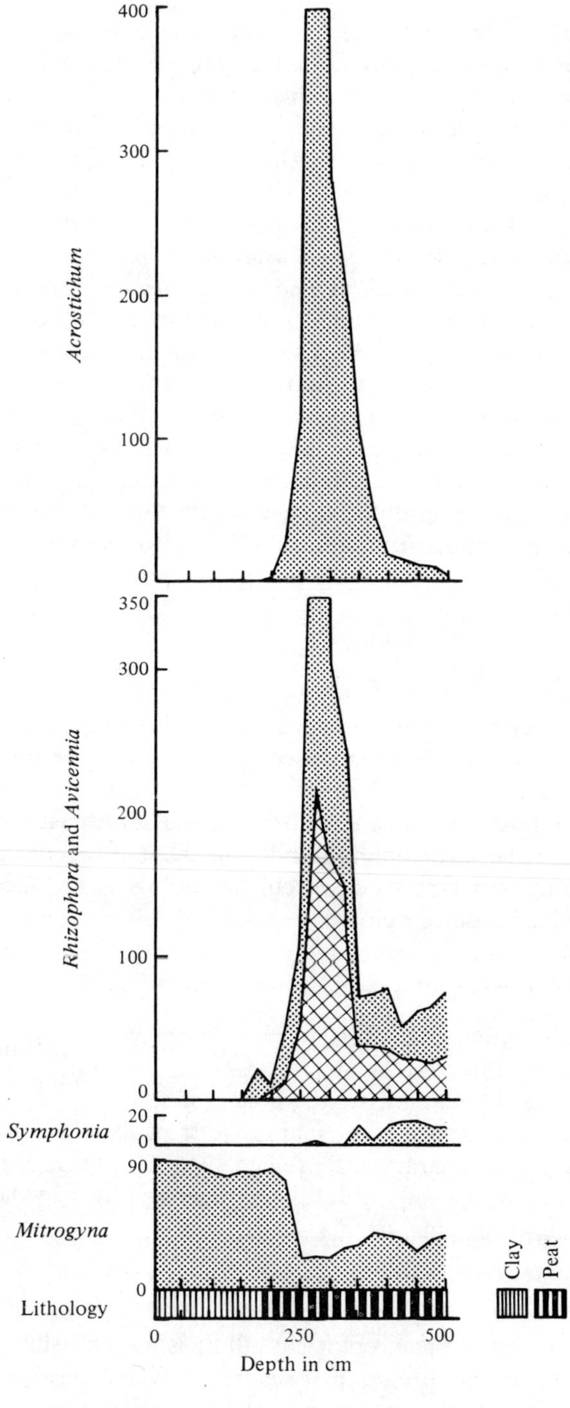

Figure 3.1
Pollen diagram at Agneby (Core No. 2) according to Assemien (1971).

was obtained with Chenopodiaceae (probably *Suaeda marina*) which are very common in the area and are equally abundantly represented in the pollen record.

Three other main categories of pollen grains were also found among the allochthonous pollens, produced mainly in the neighbouring vegetation types, within a radius of less than 10 km from the sites studied. This category includes the Gramineae and Cyperaceae which are relatively rare in the intertidal mangrove areas. Miscellaneous pollen grains are those produced either by cultivated or planted species (*Cocos*, *Casuarina*, *Borassus*, etc.) or by quasi-pantropical terrestrial weeds such as *Boerhavia*, *Jatropha*, *Tribulus*, and members of the Compositae. Unidentified pollens, presumably of allochthonous origin, represented about 8 per cent of the pollen grains found in the sediment.

An attempt by Tissot (1979) to study the evolutionary trends of mangrove vegetation in the same delta has reported two distinct stratigraphic trends, according to the location of the cores (see Fig. 3.2). Although for technical reasons no radio-carbon age determinations were obtained, theoretical considerations based on stratigraphic data indicate that the contrasting trends relate to vegetation changes from about 2000 B.P. to the present.

Comments and conclusions

Few palynological contributions to the history and evolution of mangrove vegetation have been published. However, due to the simplified and somewhat atypical flora, mangrove vegetation is certainly suitable for intensive palaeopalynological research. One of the best examples is provided by Figure 3.3. in which a deep drill hole provides an excellent picture of the past evolution as well as the present and the forthcoming trend(s) in the natural coastal ecosystem.

In conducting this type of research, palynologists are faced with three main difficulties, as outlined below.

IDENTIFICATION

In most instances, either each fossil type matches exactly the pollen of only one living species (in this case the identification can be considered as absolutely reliable), or the fossil type matches the pollen of several modern species. This is a frequent problem even if some mangrove trees like *Rhizophora* and *Sonneratia* produce a very characteristic pollen type. However, a good reference collection and knowledge of the local phytogeography may help in solving most identification problems.

METHODOLOGY

Since plant communities have ecological affinities, it is possible to classify the pollen assemblages in the core samples according to ecological groups of pollen to form the basis for a reconstruction of the past vegetation. This requires an adequate knowledge of the local phytogeography based on field observations

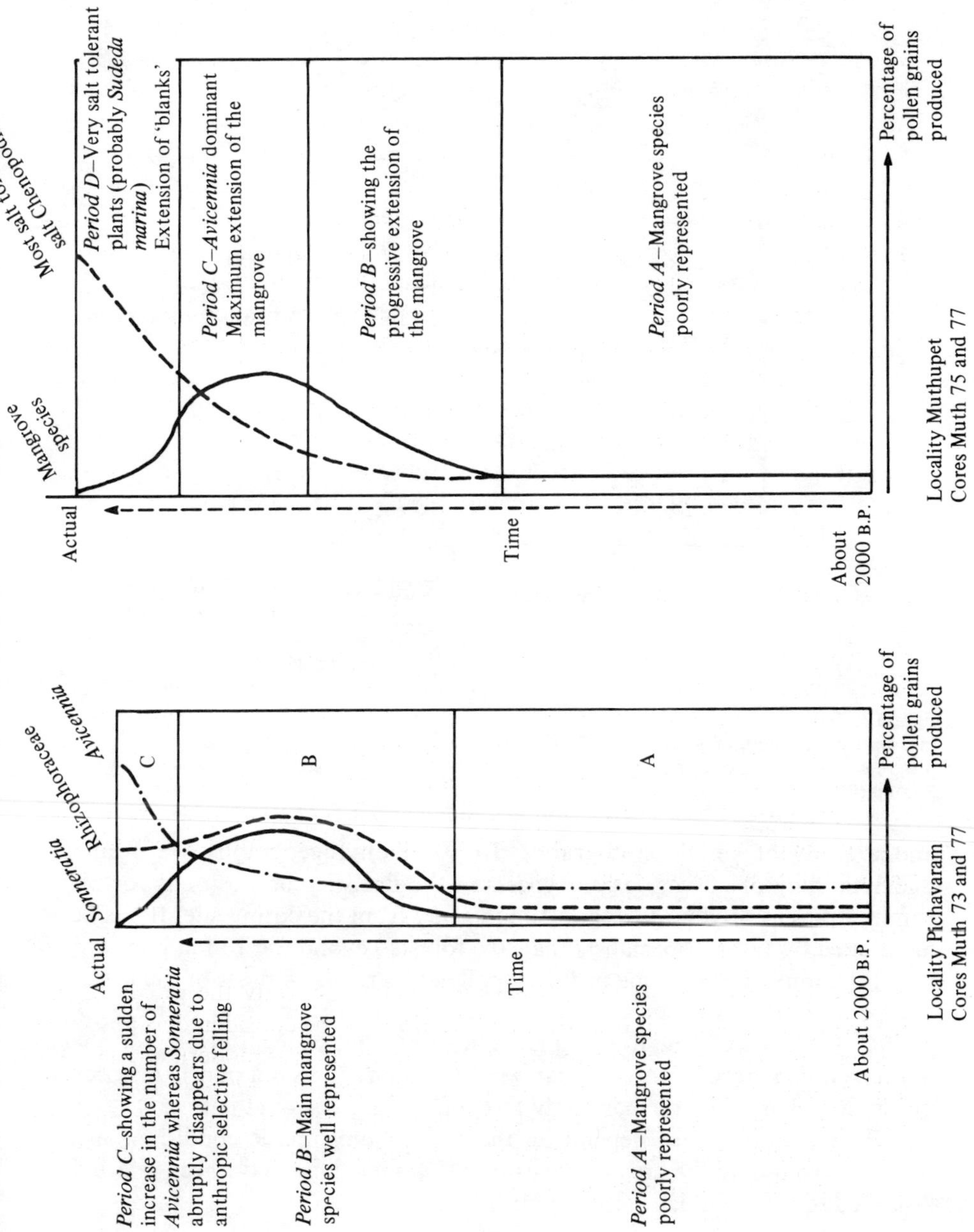

Figure 3.2
The evolution of mangrove vegetation in the Cauvery delta (southern India), according to Tissot (1979).

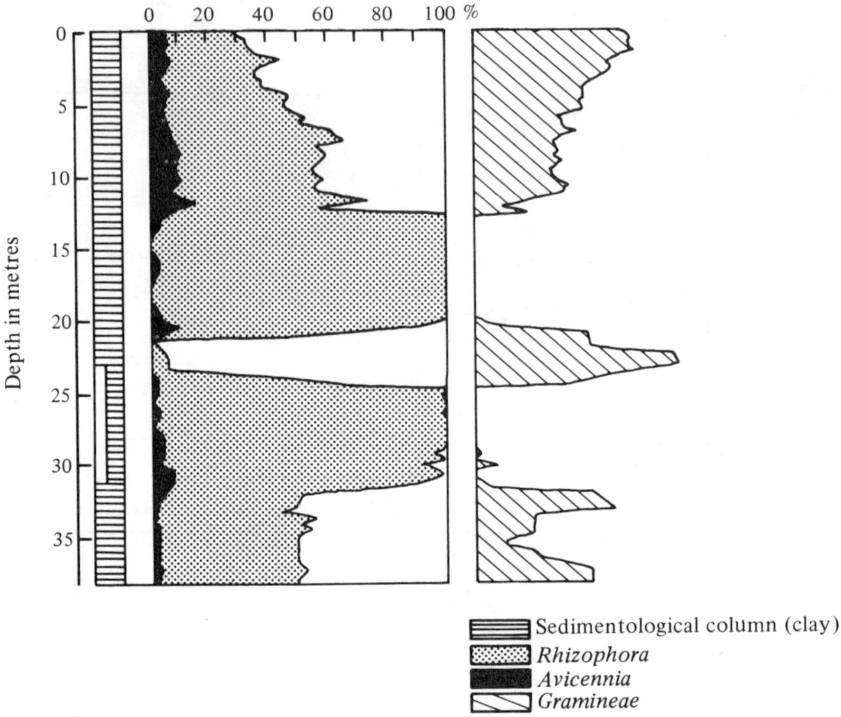

Figure 3.3
Part of a pollen diagram (Core No. T28A), extracted from Wijmstra (1971) showing the progressive vanishing of a mangrove ecosystem in the Guiana Coastal Basin (west-north-west Paramaribo)

and a study of aerial photographs. By local phytogeography we mean an understanding of all vegetation types and the floristic compositions for plant communities located no less than 10 km away from the coring site. It has been mentioned by several specialists that 'the forests beyond the 10-km limit are of very little importance to the ordinary pollen diagram and most of the material will be derived from sources much nearer the site' (Faegri and Iversen, 1964, p. 104). In the case of mangrove plants, it seems that water transport of pollen grains is much more important than wind transport. This is a poorly researched question which needs further study to explain distributional patterns.

Pollen source areas depend on the present bioclimatic conditions in the region surrounding the area of deposition. Two instances have been selected in Asian countries to illustrate this point.

Delta of the Cauvery

In the Cauvery delta of south-eastern India, the climatic conditions are dry and the pollen spectra of Pichavaram mangrove areas reveal three main groups in the stratigraphic sedimentary record (Table 3.1).

TABLE 3.1

Pollen type	Main botanical group	Ecological group of pollen
Autochthonous pollens	Rhizophoraceae, *Avicennia* *Sonneratia*, *Excoecaria*	1. Mangrove group
Allochthonous pollens	Cyperaceae (fresh water herbaceous swamp and weeds in irrigated fields)	2. Herbaceous secondary vegetation types
	Gramineae (grasslands or crops: *Oriza*, *Saccharum*, *Cymbopogon*, etc.)	3. Cultivated plants and weeds of cultivated fields
	Cultivated plants and weeds (*Cocos*, *Borassus*, Compositae, *Tribulus*, *Justicia*, *Jatropha*)	

In this case, pollen diagrams reflect the total absence of any kind of terrestrial forest in the study area. Such a distribution of pollen grains, showing an impressive number of weeds mixed with cultivated plants, must be a general feature on Indian shores due to the fact that the country is densely populated and all arable land is cultivated, particularly in the deltas. In this case, the interpretation of pollen diagrams is rather simple and the evolution of mangrove vegetation in the course of time can be followed without much difficulty, both in floristic diversity and in density of species.

Delta of the Musi River

In the delta of the Musi River (south-eastern Sumatra) where human interference is minimal, the natural evolution of the mangrove vegetation is a slow process. Present research and field observations have shown that main sources of pollen which influence the distribution include at least six ecological groups, namely:

1. Mangrove group (*Rhizophora*, *Sonneratia*, *Excoecaria*, etc.)
2. Black mangrove group (*Nypa*, *Pandanus*, *Gluta*, *Oncosperma*, etc.)
3. Swamp forest (*Shorea*, *Campnosperma*, *Dyera*, *Gonistylus*, etc.)
4. *Melaleuca* swamps (*Melaleuca*, Melastomaceae, Verbenaceae, Dilleniaceae, etc.)
5. Secondary open swamp forests (*Macaranga*, *Licuala*, *Staenochloena*, etc.)
6. Cultivated plants and weeds

Thanks to these ecological groups of pollen grains, and in spite of the floristic complexity of the area, the evolution of each vegetation type presently found in the delta, particularly the mangroves, can be easily determined.

TECHNICAL

Technical difficulties are often due to the fact that pollen diagrams represent the vegetation of a small area near the site of deposition. Moreover, according to the type of sediment and its exact location, the rate of deposition and pollen concentration fluctuate significantly. Another cause of pollen diversity from core to core is that mangrove pollen grains are mainly transported by streams and subsequently by coastal currents. In some sites, pollen and spores reworked from older deposits may also affect the interpretation of a pollen diagram.

Under these circumstances and in order to avoid a wrong interpretation of pollen diagrams, it is absolutely necessary to evaluate local pollen transport by water and to utilize a large number of cores for the study of each delta. The interpretation must include data concerning the relationship between the pollen spectra of surface soil samples and those found in stratigraphic core samples.

Finally, the term 'microfossil' involves a large variety of micro-organisms. Until now, the identification of the pollen itself has been the main concern of palynologists. No doubt, however, that several other taxonomic groups such as fern spores, fungal spores, algal remains and animal microfossils (e.g. smaller foraminiferans) are of practical interest in the study of fossil sediments. Research concerning these groups is at present carried out by only a few specialists and the research completed is exceptional on its own merits. However, it frequently fails to take into consideration the associated pollen stratigraphy, a problem that identifies a promising area of expanded research development in mangrove palynology.

References

AOUTIN, J. 1967. Les pollens et les spores des sédiments du delta de l'Ogoue (Gabon). *Mem. E.P.H.E.* (Montpellier). 160 pp.

ASSEMIEN, P. 1969. Pollens fossiles de *Rhizophora* à Bogue (Basse vallée du Sénégal). *Pollen et Spores*, Vol. XI, No. 1, pp. 73–81.

——. 1971. Étude comparative des flores actuelles et quaternaires récentes de quelques paysages végétaux d'Afrique de l'Ouest. Abidjan, Université d'Abidjan. 257 pp. (Thesis.)

BARTLETT, A. S.; BARGHOORN, E. S. 1973. Phytogeographic History of the Isthmus of Panama during the Past 12,000 Years. In: Alan Graham (ed.) *Vegetation and Vegetational History of Northern Latin America*, pp. 203–99. Amsterdam, Elsevier.

BLASCO, F.; CARATINI, C. 1973. *Mangrove de Pichavaram, phytogéographie et palynologie*. Talance, CEGET. (Travaux et documents de géographie tropicale, No. 8.)

BONNEFILLE, R. 1971. Atlas des pollens d'Ethiopie. Pollens actuels de la basse vallée de l'Omo récoltes botaniques, 1968. *Adansonia*, Vol. 11, pp. 463–518.

CARATINI, C. 1974. Leguminosae, Sapotaceae, Verbénaceae. *Pollen et spores d'Afrique tropicale*, pp. 178–9, 186, 189, 241, 243, 264–5. Talence, Association des Palynologues de Langue Française/CEGET. (Travaux et documents de géographie tropicale, No. 16.)

CARATINI, C.; BLASCO, F.; THANIKAIMONI, G. 1973. Relation Between the Pollen Spectra and the Vegetation of a South Indian Mangrove. *Pollen et Spores*, XV, pp. 281–92.

DUNBAR, A.; ERDTMAN, G. 1969. On the Pollen Grains of *Aegiceras corniculatum*. *Grana palynologica*, Vol. 9, pp. 63–71.

FAEGRI, K. 1966. Some Problems of Representativity in Pollen Analysis. *The Palaeobotanist*, Vol. 15, pp. 135–40.

FAEGRI, K.; IVERSEN, J. 1964. *Text Book of Modern Pollen Analysis*, 2nd rev. ed. Munksgaard.

FAGERLING, F. 1952. The Real Signification of Pollen Diagrams. *Bot. netiser*, pp. 185–224.

FRIEDRICH, H. C. 1956. Studien über die natürliche Verwandtschaft der Plumbaginales und Centrospermae. *Phyton*, Vol. 6, pp. 220–63.

GERMERAAD, J. H.; HOPPING, C.; MULLER, J. 1968. (Cited in Muller and Caratini, 1977.) Palynology of Tertiary Sediments from Tropical Areas. *Review Palaeobot., Palynol.*, Vol. 6, pp. 189–348.

GROOT, J. J.; GROOT, C. R. 1966. Marine Palynology: Possibilities, Limitations, Problems. *Marine Geology*, Vol. 4, pp. 387–95.

HUANG, T. C. 1972. *Pollen Flora of Taiwan*. Taipeh, National Taiwan University Botany Department Press. 297 pp.

LEOPOLD, F. B. 1969. *Miocene Pollen and Spore Flora of Eniwetok Atoll, Marshall Islands*, pp. 1133–85, pl. 304–11. Washington D.C., US Government Printing Office. (Geological Survey Professional Paper, No. 260.)

MAHESWARI DEVI, H. 1972. Salvadoraceae. A Study of its Embryology and Systematics. *J. Indian Bot. Soc.*, Vol. 51, pp. 56–62.

MALLIK, H.; CHAUDHURI, S. L. 1968. Palynological Studies of the Sunderban Flora for Identification of Peat Pollen. *Bull. Bot. Soc. Bengal*, Vol. 22, No. 1, pp. 105–8.

MITRA, K. 1969. Pollen Morphology and Identification of Gardenieae. In: J. Sen (ed.) *Memorial Volume*, pp. 223–8. Botanical Society of Bengal.

MOLDENKE, H. N. 1960. Materials toward a Monograph of the Genus *Avicennia*. *Phytologia*, Vol. 7, pp. 123–293.

MULLER, J. 1959. Palynology of Recent Orinoco Delta and Shelf Sediments. *Micropaleontology*, Vol. 5, No. 1, pp. 1–32.

——. 1961. A Palynological Contribution to the History of the Mangrove Vegetation. *Abstract of Symposium Papers*, pp. 146–7. Honolulu, Tenth Pacific Science Congress.

——. 1969. A Palynological Study of the Genus *Sonneratia*. *Pollen et Spores*, Vol. XI, No. 2, pp. 223–98.

MULLER, J.; CARATINI, C. 1977. Pollen of *Rhizophora* (Rhizophoraceae) as a Guide Fossil. *Pollen et Spores*, Vol. XIX, No. 3, pp. 361–89.

PENNINGTON, T. D.; STYLES, B. T. 1975. A Generic Monograph of the Meliaceae. *Blumea*, Vol. 22, pp. 419–540.

RAO, A. N.; TIAN, O. E. 1974. Pollen Morphology of Certain Tropical Plants. *J. Palynology*, Vol. X, pp. 1–37.

RIEGEL, W. 1965. *Palynology of Environments of Peat Formation in South Western Florida*. Ann Arbor, Mich., University Microfilms Inc. 190 pp.

SHARMA, B. D. 1969. Studies of Indian Pollen Grains in Relation to Plant Taxonomy. Sterculiaceae. *Proc. natn. Inst. Sci. India*, Vol. B.35, pp. 320–59.

SOWUNMI, M. A. 1974. Pollen Grains of Nigerian Plants. I: Woody species. *Grana*, Vol. 13, pp. 145–86.

THANIKAIMONI, G. 1970. Les palmiers: palynologie et systèmatique. *Tr. Sect. Sci. Tech. Inst. Fr. Pondichery T.*, Vol. 11. 286 pp.

THANIKAIMONI, G.; CARATINI, C.; LUGARDON, B. 1974. Palmae. *Pollen et spores d'Afrique tropicale*, pp. 218–19. Talence, Association des Palynologues de Langue Française/CEGET. (Travaux et documents de géographie tropicale, No. 16.)

TISSOT, C. 1979 *Palynologie et évolution récente de deux mangroves du Tamil Nadu (Inde)*. Montpellier, École Pratique des Hautes Études, Palynologie, Université des Sciences et Techniques du Languedoc. 118 pp.

WIJMSTRA, T. A. 1971. *The Palynology of the Guiana Coastal Basin*. Oegstgeest., Drukkevrij de Krempenaer. 62 pp.

Bibliography

GRAHAM, A. (ed.) *Vegetation and Vegetational History of Northern Latin America*. Elsevier, Amsterdam, 1973. 393 pp.

MULLER, J. A Palynological Contribution to the History of the Mangrove Vegetation in Borneo. In: L. M. Cranwell (ed.) *Ancient Pacific Floras*, pp. 33–42. Honolulu, University of Hawaii Press, 1964.

SOLOMON, A. M.; KING, J. E.; MARTIN, P. S.; THOMAS, J. Further Scanning Electron Photomicrographs of Southeastern Pollen Grains. *Arizona Acad. Sci.*, Vol. 8, 1973, pp. 135–57.

THANIKAIMONI, G. Index bibliographique sur la morphologie des pollens d'Angiospermes. Supplément 2. *Inst. Fr. Pondichery. Trav. Sec. Sci. Tech.*, Vol. 13, 1976, pp. 1–386.

Part II

Community structure and description

Part II

4

Botanical surveys in mangrove communities

Valentine J. Chapman
In memoriam

Species composition

The predominant plants in a mangrove swamp are the trees, and it is these species that are generally listed. However, phanerogamic epiphytes may occur as well as some phanerogams appearing as ground flora, especially towards high-tide mark. Any study of a mangrove swamp should take cognizance of all these plants. In addition, there is the occasional fern, e.g. *Acrostichum aurem*, *A. speciosum* and a range of marine algae, lichens and fungi. The former are usually found on the pneumatophores and trunk bases and the latter two on the leaves and twigs. If at all possible, a list of all species involved should be compiled together with a note on their habitat.

Table 4.1 is a review of the distribution of mangrove phanerogams from mangroves in the major regions where they occur. Workers in the past have varied in their interpretation of what can be included as a mangrove (these are the tree species) when one is considering the landward edge of a swamp or the transition to freshwater conditions.

Table 4.2 is a review of the marine algae that have been recorded from mangrove swamps in various parts of the world. This is an area of study that, together with the lichens and fungi, has not as yet received much study and data are wanted from all parts of the world. Table 4.3 reviews the lichen flora and Table 4.4, the fungal flora.

In order to establish a complete species list, a survey must be made of the entire swamp under study. Any species not readily identifiable in the field should be collected; this applies especially to fungi, algae and lichens. On return to the laboratory, species should be determined by reference to appropriate keys and by checks with herbarium material, if available. Species that belong to difficult genera are best referred to an authority in the relevant field. This generally will involve sending adequate material overseas. If mud samples are taken at different depths, they should be analysed for pollen (see Chapter 3) as such information can lead to a past history of the swamp.

TABLE 4.1. Distribution of mangrove phanerogams

Species	Family	Pacific America	Atlantic United States	West Africa	East Africa	Madagascar	Red Sea	India	Sri Lanka	Burma	Malaysia	Indonesia	Borneo	Papua New Guinea	Thailand and Viet Nam	Australia	Ryukyus	Philippines	Taiwan	China	Pacific Islands	New Zealand
Rhizophora mucronata	Rhizophoraceae				+	+	+	+	+	+	+	+	+	+		+	+	+			+	
apiculata								+	+	+	+	+	+	+	+	+		+		+	+	
stylosa											+	+	–	+		+		+	+		+	
mangle		+	+	+																	+	
harrisonii		+	+	+																		
(includes *brevistylosa*)																						
Rhizophora racemosa			+	+																		
lamarckii																					+	
Bruguiera gymnorrhiza					+	+	+	+	+	+	+	+	+	+	+	+	+	+			+	
cylindrica								+	+	+	+	+	+	+	+			+				
exaristata														+		+						
parviflora								+		+	+	+	+	+	+	+		+				
sexangula								+			+	+	+	+	+	+		+	+	+		
hainesii											+											
Ceriops tagal					+	+	+	+	+	+	+	+	+	+	+	+		+				
decandra								+	+	+	+	+	+	+	+			+			+	
Kandelia candel								+	+	+	+	+	+	+			+		+	+	+	
Avicennia marina	Avicenniaceae				+	+	+	+	+	+	+	+	+	+	+	+	+	+	+	+	+	
officinalis										+			+					+				
lanata											+											
alba											+	+										
balanophora															+							
eucalyptifolia												+				+		+				
tonduzii		+																				
bicolor		+																				
germinans				+																		
africans		+	+																			
schaueriana			+																			
Xylocarpus mekongensis	Meliaceae							+		+	+	+	+									
granatum					+	+		+	+	+	+	+	+	+	+	+	+	+			+	
moluccensis					+	+		+	+	+	+	+	+	+	+	+	+	+			+	
Amoora cucullata								+		+												
Laguncularia racemosa	Combretaceae	+	+	+																		
Conocarpus erecta		+	+	+																		
Lumnitzera racemosa					+	+		+	+	+		+		+		+	+	+	+			
var. *lutea*												+										
littorea								+		+	+	+	+	+	+	+	+	+			+	
Terminalia catappa										+												
sertocarpa															+							
Camptostemon schultzii	Bombacaceae													+		+						
philippinense													+					+				
Aegialites annulata	Plumbaginaceae											+		+		+						
rotundifolia								+		+	+											
Aegiceras corniculatum								+	+	+	+	+	+	+	+	+		+		+		
floridum												+		+			+	+			+?	
Sonneratia caseolaris	Sonneratiaceae							+	+	+	+	+	+	+	+	+		+			+	
ovata											+	+		+	+							
alba					+	+		+	+	+	+	+	+	+		+	+	+			+	
apetala								+	+													
griffithii											+	+										
S. alba × *S. caseolaris*													+			+						
× *S. ovata*													+									
Scyphiphora hydrophyllacea	Rubiaceae							+	+	+	+	+	+	+	+	+	+	+			+	
Dolochandrone spathacea	Bignoniaceae							+	+	+	+	+	+	+		+					+	
Cydista equinoctialis			+																			

TABLE 4.1.—*continued*

Species	Family	Pacific America	Atlantic United States	West Africa	East Africa	Madagascar	Red Sea	India	Sri Lanka	Burma	Malaysia	Indonesia	Borneo	Papua New Guinea	Thailand and Viet Nam	Australia	Ryukyus	Philippines	Taiwan	China	Pacific Islands	New Zealand
Excoecaria agallocha	Euphorbiaceae							+	+	+	+	+			+	+		+			+	
Pelliciera rhizophorae	Theaceae	+																				
Heritiera fomes	Sterculiaceae							+	+	+												
littoralis					+	+			+			+										
Osbornia octodonta	Myrtaceae															+						
Derris heptaphylla	Leguminosae							+								+		+			+	
heterophylla										+		+				+						
Cyanometra ramiflora								+	+	+												
iripa												+										
Drepanocarpus lunatus			+	?																		
Intsia bijuga											+	+									+	
Nypa fructicans	Palmae							+	+	+	+	+	+	+	+	+		+			+	
Oncosperma filamentosa											+	+	+		+			+				
horrida												+										
Euterpe cuatrecasana		+																				
Phoenix reclinata					+																	
spinosa				+																		
paludosa								+														
Raphia vinifera			+	+																		
Pemphis acidula	Lythraceae																				+	
Sarcolobus carinatus	Asclepiadaceae							+														
Cynanchium carnosum																+						
Dischidia saccata													+									
Acrostichum aureum	Polypodiaceae	+	+	+	+			+		+	+							+			+	
speciosum														+		+					+	
Dendrobium rhizophoreti	Orchidaceae										+											
Bulbophyllum xylocarpi											+											
Clerodencron inerme	Verbenaceae							+		+	+	+						+			+	
Acanthus ilicifolius	Acanthaceae		+			+		+	+	+	+	+	+	+	+	+	+	+		+		
ebracteatus																		+			+	
speciosum																		+				
Cerbera odollam	Apocynaceae							+		+	+											
manghas											+											
Brownlowia argentata	Tiliaceae												+					+				
lanceolata													+								+	
Rhabdadenia biflora	Apocynaceae		+																			
Pavonia scabra	Malvaceae		+																			
rhizophorae		+																				
Thespesia acutissima					+	+																
populnea		+	?															+?			+?	
Hibiscus tiliaceus		+	+	+	+	+																
Brachypteris ovata	Malpighiaceae		+																			
Tillandsia spp.	Bromeliaceae	+	+																			
Conostegia polyandra	Melastomaceae	+																				
Rustia occidentalis	Rubiaceae	+																				
Mymecodia tuberosa												+										
Muelleria frutescens	Leguminosae	+																				
Ardisia granatensis	Myrsinaceae	+																				
Tubelostylis rhizophorae	Compositae	+																				
Pandanus candelabrum	Pandanaceae			+																		
Dodonaea viscosa	Sapindaceae		+	+																		
Amyema gravis	Loranthaceae										+											
Loranthus quinquenervis											+		+									
Viscum ovalifolium												+										

Source: Chapman (1975).

TABLE 4.2. Marine algae in mangroves

Family and species	Sumatra	Singapore	Celebes	Fiji	Samoa	Australia	United States	Brazil	Venezuela	West Africa	New Zealand
Cyanophyceae											
Microcoleus tenerrimus							+				
Lyngbya maiuscula							+				+
Sirocoleum guyanense							+				
Chlorophyceae											
Enteromorpha sp.	+										+
Ulva sp.	+										
Rhizoclonium sp.	+										
Lola tortuosa											+
Cladophora limicola					+	+				+	
fuliginosa									+		
Boodlea sp.	+	+								+	
Chaetomorpha sp.										+	
Vaucheria sp.											+
Caulerpa fastigiata								+	+		
racemosa									+		
sertularioides							+				
taxifolia							+				
verticillata							+				
Batophora oerstedii							+		+		
Valonia ventricosa									+		
Phaeophyceae											
Ectocarpus mitchellae								+			
Dictyota sp.	+		+								
Dictyopteris propagulifera	+	+	+	+							
Hotmosira banksii var. *limicola*											+
Rhodophyceae											
Acanthophora spicifera									+		
Bostrychia radicans	+	+		+				+	+	+	
tenella	+	+									
calliptera	+	+							+		+
mita f. inermis	+										
harveyi						+	+				+
kelanensis	+	+		+							
moritziana		+	+								
tenuis		+									
simpliuscula									+	+	
binderi		+									
scorpioides											+
var. *montagnei*											
Caloglossa lepriewrii	+			+		+	+	+	+		+
var. *hookeri*	+	+									
bombayensis	+	+									
stipitata	+	+									
beccarii	+	+									
adnata	+	+		+							
agasawarensis				+							
Catenella nipae	+	+				+					+
impudica	+	+							+		
repens								+		+	
Centroceras clavulatum								+			
Caulacanthus indicus		+									
Chondria riparia	+	+									
Gelidium caulacantheum											+
Gracilaria secundata f. pseudoflagellifera							+				+
Lophosiphonia sp.	+	+									
Murrayella periclados		+						+			

TABLE 4.3. Mangrove lichens

Bulbothrix apophysata
goebelii
isidiza
tabacina
Caldonia macilenta
Coccocarpia cronia
erythroxyli
pellita
Collema laeve
glaucophthalmum
var. *implicatum*
rugosum
Dirinaria applanata
aspera
confluens
consimilis
melanoclina
picta
Heterodermia tremulans
obscurata
speciosa
Hypotrachyna immaculata
Leptoginum phyllocarpum var. *isidiosum*
Menegazzia circumsorediata
globulifera
Normandia pulchella
Pannaria elatior
lurida
mariana
Parmelia erumpens
borreri var. *coralloides*
subrudecia
Parmelina aurelenta
damaziana
spumosa
aff. tiliacea
Parmotrema austrosinense
crinita
cristifera
disparile
perlata
permutatum
rampoddense
reticulata
robustum
saccatilobum
subtinctorium
tinctorum
Physcia adscendens
caesia
Physcia dipolia
tribacoides
Physciopsis adglutinata
Physma byrsinum
Pseudoparmelia caperata
carneopruinata
aff. crozalsiana
rahengensis
rutidota
soredianus
texana
Pyxine berteriana
caestropruinosa
cocoes
aff. physciaeformis
retirugella
subcinerea
Ramalina celestri
disparata
duriaei
exiguella
aff. fecundata
inflata
aff. leiodea
myrioclada
peipusilla
peruviana
reagens
subpusilla
tenella
Relicina amphithrix
circumnodata
limbata
samoensis
subabstrusa
sublanca
sydneyensis
Teloschistes chrysophthalmus
flavicans
steberianus
Ushea baileyi
leprosa
propinqua
ramulossissima
rubescens
rubicunda
indulata
Xanthoria ectanea

Source: Stevens (1979); Stevens and Rogers (1979).

TABLE 4.4. Fungi associated with mangroves

MARINE FUNGI		
Phycomycetes	*Phytophthora* spp.	*A. marina* var. *resinifera* (N.Z.).
Ascomycetes	*Buellia haliotrepha*	*R. mangle, A. germinans, Hibiscus tiliaceus*
	Didymosphaeria enalia	*R. mangle, A. germinans, A. africana*
	Didymosphaeria rhizophorae	*R. mangle*
	Gnomonia longirostris	*A. marina* var. *resinifera*
	G. marina	*A. marina* var. *resinifera*
	Haligena viscidula	*R. racemosa*
	Halosphaeria quadricornuta	*R. mangle, A. marina* var. *resinifera, Hibiscus tiliaceus*
	Heliascus kanaloanus	*R. mangle*
	Hydronectria tethys	*R. mangle, A. germinans, Hibiscus tiliaceus*
	Keissleriella blepharospora	*R. mangle*
	Leptosphaeria avicenniae	*A. germinans, A. africana*
	Lignincola laevis	*R. mangle, Hibiscus tiliaceus*
	Lulworthia sp.	*R. mangle, R. racemosa, A. germinans, A. africana, A. marina* var. *resinifera, Hibiscus tiliaceus*
	Metasphaeria australiensis	*A. germinans, A. marina* var. *resinifera, Hibiscus tiliaceus*
	Mycosphaerella pneumatophorae	*A. germinans, A. africana*
	Ophiobolus australiensis	*A. marina* var. *resinifera*
	Paraliomyces lentiferus	*R. mangle, A. germinans*
	Torpedospora radiata	*R. mangle, A. germinans*
	Trematosphaeria mangrovis	*R. racemosa*
Deuteromycetes	*Cirrenalia pygmea*	*R. racemosa*
	C. tropicalis	*R. racemosa*
	Culcitalna achraspora	*R. racemosa, A. germinans*
	Cytospora sp.	*R. mangle, R. racemosa*
	Halocyphina villosa	*R. mangle*
	Phialophorophoma litoralis	*A. marina* var. *resinifera*
	Phoma sp.	*R. mangle, R. racemosa, A germinans, A. africana, Hibiscus tiliaceus*
	Robillarda rhizophorae	*R. mangle*
Basidiomycetes	*Nia vibrissa*	*R. mangle*
TERRESTRIAL FUNGI		
Myxomycetes	*Arcyria cinerea*	*R. mangle*
Ascomycetes	*Anthostomella rhizomorphae*	*R. mangle*
	A. rhizophorae	*R. mangle*
	Botryosphaeria ribis	*Laguncularia racemosa*
	B. ribis var. *chromogena*	*L. racemosa, Hibiscus tiliaceus*
	Eudimeriolum avicenniae	*Avicennia* sp.
	*Irene laguncularia*e	*L. racemosa*
	I. sepulta	*Avicennia*
	Meliola nigra	*L. racemosa*
	M. triumfettae	*H. tiliaceus*
	*Micropeltis laguncularia*e	*L. racemosa*
	Mollisia petiolarum	*H. tiliaceus*
	Patellaria atrata	*H. tiliaceus*
	Phyllachora minuta	*H. tiliaceus*
	Physalospora fusca	*H. tiliaceus*
	P. hibisci	*H. tiliaceus*
	*P. laguncularia*e	*L. racemosa*
	P. rhizophorae	*R. mangle*
	Physalosporopsis rhizophoricola	*R. mangle*
	*Schizothyrium laguncularia*e	*L. racemosa*
Deuteromycetes	*Ascochytella rhizophoropsis*	*R. mangle*

TABLE 4.4—*continued*

	Botrytis argillacea var. *avicenniae*	*Avicennia*
	Cercospora hibisci	*H. tiliaceus*
	C. hibiscina	*H. tiliaceus*
	C. rhizophorae	*R. mangle*
	Colletotrichum hibiscicola	*H. tiliaceus*
	'Diplodia natalensis'	*H. tiliaceus*
	Helminthosporium glabroides	*L. racemosa*
	Leptothyrium rhizophorae	*R. mangle*
	Pestalotia disseminata	*R. mangle*
	P. guepini	*R. mangle*
	P. longiaristata	*R. mangle*
	P. versicolor	*R. mangle*
	P. Zahlbruckneriana	*R. mangle*
	Phoma rhizophorae	*R. mangle*
	Phomopsis rhizophorae	*R. mangle*
	Sphaeronaema avicenniae	*Avicennia*
Basidiomycetes	*Fomes avicenniae*	*Avicennia*
	Phellinus gilvus	*R. mangle*
	Psathyrella sp.	*R. mangle*
	Schizophyllum commune	*R. mangle, Avicennia*
	Trametes rhizophorae	*R. mangle*
	Tulasnella bifrons	*Hl tiliaceus*
	T. pacifica	*H. tiliaceus*
	T. violea	*H. tiliaceus*

Source: Chapman (1976)

In the final account of any swamp, it is recommended that all species be listed and placed in the relevant genera and families.

Where the swamp abuts on to elevated land, all species that occur up to high-water mark of spring tides should be included. In places where the mangroves give way to riverine forest or swamp, all species should be recorded up to the point where distinctly saline water (1 per cent NaCl) ceases to be found in the lower water strata of the main channel at high water of spring tides. In both cases, it is espically desirable that species occurring in the zone between mean low-water mark and high-water mark of spring tides should be recorded separately.

Local floras should be used for identification of macroalgae and fungi. A key to the Australian lichens (Table 4.5) may serve others in this region.

COLLECTING AND PRESERVING SPECIMENS

Past workers have generally collected specimens of the higher plants and after drying and pressing, they have been deposited in a local or overseas herbarium. There may be cases where macroalgae have been collected and preserved as herbarium specimens.

In the case of the angiosperms, it is most desirable that samples of leaves, flowers and fruits/seedlings should be collected and preserved. If trees or plants are not in flower, visits should be made at intervals until they can be collected. Twigs with leaves and flowers should be dried between sheets of newspaper or blotting paper, either placed under weights or contained within a plant press.

TABLE 4.5. Key to Australian macrolichens

1.	Thallus fruticose, i.e. without dorsiventral differentiation, round or ribbon like	Section I
	Thallus foliose, i.e. with dorsiventral differentiation, leaflike, flat with obvious lobes	2
2.	Thallus containing green algae	Section II
	Thallus containing blue-green algae	Section III

Section I. Fruticose lichens

1.	Thallus branches with a central axis	2
	Thallus branches without a central axis	8
2.	Axis hollow	*Usnea baileyi*
	Axis solid	3
3.	Thallus tinged with red	4
	Thallus green, yellow-green or yellow-orange	5
4.	Medullary reaction K + yellow	*Usnea rubicunda*
	Medullary reaction K + yellow turning red	*Usnea rubescens*
5.	Medullary reaction K + yellow turning red	6
	Medullary reaction K + yellow or K−	7
6.	Fibrils dense on branches, simple	*Usnea ramulosissima*
	Fibrils dense and branched	*Usnea undulata*
	Fibrils absent or few	*Usnea propinqua*
7.	Medulla lax and twice as wide as axis	*Usnea leprosa*
	Medulla not as above	8
8.	Thallus colour yellowish-orange or orange	*Teloschistes flavicans*
	Thallus colour green, or yellow-green	9
9.	Thallus inflated and perforate	*Ramalina perpusilla*
	Thallus not inflated and perforate	10
10.	Thallus sorediate	11
	Thallus not sorediate	13
11.	Soredia bursting from stems and apices	*Ramalina peruviana*
	Soredia not bursting from apices	12
12.	Thallus rarely with apothecia but always with marginal and laminal soredia	*Ramalina reagans*
	Thallus regularly bearing apothecia	13
13.	Thallus branches subterate, narrow, ending in pointed tips, often blackened	*Ramalina exiguella*
	Thallus branches canaliculate or broad and slightly canaliculate	*Ramalina* aff. *leiodea*

Section II: Foliose lichens with green algae

1.	Cilia present	2
	Cilia absent	13
2.	Thallus with ciliate isidia	*Parmotrema crinita*
	Thallus without ciliate isidia	3
3.	Soredia present	4
	Soredia absent	5
4.	Thallus markedly maculate	*Parmotrema reticulata*
	Thallus not maculate, or only slightly	*Parmotrema rampoddense*
5.	Cilia simple	6
	Cilia bulbate	11
6.	Medulla pale yellow	*Parmelina aurulenta*
	Medulla white	7

TABLE 4.5.—*continued*

7.	Isidia present	8
	Isidia absent	10
8.	Isidia pustulate	*Parmelina spumosa*
	Isidia cylindrical	9
9.	Lobes narrow and very small	*Parmelina* aff. *tiliaceae*
	Lobes broad and large	*Parmotrema subtinctoria*
10.	Medullary reactions K –, C –, KC + pink, P –	*Parmelina damaziana*
	Medullary reactions K + yellow turning red, P + orange	*Parmotrema parahypotropum*
11.	Thallus yellow green	*Relicina sydneyensis*
	Thallus greenish-grey, green or whiteish-grey	12
12.	Medullary reaction K + yellow turning red	*Bulbothrix tabacina*
	Medullary reaction K –	*Bulbothrix goebelii*
13.	Isidia present	14
	Isidia absent	17
14.	Medulla yellow in upper part, white below	*Pyxime physciaeformis*
	Medulla not as above	15
15.	Medullary reactions K –, C –, KC + rose, P + orange	*Parmotrema saccatilobum*
	Medullary reactions not as above	16
16.	Lobes narrow and very small	*Parmelina* aff. *tiliaceae*
	Lobes broad and large	*Parmotrema tinctorum*
	Lobes not as above	17
17.	Soralia absent	18
	Soralia present	19
18.	Apothecia present, black with black margin	*Pyxine berteriana*
	Apothecia present, black with white margin	*Dirinaria confluens*
19.	Rhizines black, densely dichotomously branched	*Hypotrachyna immaculata*
	Rhizines pale or dark, simple or squarrously branched	21
20.	Thallus maculate	*Pyxine retirugella*
	Thallus yellow-green	*Pseudoparmelia caperata*
	Thallus greenish-grey or greyish-white	*Heterodermia obscurata*
21.	Soralia marginal on broad sinuate lobes	*Parmotrema austrosinense*
	Soralia marginal	*Heterodermia tremulans*
	Soralia on thallus ridges, coalescing	*Pseudoparmelia* aff. *crozalsiana*
	Soralia not as above	22
22.	Cortex K + yellow	23
	Cortex K –	27
23.	Thallus maculate and reticulate	*Parmelia erumpens*
	Thallus not maculate	24
24.	Medullary reaction P + orange	*Parmotrema robustum*
	Medullary reaction P –	25
25.	Sekikaic acid present	*Dirinaria consimilis*
	Sekikaic acid absent	26
26.	Apothecia present, disc red-pruinose	*Dirinaria melanoclina*
	Apothecia absent	*Pseudoparmelia texana*
27.	Lower surface pale	*Physcia tribacoides*
	Lower surface black at centre	28
28.	Thallus lobes contiguous, apices discrete	*Dirinaria picta*
	Thallus lobes confluent to apices	29

TABLE 4.5.—*continued*

29.	Soralia laminal, initially as minute verrucae becoming small crateriform	*Dirinaria aspera*
	Soralia laminal, globose-capitate, sometimes coalescing	*Dirinaria applanata*
	Soralia not as above	30
30.	Medulla pale yellow	*Pyxine subcinerea*
	Medulla white	31
31.	Cortex UV + yellow	*Pyxine cocoes*
	Cortex UV−	*Physciopsis adglutinata*
Section III. Foliose lichens with blue-green algae		
1.	Thallus gelatinous when wet	2
	Thallus not gelatinous when wet	5
2.	Thallus lead-grey	*Leptogium phyllocarpum* var. *isidiosum*
	Thallus dark green or black	3
3.	Thallus isidiate	*Collema rugosum*
	Thallus not isidiate	4
4.	Apothecia with white pruinose disc	*Collema laeve* var. *laeve*
	Apothecia with red to black disc	*Collema glaucophthalmum* var. *implicatum*
5.	Thallus dark brown mottled grey, very thick	*Physma byrsinum*
	Thallus pale brown to light grey, thin	6
	Thallus lead-grey	7
6.	Thallus with isidia-like soredia	*Pannaria elatior*
	Thallus without isidia-like soredia	*Pannaria lurida*
7.	Thallus isidate	*Coccocarpia cronia*
	Thallus without isidia	8
8.	Thallus lobes with zoned rings	*Coccocarpia erythroxili*
	Thallus lobes wedge-shaped	*Coccocarpia pellita*

Source: Stevens and Rogers (1975).

The drying paper needs to be changed daily at first and less frequently later. When the specimens are completely dry they should be mounted on herbarium sheets and properly labelled. Where small fruits or seedlings (viviparous species) are produced they can be dried as above and mounted on the herbarium sheet. With large fruits (e.g. *Sonneratia*) and seedlings (e.g. *Rhizophora, Bruguiera*) preservation is best achieved by using 3 per cent formalin in bottles. In the case of herbarium sheets, protection from future damage by insects or fungi can be achieved by exposing the sheets to mercuric chloride. Any ferns collected can be dried as above and mounted. Wherever possible, sporulating and vegetative fronds should be collected. Lichens, after collection, should be dried and then stored in small packets attached to herbarium sheets and properly labelled. Macroalgae found on the mud or on pneumatophores are preserved by floating out in a bowl of fresh water and then allowed to subside on to a herbarium sheet placed in the water under the material and gradually lifted out. A piece of muslin is placed over the specimen which is then dried and pressed, as for the higher plants. Fungi are best preserved in liquid in bottles, but some may need to be cultured in order to identify them properly.

The procedure is to scrape some of the material off on to agar plates in petri dishes and incubate them. From the above, it can be appreciated that future studies of mangrove swamps involve more than just the higher plants; full collections of all plants need to be made.

MICROFLORA

Very little work has been done on the bacterial flora of mangrove muds. Troll and Dragendorff (1931) considered that the black colour of many mangrove muds is produced by anaerobic bacteria reducing sulphates to sulphides. They suggested that the species involved could be *Microspora aestuarii*. Similarly, very little work has been done upon the diatoms that must be associated with both the muds and the water in the channels. Bourelly and Manguin (1952) listed the following species from mangrove muds in Guadeloupe: *Thalassiosira fluviatiles* var. *mangrovii*, *Propidoneis lepidoptera*, *Hautschira sigma*, *Nitzschia affinis* and *N. sigma f. longissima*. Mud samples also need to be examined for microscopic algae which can play an important role in the habitat; especially those Cyanophyceae that are nitrogen fixers, e.g. *Anabaena* spp.

In order to study the microflora, scrapings of the mud surface and of decaying leaves should be taken at intervals across the swamp, and small amounts examined in water under a microscope. This will enable identification of diatoms and microscopic Cyanophyceae. Another portion should be put on an agar plate and incubated; this will provide growth of any bacteria with their subsequent identification. The soil microflora may extend down into the soil, so pits should be dug across a transect and soil samples taken at 5, 10 and 15 cm. The nature of the soil should be recorded. These depth samples should be examined in the same manner as suface samplings.

Muslin nets should be used to collect diatom samples from creeks or other standing waters in the swamp. Such samples can be preserved in 3 per cent formalin/sea-water and examined under a microscope.

The microflora is a very important aspect of the mangrove environment because of the part played in the food chain. It has been grossly neglected in the past, and information is urgently required from mangal all over the world.

PNEUMATOPHORES AND PROP ROOTS

One of the morphological features that enables mangrove trees to grow and survive in the adverse habitat is the existence of aerating roots, which either emerge from the soil or which take the form of prop roots, as in the genus *Rhizophora* (Fig. 4.1). Erect vertical aerial roots emerging from the soil are found in the genera *Sonneratia*, *Avicennia* and *Laguncularia* (Fig. 4.2 and 4.3). In the latter genus and *Xylocarpus mekongensis*, the pneumatophores tend to have a bulbous extremity, while in *Camptostemon*, outgrowths are characteristic (Fig. 4.4(a)). In the genera *Bruguiera* and *Lumnitzera*, the horizontal roots grow up, out of the soil at intervals and then go down again producing a knee-like bend (Fig. 4.4(b)). In *Heritiera littoralis*, *Carapa moluccensis* and *Xylocarpus granatum*, the upper portion of the horizontal roots emerges vertically out of the soil as a sharp ridge that becomes flattened (Fig. 4.4(c)). Prop roots have more recently been reported from *Avicennia* but these are abnormal. For details of the various types of pneumatophores, see Chapman (1976).

Figure 4.1
Prop roots of *Rhizophora apiculata*
(*Photo:* Forest Research Institute, Kepong, Selangor, Malaysia).

In any survey, notes should be made of the various types of pneumatophores to be found and samples collected for drying and mounting.

While some genera, e.g. *Rhizophora*, *Sonneratia*, *Avicennia*, have had their pneumatophores and prop roots studied quite thoroughly, there are other genera for which information is lacking. Physical measurements are needed and soil should be excavated to determine the relationship of the absorbing rootlets

Figure 4.2
Pneumatophores of *Sonneratia*.

Figure 4.3
Pneumatophores of *Avicennia*.

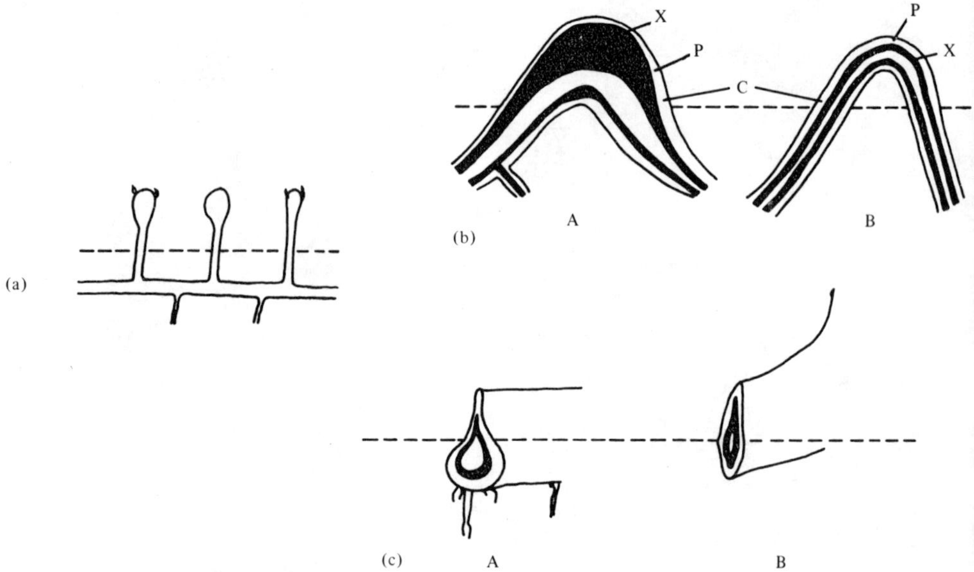

Figure 4.4
(a) Bulbous tops of pneumatophores typical of *Laguncularia*.
(b) Horizontal roots with knee-like bend typical of (A) *Bruguiera* and (B) *Lumnitzera*.
(c) Upper portion of horizontal roots are sharply flattened in the vertical plane in such species of mangrove as (*A*) *Camptostemon* and (B) *Heritiera littoralis*.

to the pneumatophores. Sections should also be cut to determine the extent of any aerenchyma and its relationship to the tissues of the below-ground roots.

DENDROCHRONOLOGY

Information about rings in mangroves is practically non-existent, except for *Avicennia* where there is considerable interest over the significance of the rings. It would seem from the information available (Chapman, 1976) that the rings do represent annual increments in arid areas but that in moist zones more than one ring may be produced annually. Control of cambial development appears to be endogenous rather than exogenous. Growth rings are generally absent in *Excoecaria* and species of *Rhizophora*.

When trees are felled, rings can be counted, but unless the date of seedling establishment is known, this only gives the age by assumption. Rings can also be counted by taking a core from the trunk as near the base as possible.

It is evident that more data are greatly desired. Seedlings should be marked and then some years later (ten is recommended) they can be cut down and rings counted. Only in this way can it be established that the rings actually represent annual increments.

SEEDLINGS

The viviparous seedlings of many mangrove genera are well known (Fig. 4.5). Abnormalities such as double seedlings and albino seedlings (*Rhizophora*) have also been recorded.

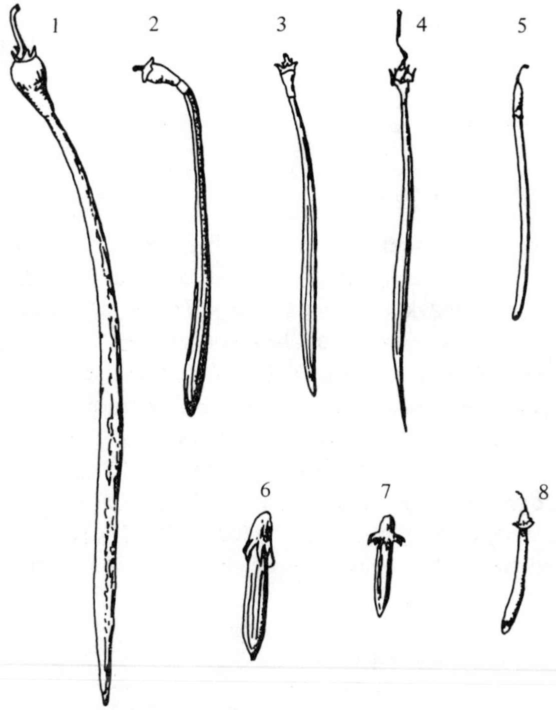

Figure 4.5
Viviparous fruits of various mangrove species: 1. *Rhizophora mucronata*; 2. *R. conjugata*; 3. *Ceriops condolleana*; 4. *Kandelia rheedii*; 5. *Bruguiera parviflora*; 6. *B. gymnorrhiza*; 7. *B. eriopetala*; 8. *B. caryophylloides*. (After Watson, 1928).

Seedling productivity can be estimated by means of the total produced annually per tree, either in relation to tree height or to estimated leaf biomass. If this figure is then compared later with the current year's seedlings that have established under the parent tree, this provides some estimate of potential seedling distribution. Actual observations at times of flooding tides is perhaps even more accurate.

Seedling productivity and distribution are both items which require much more information. The distribution must be considered in relation to hydrology and geomorphology because only then can distribution patterns be understood. If a species zone is very extensive, it is obviously desirable to study productivity and distribution at both the seaward and landward limits. The kind of information required is best exemplified by the work of Yamashiro (1961) on seedlings of *Kandelia candel* (Table 4.6).

TABLE 4.6. Distribution of marked seedlings of *Kandelia*

Station of mother trees	Found under mother tree	Within 10 m of mother tree	Within 50 m of mother tree	Unknown	Total
Near flood-tide line	24	76	58	717	875
Swamp centre	7	46	8	336	397
Near ebb-tide line	—	7	1	574	582
TOTAL	31	129	67	1 627	1 854

Source: Yamashiro (1961).

SPECIES AND SEA-LEVEL

On a broad basis, species and sea-level are interrelated by the zonations that have been recorded in transects from the seaward edge to the landward edge at extreme high-water mark (Figs. 4.6 and 4.7). In most cases, however, the exact relationship to local tidal levels has not been established. Early work in this field (Watson, 1928; de Haan, 1931) placed the different species within inundation classes. The first worker proposed five inundation classes and the latter, six (Table 4.7). It can be noted that there is not complete agreement between the two workers. Table 4.8 (Chapman, 1976) shows the requirements of species on the west coast of Malaysia and Table 4.9 adds in New World species. As pointed out by Chapman (1976), neither scheme gives any indication of the extent of the belts and it is better to use diagrams that relate belts to the tides (Figs. 4.8 and 4.9).

TABLE 4.7. Inundation classes for Malayan mangroves

Class	Flooded by	Hight above admiralty datum (feet)	Times flooded per month (Watson)	Days flooding per month (de Haan)
1.	All high tides	0– 8	56–62	20+
2.	Medium high tides	8–11	45–59	10–19
3.	Normal high tides	11–13	20–45	4– 9
4.	Spring high tides	13–15	2–20	2– 4
5.	Abnormal or equinoctial tides	15	2	2
6.	Seasonal in wet season only			Seasonal flood

Two methods are readily available to establish the relationship of various species to tide levels: (a) a levelling survey based on a bench-mark above high-tide level—the bench mark can be related to a tide level in the nearest harbour; (b) a levelling survey based on a tide pole established in the swamp and related to the nearest harbour tide machine.

Tide tables or actual tide records can be used to show the relationships as in Figures 4.8 and 4.9.

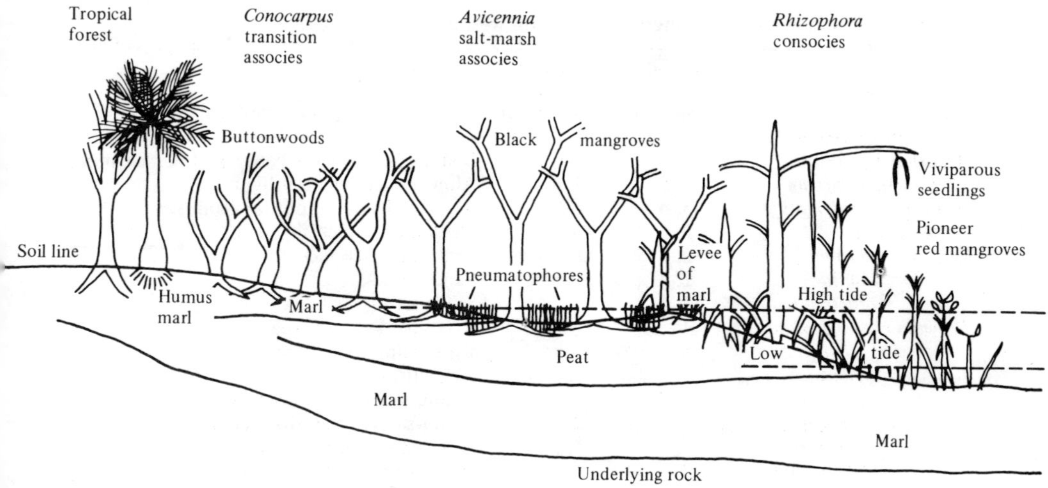

Figure 4.6
Diagrammatic transect of Florida mangrove communities from the pioneer *Rhizophora* family to the tropical hammock forest (after Davis, 1940).

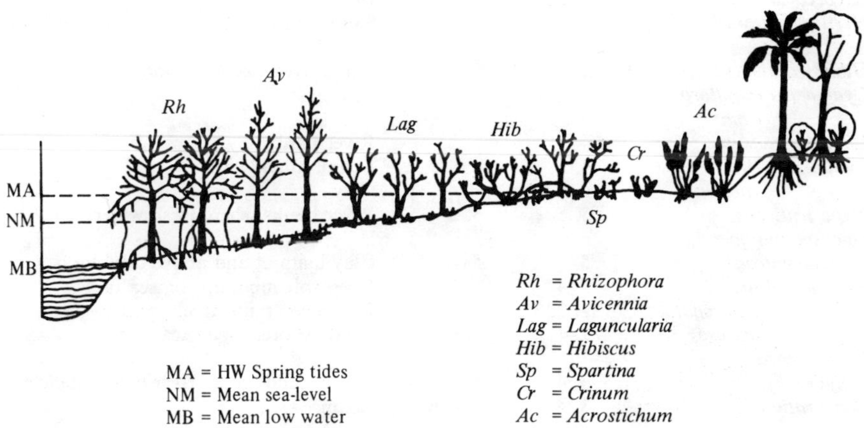

Figure 4.7
Schematic profile of mangrove vegetation near Itanhaem, southern Brazil (after Lamberti, 1969).

TABLE 4.8. Requirements of mangrove species on west coast of Malaysia

Species	Inundation class (Watson)	Inundation class (de Haan)	Soil and position
Acanthus ilicifolius	4, 5	1(?)	Loam, clay on river banks and in clearings
Acrostichum aureum	3–5	3, 4	Almost everywhere if light
Aegiceras corniculatum	3, 4	1(?)	Riverbanks, loam not far from sea
Avicennia alba	2	—	Deep mud, brackish
marina	2, 3	1	Firm mud, sea face
lanata[1]	2, 3	—	Sandy mud
officinalis	3, 4	—	Stiff mud, river banks
Brownlowia lanceolata	4	—	River banks, open spaces
riedelii	5	—	River banks, sandy mud
Bruguiera cylindrica	4	3, 4	New stiff clay
sexangula	3, 4	5	Loam, wetter soils near back
gymnorrhiza	3, 5	2, 3	Loam-sandy mud, drier areas
parviflora	3, 4	1, 2	Anywhere if well drained
exaristata[2]	4, 5	—	Rear swamps and along creeks
hainesii[2]	5	—	Dry, very high spring tides only
Calophyllum inophyllum[1]	6	—	Loam, inland river banks
Xylocarpus moluccensis	4, 5	3, 3	Loam near tidal limit
granatum[1]	3–5	3	Sandy mud, near river banks
Cerbera manghas (= C. lactaria)	4, 5	5	River banks near tidal limit
Ceriops tagal	3, 4	3	Loam brackish
decandra[2]	5	3, 4	Land fringe
Derris heterophylla	4, 5	3	Loam on river banks
Excoecaria agallocha	4, 5	3	Clay
Heritiera littoralis	5	3	Sandy loam; river banks (brackish) and in land areas
Hibiscus tiliaceus	5	6	Loam, river banks inland
Cynometra ramiflora	5	5	
iripa	5	6	
Kandelia candel	4	—	River banks
Lumnitzera littorea	4, 5	4	Loam, inland edge
racemosa	4, 5	—	Clay
Nypa fruticans	3, 5	3, 5	River banks where fresh water
Oncosperma horrida (filamentosa)	5	5	Clay/loam at and above tidal limit
Rhizophora apiculata	3, 4	1, 2	Deep soft mud, not on sea face
mucronata	1, 2, 3	1, 2	Firm deep mud, creeks and rivers
stylosa[2, 3]	2–4		Sandy shores, open sea face
Scyphiphora hydrophyllacea	3, 4	4	Loam, sandy mud, river banks and clearings
Sonneratia ovata	3, 4	Loam	Loam
caseolaris (= S. acida)	4, 5	2, 3, 4, 5	Loam, near river water
alba	2, 3	1	Rich mud—sea face, river banks
Terminalia catappa	—	4	Loam, river banks

1. *X mekongensis* also occurs in this area and the requirements of the three species need to be worked out.
2. Species not in Watson's original list; data derived from Van Steenis (1958).
3. Species peculiar to east coast.

TABLE 4.9. Inundation classes of mangrove species in New and Old Worlds

Watson (1928)	de Haan (1931)	Dominant in Old World	Chapman (1944)	Dominant in New World
	A. Brackish to saline; salinity at H.T. 1–3%			
1. Land flood at all high tides	A1. Areas flooded 1–2 times daily 20 days minimum per month	*Sonneratia alba* *S. apetala* *Avicennia marina*	530–700 + subs. per annum	*Rhizophora mangle*
2. Flooded by medium high tides	A2. Areas flooded 10–19 days per month	*Rhizophora*		
3. Flooded by normal high tides		*Bruguiera*	400–530 subs. per annum	*Avicennia germinans*
	A3. Flooded 9 days per month	landward fringe *Xylocarpus granatum*		
4. Flooded by Spring tides only	A4. Flooded only a few days per month	*Lumnitzera littorea* *Bruguiera sexangula*	150–250 subs. per annum	*Laguncularia salina*
5. Flooded by storm high tides only		Halophytes or salt flats	4–100 subs. per annum.	*Salina* or *Laguncularia–Conocarpus*
	B. Fresh to brackish water; salinity 0–10%			
	B1. More or less under tidal influence	*Nypa*		

Source: Chapman (1975).

Much more work of this nature is required because only then can the overall world picture of individual species, relative to tides, be obtained. Sea-level and tidal inundations are not the only major factors involved. Soil-drainage pattern, salinity and actual location (e.g. mouth of river, lagoon, etc.) can be significant and should not be ignored.

From the point of view of seedling survival, the recording of the occurrence of days of continuous exposure at neap-tide periods is highly significant. It is important that studies should be made of seedling fall in relation to these periods of continuous exposure associated with the incidence of precipitation (Chapman, 1976).

PHENOLOGY, LIFE HISTORY

Items included here are, for example: (a) time of flowering; (b) time of fruit production; (c) time of seedling fall; (d) time of seed germination; and (e) time

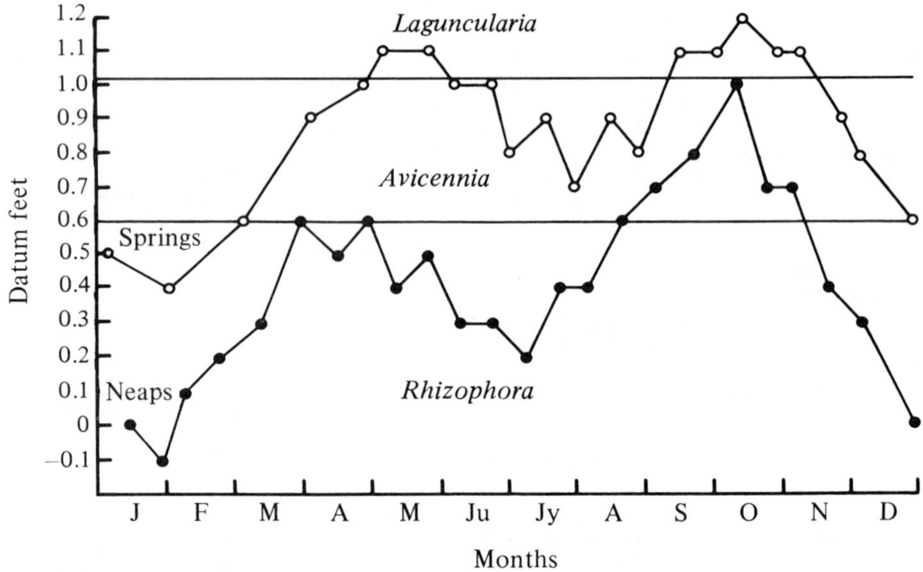

Figure 4.8
The relation of the mangroves in Kingston Harbour to neap and spring tides.

of new leaf production. Where possible, attempts should be made to see if any of these items are related to either seasonal temperatures or to seasonal (if any) rainfall. Very little information, even on items (a) and (b), has been published but it may well be available in some areas. Continuous observations over a number of years is the only effective method. This is an area in which much information is required from a range of regions in order to establish the extent to which individual species may or may not exhibit any uniformity throughout their geographic range.

Vegetation studies

COMMUNITY LISTS

An overall view of mangrove communities has been set out by Chapman (1976, 1977) and Table 4.10 sets out the list of those generally recognized. Chapman (1976) and authors in *Ecosystems of the World*, Vol. 1, should be consulted for species recorded from these communities or for papers that give such lists. In the lists it is sometimes possible to ascertain those species that are regular components of a community and those which can be regarded as casuals.

In any region or area, the first step by observation is to establish the associations that can be recognized. When this has been done the species

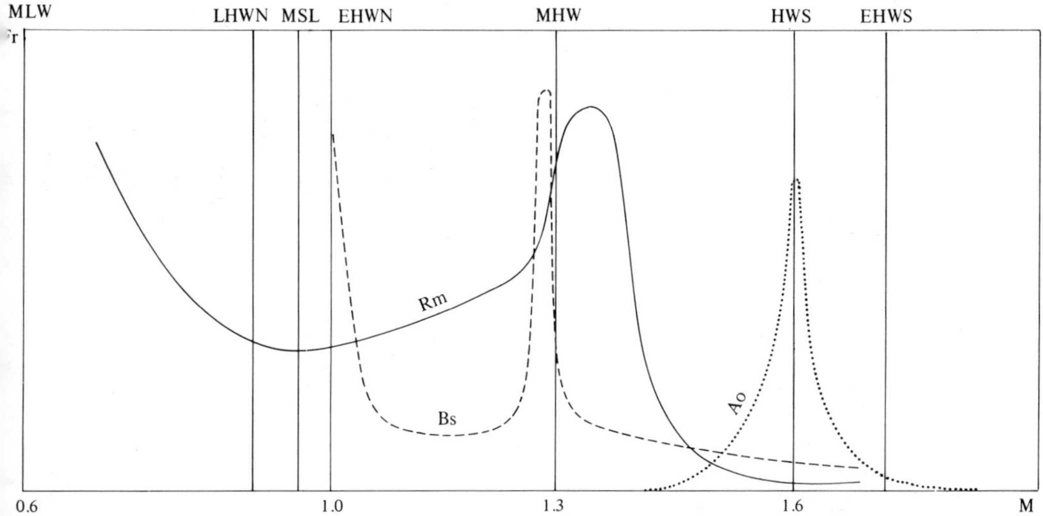

Figure 4.9
Curves representing frequency (Fr) of *Rhizophora mucronata* (Rm), *Bruguiera sexangula* (Bs), and *Avicennia offiicinalis* (Ao) in relation to tidal features. MLW = mean low water; LHWN = lowest high-water neap tides; EHWN = highest high-water neap tides; MSL = mean sea-level; MHW = mean high water; HWS = high-water spring tides; EHWS = extreme high-water spring tides. Height in metres above sea-level datum.

present in each association can be established either by observation or by the use of quadrats. In studies such as this, it must be remembered that where one association abuts on to another one, there will be a zone, the ecotone, where a mixture of species from both associations can be present. Ecotones should not be included in either association but treated separately. Species lists should be restricted solely to areas clearly within the association being studied.

It is desirable that all lists should be divided into (a) regular components of the association, and (b) casual components of the association.

If any association is utilized commercially in any way (e.g. selective felling of a particular species or felling leaving only nursery specimens), it is essential that the effect on the species list should be recorded and, in such cases, the reinstatement of the community, if it occurs, should be followed annually by listing species present and newly arrived.

DENSITY

This is (a) number of plants of a single species per unit area, and (b) total number of plants per unit area.

For convenience, it is probably best to divide the plants into strata (synusiae) and count the plants for each stratum. It is recommended that the

TABLE 4.10. Distribution of principal mangal associations

	Lower California	Central America	West Indies/Florida	Brazil/Uruguay	West Africa	East Africa	India	Burma/Indochina	North Australia	Pacific Islands	Philippines
RHIZOPHORETEA											
A. Rhizophoretalia											
1. Rhizophorion occidentale											
Rhizophoretum manglae	+	+	+	+	+					+	
Rhizophoretum racemosae		+		+	+						
Rhizophoretum harrisonii		+			+						
2. Rhizophorion orientale											
Rhizophoretum mucronatae						+	+	+	+	+	+
Rhizophoretum apiculatae							+	+		+	+
Rhizophoretum stylosae								+	+	+	+
3. Bruguierion											
Bruguieretum gymnorrhizae						+	+	+	+	+	+
Bruguieretum parviflorae								+	+		+
Bruguieretum cylindricae								+	+		+
Bruguieretum sexangulae							+	+			
Bruguiereto–Xylocarpetum								+	+		
4. Ceriopion											
Ceriopetum tagalae						+	+	+	+	+	+
Ceriopeto–Aegiceretum corniculatae							+	+			
5. Kandelion											
Kandelietum candeli							+	+		+[1]	+
6. Avicennion occidentalis											
Avicennietum germinansae	+	+	+	+							
Avicennietum africanae					+						
Avicennietum germinans/A. schauerianae			+	+							
7. Avicennion orientalis							+	+			
Avicennietum albae							+	+	+	+	+
Avicennietum officinalis							+	+			+
Avicennietum marinae						+	+	+	+[2]	+	+
Avicennietum albae–marinae								+	+		
Avicennieto–Excoecarietum										+	+
B. Sonneratietalia											
1. Sonneration											
Sonneratietum albae						+	+	+	+	+	+
Sonneratietum caseolariae							+		+	+	+
Sonneratietum apetalae							+				
Sonneratio–Camptostemonetum										+	+
COMBRETETEA											
A. Combretalia											
1. Conocarpion											
Conocarpetum erectae	+	+	+		+						

TABLE 4.10.—*continued*

	Lower California	Central America	West Indies/Florida	Brazil/Uruguay	West Africa	East Africa	India	Burma/Indochina	North Australia	Pacific Islands	Philippines
2. Laguncularion											
Laguncularietum racemosae		+	+	+	+						
3. Lumnitzion											
Lumnitzeretum racemosae							+	+	+		+
Lumnitzeretum littorale								+	+	+	+
Lumnitzereto–Xylocarpetum obovatae										+	+
B. Xylocarpetalia											
1. Xylocarpion											
Xylocarpetum granatae						+	+	+	+	+	+
Xylocarpetum moluccensis						+[3]	+	+	+	+	+
Xylocarpetum australasicae									+		
Xylocarpetum benadirensae						+					
C. Pellicieretalia											
1. Pellicierion											
Pellicieretum rhizophorae	+										
D. Acrostichetea											
1. Halo–Acrostichion											
Acrostichetum aureae		+			+		+				+
Acrostichetum speciosae									+		
E. Excoecarietalia											
1. Exoccarion											
Excoecarietum agallochae							+	+	+	+	
Excoecarieto–Xylocarpetum australasicae									+		
2. Acanthion											
Acanthetum ilicifoliae							+	+			+
F. Heritietalia											
1. Herition											
Heritieretum minori							+	+			
Heritieretum littorali										+	+
G. Aegiceretalia											
1. Aegicerion											
Aegiceretum corniculatae									+	+	+

1. Japan.
2. South Australia also.
3. Madagascar.

Source: Chapman (1977).

following synusiae be recognized: (a) canopy (density in number per hectare); (b) sub-canopy, shrubs and large ferns (e.g. *Acrostichum*) (density in number per 100 m^2); and (c) herb layer (density in number per 4 m^2).

Under special circumstances, it may be necessary to use areas other than the above, though these are generally recognized as the standard ones. If the association does not occupy a substantial area, then a quadrat of less than one hectare will be necessary. The smaller quadrats (e.g. 100 m^2 and 4 m^2) should be selected at random and if the association occupies a substantial area then a number of these smaller quadrats should be established in order to obtain a satisfactory answer.

DOMINANCE

Typically, this is determined by the dominant species in the canopy layer even if a species in the shrub layer is more abundant. The dominant is that species with the greatest density per unit area. The dominants give their name to the various associations that have been recognized (Table 4.11).

It is recommended that in any study of an association, dominance in the

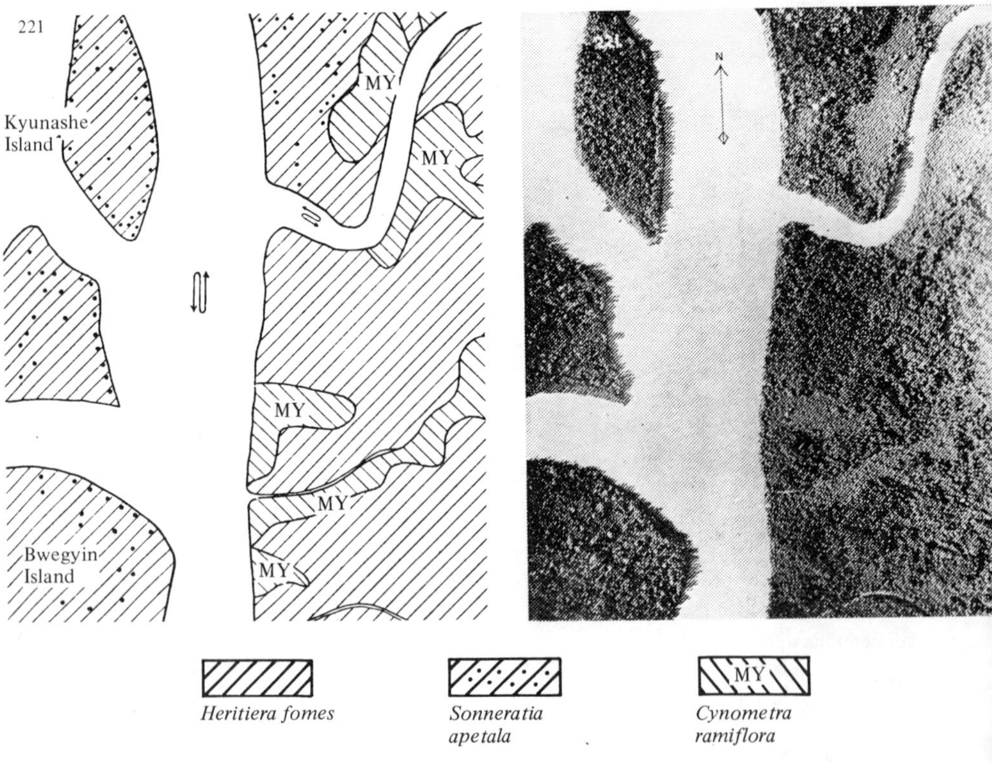

Figure 4.10
Mangrove vegetation map based on aerial photography of the Kadonkani Reserve, Irrawaddy River, Burma.

canopy should be regarded as dominance for the association. If there is great abundance of a subsidiary species, this can be recognized by adding its name to that of the dominant, e.g. *Ceriops tagal–Aegiceras corniculatum*.

If the synusiae are readily identifiable, it is desirable to record the dominance in the various strata. In the case of the herb layer this can be important in the landward associations.

TECHNIQUES

A variety of techniques have been used in plant sociological studies, for example: (a) maps based on aerial photographs (Fig. 4.10); (b) quadrats as described in the above statements; (c) belt transects through a homogeneous habitat; (d) line transects across environmental gradients as in Figures 4.8 and 4.9; and (e) type maps prepared from aerial photographs together with intensive ground truthing. (Fig. 4.11.)

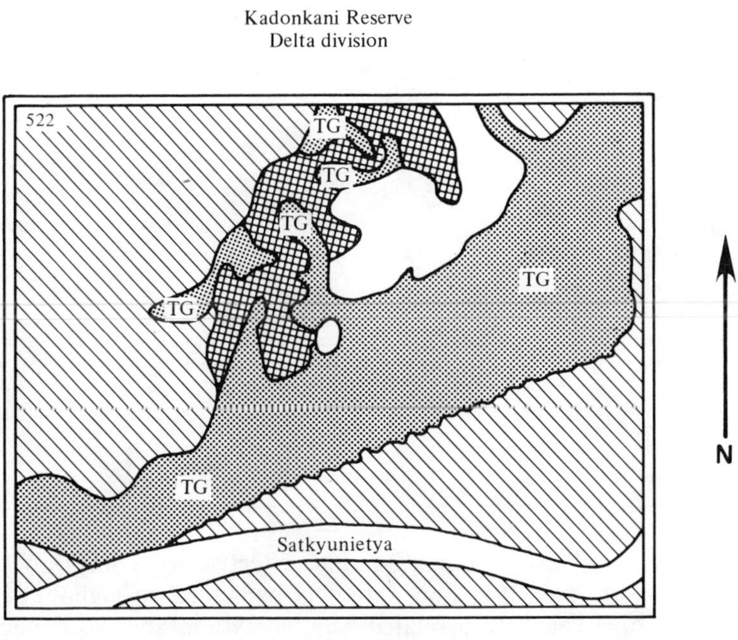

Figure 4.11
Detailed mangrove vegetation map based on aerial photography.

Stratification pattern: Very little information is available here. Research elsewhere (see Chapter 6) has shown that there is a clear distinction between numbers and diameters of young and old trees in the swamps.

Until information on stratification in the different successional zones becomes available there is no possibility of considering physionomic types. Future studies should aim to investigate the number of strata that can be recognized; it is likely to be main canopy, understorey and ground vegetation—their composition and degree of foliage cover (see Table 4.11).

TABLE 4.11. Presence and relative abundance (scale 1–10) of species in a mature mangrove forest associes in Florida

	Stations		
Species	BR1–4	SR1–4	CR1–5
Trees			
Rhizophora mangle	5	6	5
Avicennia germinans	3	3	4
Laguncularia racemosa	1	1	2
Conocarpus erecta	2	1	1
Sabal palmetto	1	1	1
Eugenia axillaris	1	1	
Eugenia buxifolia	1	1	
Coccolobis uvifera			1
Shrubs			
Batis maritima	7	6	8
Salicornia perennis	2	1	2
Borrichia frutescens	1	1	2
Borrichia arborescens		1	
Bumelia augustifoli		1	
Herbs and ferns			
Acrostichum aureum	2	3	1
Spartina alterniflora	1	1	1
Juncus roemeriancus	1	1	
Sporobolus virginicus	1		1
Epiphyte			
Tillandsia circinata		1	1

Source: Davis (1940).

Complexity indices: This is an area in which effective work has only recently been initiated (Pool et al., 1977), but should be expanded throughout the world for its comparative value. In a related way, ecosystem analysis (Lugo et al., 1976) deals with an individual system and shows how models and computer programs can be established for ecological factors but no programs have so far been worked out or proposed that would deal with species complexity in mangroves. Changes in pattern can be worked out by determining species change with latitude. This needs to be done for all members of the flora and then put together.

Vegetation maps: Any study of a specific area is greatly enhanced if a vegetation map is prepared as a result. Depending on the size of the region studied these can be at various scales, but it is recommended that no map should be on a smaller scale than 1:50,000 (one mile to one inch).

In general, it may be stated that for any region a combination of quadrats, belt transects and air photographs is likely to provide the best overall picture.

In the case of vegetation maps, the scale will depend upon the area involved. It is unlikely to be of great value if the scale is smaller than 1:50,000. The size of quadrat depends on the vegetation and ranges from 1 m^2 to 1 hectare. Transects are desirable in all cases because they demonstrate any changes from one association to another, and also indicate the extent of any ecotones. Depending upon the size and distribution of canopy species, transects should be either 5 or 10 metres wide. Aerial photographs may not always show zonation and, hence, the necessity for ground checks. Such photographs will, however, give an indication of density of the canopy species.

Since communities do change as accretion of mud reduces frequency of flooding, arrangements need to be made for permanent monitoring of communities by means of the techniques above. This monitoring process is even more important where there is human pressure and utilization of the forests.

Figure 4.12
Oblique aerial photograph of tidal Nipa palm west of Kerema, Gulf District, Papua New Guinea.

SOIL SAMPLING

All communities should have their soils sampled and described down to a standard depth. The basic techniques is to acquire air photographs of the area being studied and then make ground-truth checks in order to establish the exact boundaries of the different communities. Oblique aerial photographs (Fig. 4.12) are not to be recommended as they are difficult to interpret. Infrared aerial photographs can also be used for this purpose, and if the area is especially large, it may be possible to secure Landsat images which will probably need to be enlarged to an appropriate scale.

Such maps can be used to calculate areas of water occupied in relation to adjacent harbour or lagoon waters and an attempt made to relate these figures to fisheries harvests. They can also be valuable in relation to descriptions of historical changes and in support of the need for conservation.

References

Bourelly, P.; Manguin, E. 1952. *Algues d'eau douce de la Guadeloupe. Paris, SEDES. 282 pp.*

Chapman, V. J. 1944. The 1939 Cambridge University Expedition to Jamaica. I: A Study of the Botanical Processes Concerned in the Development of the Jamaican Shore-line. *J. Linnean Soc.* (Botany), Vol. 52, pp. 407–47.

——. 1975. Mangrove Biogeography. In: G. Walsh, S. Snedaker and H. Teas (eds.), *Proceedings of the International Symposium on Biology and Management of Mangroves* (held in Honolulu, 8–11 October 1974), pp. 3–22. Gainesville, Fla., Institute of Food Agricultural Science, University of Florida. 2 vols. 846 pp.

——. 1976. *Mangrove Vegetation.* Strauss & Cramer. 499 pp. — 581.52632C

——. (ed.) 1977. *Ecosystems of the World.* Vol. I: *Wet Coastal Ecosystems.* New York, Elsevier. 428 pp.

Davis, J. H., Jr. 1940. *The Ecology and Geologic role of Mangroves in Florida.* (Carnegie Institute, Washington Publication, 517, pp. 303–412. Papers from the Tortugas Lab, Vol. 32.)

de Haan, J. D. 1931. Het een en auder over de Tjilatjapsche vloedbosschen. *Tectona*, Vol. 24, pp. 39–76. (Quoted in: Chapman, V. J., *Mangrove Vegetation.* Strauss & Cramer. 1976. 447 pp.)

Lamberti, A. 1969. Contribuicão a Conhecimento da Ecologia das Plantas do Manquezal de Itanhaem. *Univ. São Paulo, Fac. Filos. Cien. e Letras Bol.*, 317; *Bot.*, Vol. 23, pp. 7–218.

Lugo, A. E.; Sell, M.; Snedaker, S. C. 1976. Mangrove Ecosystem Analysis. In: B. C. Patten (ed.), *Systems Analysis and Simulation Ecology*, Vol. 4, pp. 113–45. New York, Academic Press. 593 pp.

Pool, D. J.; Snedaker, S. C.; Lugo, A. E. 1977. Structure of Mangrove Forests in Florida, Puerto Rico, Mexico and Costa Rica. *Biotropica*, Vol. 9, No. 3, pp. 195–212.

Stevens, G. N. 1979. Distribution and Related Ecology of Macrolichens on Mangroves on the East Australian Coast. *Lichenologist*, Vol. 11, No. 3, pp. 293–305.

Stevens, G. N.; Rogers, R. W. 1979. The Macrolichen Flora from the Mangroves of Moreton Bay. *Proc. R. Soc. Qd*, Vol. 90, pp. 33–49.

Troll, W.; Dragendorff, O. 1931. Uber die luftwürzeln von *Sonneratia* L. und ihre biologische Bedeutung. *Planta*, Vol. 13, pp. 311–473.

van Steenis, C. G. G. J. 1958. Ecology of Mangroves. *Flora Malesiana*, Vol. I. Ser. 5, pp. 431–41.

Yamashiro, M. 1961. Ecological Study on *Kandelia candel* (L.) Druce, with Special Reference to the Structure and Falling of the Seedlings. *Hikobia*, Vol. 2, No. 3, pp. 209–14.

Watson, J. G. 1928. *Mangrove forests of the Malay Peninsula.* Singapore, Fraser and Neave. 275 pp. (Malayan Forest. Rec., No. 6.)

5

Taxonomic considerations of the mangrove species

François Blasco
Institut de la Carte Internationale du Tapis Végétal
39 allée Jules-Guesde
31400 Toulouse, France.

General considerations

The total flora of woody plants usually found in the mangroves of the world includes twenty-three genera and fifty-three species belonging to sixteen families. A few varieties have been described mainly in the *Avicennia* group. Most of the taxa listed in Table 5.1 are either trees or shrubs; climbers are exceptional (e.g. *Derris uliginosa*, *D. heterophylla*, *Dalbergia spinosa*) as are several of the Asclepiadaceae.

Some other halotolerant taxa occasionally encountered in subtropical mangrove forests are considered marginal and atypical, and are not listed in Table 5.1; this group includes oaks (*Quercus glauca* var. *amamiana*), conifers (*Taxodium distichum*), *Drypetes karapinensis*, *Hibicus tiliaceus*, several *Pandanus* and palms (*Oncosperma*).

Many herbaceous or semi-woody halophytes are often found in mangroves, generally wherever there has been a habitat disturbance as a result of human interference. Members of this large group of plants may be considered to be more characteristic of tropical salt marshes and their presence in mangrove stands reflects a marginal distribution. Most of these herbs and understorey shrubs have a very broad geographical distribution and are often pantropical. Most of them belong to one family, that of Chenopodiaceae which has more than 100 *Suaeda* species, about 200 *Atriplex*, some 150 *Salsola* and 30 species of *Salicornia*. Some grasses (e.g. *Porteresia*, *Aeluropus*) and members of the Cyperaceae (e.g. *Cyperus*, *Carex*, *Scirpus*) may incidentally be found in mangroves as well as representatives of smaller groups such as Aizoaceae (eight species of *Sesuvium*), Pteridophytes (*Acrostichum aureum*) and Boraginaceae (*Heliotropium curassavicum*) among others.

Orchidaceae such as *Bulbophyllum dixoni* Roffe (found on *Sonneratia alba*), *Deudrobium crumenatum* S. (found on *Sonneratia* and *Rhizophora*), *Grammatophyllum speciosum* Bl. (found on *Rhizophora*) etc. are not rare locally in Asian mangroves. However, until recently, very few of the recorded epiphytic orchids were thought to be restricted to mangrove areas.[1] According to Sahavaharin et al. (1979), *Grammatophyllum speciosum*, *Paphiopedium exul* and *Deudrobium cruentum* are extremely beautiful, but rare and therefore worthy of immediate protection. Several epiphytic Asclepiadaceae belonging to the genera *Hoya*, *Dischidia*, etc., and the Melastomaceae (*Plethiandra*) may also be found in mangroves but are usually restricted to the more inland areas.

Regarding parasitic plants, it seems that only Loranthaceae are reported (several each of *Dendrophthoe*, *Viscum*); *Amyema gravis* seems to be confined to mangrove habitats.

The flora of the mangrove is therefore mainly composed of the woody species (Table 5.1) and the majority are found in tropical Asia and Australia.

TABLE 5.1. Main woody species of mangrove communities

Acanthus ilicifolius L.	Acanthaceae
Aegialitis annulata R. Br.	
A. rotundifolia Roxb.	Plumbaginaceae
Aegiceras corniculatum Blanco	Myrsinaceae
A. floridum	
Avicennia alba Blume	Avicenniaceae
A. alba var. *latifolia* Moldenke	
A. marina (Forsk.) Vierh.	
A. marina var. *acutissima* Stapf and Moldenke ex Moldenke	
A. marina var. *anomala* Moldenke	
A. marina var. *resinifera* (Forst.) Bakh.	
A. marina var. *rumphiana* (H. Hallier) Bakh.	
A. lanata Ridl.	
A. eucalyptifolia Zipp. ex Miq.	
A. tonduzii Moldenke	
A. schaueriana Stapf and Leechman ex Moldenke	
A. bicolor Standl.	
A. balanophora Stapf and Moldenke ex Moldenke	
Aviciennia officinalis L.	
A. africana P. Beauv.	
A. germinans (L.) Stearn	
Brownlowia lanceolata Benth.	Tiliaceae
Bruguiera	Rhizophoraceae
B. cylindrica (L.) Bl.	
B. exaristata Ding Hou	
B. gymnorrhiza (L.) Lamk.	
B. hainesii C. G. Rogers	
B. parviflora (Toxb.) W. and A. ex Griff.	
B. sexangula (Lour.) Poir.	
Ceriops	Rhizophoraceae
C. decandra (Griff.) Ding Hou	
C. tagal (Perr.) C. B. Rob.	
Clerodendrum inerme Gaertn.	Verbenaceae
Conocarpus erectus L.	Combretaceae
Dalbergia candenatensis Prain	Papilionaceae
D. spinosa Roxb.	
Derris heterophylla (Willd.) Back.	Papilionaceae
D. uliginosa Benth.	

1. Ding Hou (1958, p. 440) has reported that '*Humata parvula* (Wall.) Mett. on "old mossy mangrove" in S. Malaya and Borneo, and orchids as *Dendrobium rhizophoreti* J.J.S., *D. prostratum* Ridl., *D. callibotrys* Ridl., and *Bulbophyllum xylocarpi* J.J.S., and a Melastomaceae *Plethiandra sessilifolia* Ridl.' are restricted to the mangrove habitat.

TABLE 5.1.—*continued*

Excoecaria agallocha L.	Euphorbiaceae
Heritiera fomes Buch. Ham.	Sterculiaceae
Kandelia candel (L.) Druce	Rhizophoraceae
Laguncularia racemosa Gaertn.	Combretaceae
Lumnitzera racemosa (L.) Gaertn.	Combretaceae
L. littorea (Jack) Voigt	
Nypa fruticans Wurmb.	Nypaceae (~Palmae)
Pelliciera rhizophora Planch. and Triana	Pelliceriaceae
Phoenic paludosa Roxb.	Palmae
Rhizophora[1]	Rhizophoraceae
R. apiculata Blume	
R. mangle L.	
R. mucronata Lamk.	
R. racemosa G. F. W. Meyer	
R. samoensis (Hochr.) Salvoza	
R. stylosa Griff.	
Salvadora persica L.	Salvadoraceae
Scyphiphora hydrophyllaceae Gaertn.	Rubiaceae
Sonneratia[2] *alba* J. Smith in Rees	Sonneratiaceae
S. apetala Buch.-Ham.	
S. caseolaris L. Engl.	
S. griffithii Kurz	
S. ovata Backer	

1. Several hybrid populations have been reported such as: *Rhizophora* × *lamarckii* (probable hybrid, *R. stylosa* × *apiculata*), *Rhizophora* × *selala* (probable hybrid, *R. stylosa* × *samoensis*), *Rhizophora* × *harrisonii* (probable hybrid, *R. mangle* × *racemosa*); *R. harrisonii* seems to be synonymous with *R. brevistyla* (see Tomlinson, 1978; Ding Hou, 1960).
2. Several hybrids exist also in this genus. They have been partly described but not named, *Sonneratia alba* × *ovata*; *Sonneratia alba* × *caseolaris* (see Muller and Hou-Liu, 1966).

General biogeography

With the exception of the Avicenniaceae, and many groups within the Rhizophoraceae and Sonneratiaceae, the majority of the mangrove genera (eighteen out of twenty-three) are represented by only one or two mangrove species. The groups of genera may be categorized as follows.

1. Taxa having several or many terrestrial species but only one or two are halophytes. There are nine genera in this class:

Acanthus L. (Acanthaceae), 50 species, mostly in tropical or subtropical Asia and Africa.

Brownlowia Roxb. (Tiliaceae) includes some 25 terrestrial species in south-east Asia, Malaysia, Pacific Islands.

Clerodendrum L. (Verbenaceae); 400 species in tropical and subtropical regions.

Dalbergia L. f. (Papilionaceae); 300 species in tropical and subtropical regions.

Derris Lour. (Papilionaceae); 80 species in tropics.

Excoecaria L. (Euphorbiaceae) has about 40 species in the tropical countries of Asia and Africa.

Heritiera Dryand (Sterculiaceae); 35 species, tropical west Africa, Indo-Malaysia, tropical Australia, Pacific Islands.

Phoenix L. (Palmae) is represented by 17 species, at least, in the countries of Asia and Africa.
Salvadora L. (Salvadoraceae): 4 species in Africa and Asia.

2. Taxa known only by one or two species exclusively found in mangrove ecosystems (seven genera):
Aegialitis R. Br. (Plumbaginaceae); 2 species in tropical Asia and Australia.
Aegiceras Gaertn. (Myrsinaceae); 2 species; palaeotropics.
Conocarpus L. (Combretaceae); 1, possibly 2 species in Florida, tropical America and Africa.
Kandelia W. and A. (Rhizophoraceae); 1 species, monospecific genus in tropical Asia and Japan.
Laguncularia Gaertn. f. (Combretaceae); 1 or 2 species in tropical America and tropical West Africa.
Lumnitera Willd. (Combretaceae) 2 species in tropical Asia, North Australia, Pacific Islands and East Africa.
Scyphiphora Gaertn. f. (Rubiaceae), monospecific genus in Indo-Malaysian coasts and Australia.

3. Two families of plants, i.e. Nypaceae (*Nypa fruticans* Wurmb.) and Pellicieraceae (*Pelliciera rhizophora*) which are monotypic (one genus, one species only), are found in mangrove habitats.

Taxonomic keys for Rhizophoraceae, Sonneratiaceae and Avicenniaceae

With the exception of three families, Avicenniaceae, Rhizophoraceae and Sonneratiaceae, the woody species of mangroves are well-differentiated and their taxonomic position is almost certain; morphological and ecological characters stand out clearly. The many ambiguities are principally synonymic. This is often the case whenever a genus is represented by a very small number of species or when one or two species of a large genus have an exceptional distribution or appear in variations of the same habitat.

The case of the Avicenniaceae, Rhizophoraceae and Sonneratiaceae is slightly different because several species belonging to the same genera are found in the same habitat or in similar ecological conditions. Nevertheless, due to the taxonomic studies of Backer and van Steenis (1951), Breteler (1969, 1977), Ding Hou (1958, 1960), Tomlinson (1978) and others, we can consider that our present knowledge on Rhizophoraceae and Sonneratiaceae is relatively good.

For both families the available keys to the genera, based on reproductive and vegetative characters, are simple. Further research at the subspecific levels will probably contribute only small additions to what has already been done.

Rhizophoraceae

SIMPLIFIED KEY TO THE GENERA OF RHIZOPHORACEAE LIVING IN MANGROVE COMMUNITIES (see Fig. 5.1)

1. Calyx always 4 lobed. Petals entire without appendages . . . *Rhizophora.*
2. Calyx 5–16 lobed. Petals 2 lobed, multifid or with apical appendages
   3. Calyx 8–16 lobed, lobes subulate[1]-lanceolate, pointed. Petals bilobed or emerginate[2] . . . *Bruguiera.*
   3. Calyx 5 or 6 lobed, lobes generally ovate,[3] acuminate[4] or obtuse. Petals fringed with apical appendages or multifid
      4. Calyx lobes ovate 2.5 to 3.5 mm long. Hypocotyl ridged . . . *Ceriops.*
      5. Calyx lobes linear-oblong,[5] about 15 mm long. Hypocotyl smooth . . . *Kandelia.*

Rhizophora—simplified key to the species

Inflorescence, 2-flowered, shorter than the petiole. Petals glabrous; leaf blade about 15 cm long . . . *R. apiculata* Blume.

Distribution. Tropical Asian countries, Pacific Islands, New Britain, north-east coast of New Caledonia, unknown further east than New Hebrides.

Synonyms. R. candelaria D.C.; *R. conjugata* (non Linne) Arn.

— Inflorescence, 2–16 flowered, longer than the petiole. Petals hairy or woolly-hairy.

• inflorescence few-flowered (2–3, occasionally 4), leaf blade usually less than 10 cm long . . . *R. mangle* L.[6]

Distribution. West African Coasts and tropical America.

• inflorescence, many branched pedunculate cymes
  — Style very short, less than 1.5 mm
    • Leaf blade usually 10 to 20 cm long . . . *R. mucronata* Lamk.
    * Petiole 2.5–6 cm long

Distribution. Palaeotropical coastal regions, from East Africa and Madagascar to Micronesia, northern Australia and South Pacific as far as New Hebrides; introduced to Hawaii; probably absent from New Caledonia and Fiji.

1. Awl-shaped; tapering to a sharp tip.
2. Deeply notched at the apex.
3. Egg-shaped.
4. Tapering to an acute end.
5. Much longer than broad and rounded tip.
6. In his description of the *Rhizophora* of Australasia, Tomlinson (1978) gives a description of *R. samoensis* which is similar to that of *R. mangle*. However, '*R. mangle*, apart from its distribution in West Africa and tropical America, is distinguished by its conspicuous bracteoles and pointed flower buds'.

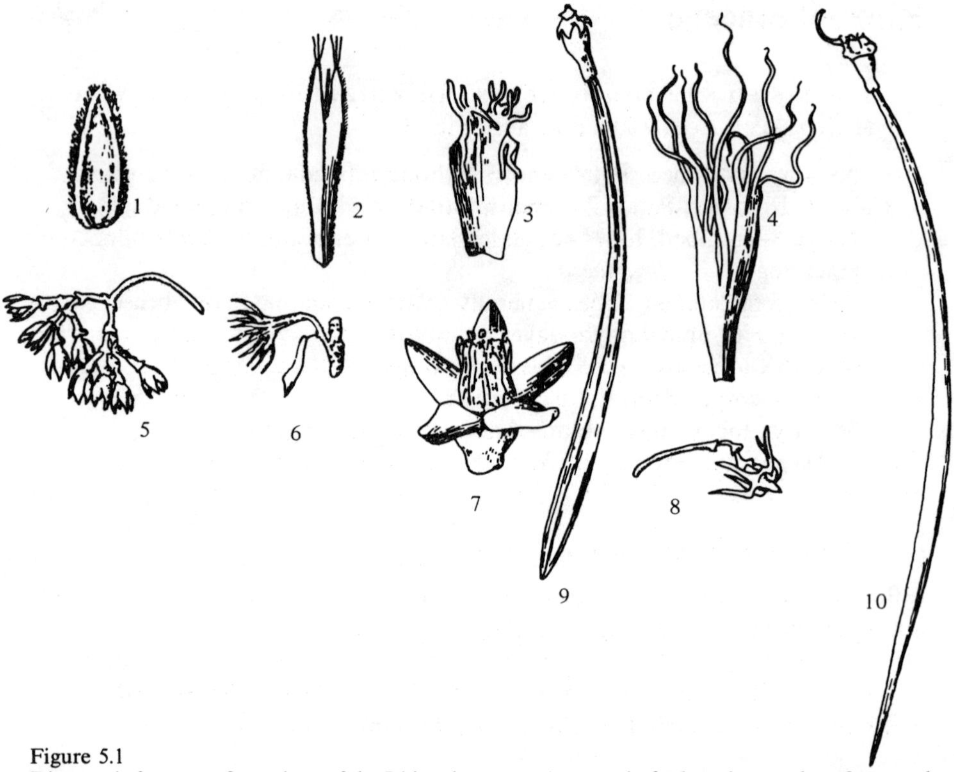

Figure 5.1
Diagnostic features of members of the Rhizophoraceae. 1 = petal of *Rhizophora stylosa*; 2 = petal of *Bruguiera gymnorrhiza*; 3 = petal of *Ceriops decandra*; 4 = petal of *Kandelia*; 5 = inflorescence of *Rhizophora stylosa*; 6 = flower of *Bruguiera hainesii*; 7 = flower of *Ceriops tagal* (corolla removed); 8 = flower of *Kandelia*; 9 = hypocotyl of *Ceriops decandra*; 10 = hypocotyl of *Kandelia*.

Synonyms. R. macrorrhiza Griff., *R. longissima* Blanco, *R. mucronata* var. *typica* Schimp.

* Petiole 1.5–3 cm long, inflorescence usually contracted; bracteole-cup irregularly lacerate or dentate . . . *R. racemosa* G. F. W. Meyer.

Distribution. West coast of tropical Africa and east coast of tropical America.

— Style filiform of about 4–6 mm long . . . *R. stylosa* Griff.

Distribution. Sandy shores of Taiwan, Philippines, Malay Peninsula, Indonesian Islands, New Guinea, New Britain, northern Australia, Melanesia; apparently unknown in Kalimantan.

Synonyms. R. mucronata var. *stylosa.*

The definitions for the following three known hybrids are the following:

1. Petals flat, sometimes slightly concave, sparsely short-hairy on the margins, sometimes also on the inside, usually covering the epipetalous stamens

only on the back. Inflorescences 2(–4) flowered. Bracteole-cup at the base of the flower with irregularly lacerate of dentate margin, not 2-lipped. Stamens usually 12–15 . . . *R.* × *lamarckii.* This taxon was considered a species (*R. lamarckii* Montr.) by Ding Hou (1960) who wrote the above definition as well as that of the following entity: *R.* × *harrisonii* Leechm.

2. Inflorescences loose, bracteole-cup at the base of the flower distinctly 2-lipped. Flower buds acute to acuminate. Pedicels (3–)6–10 mm long . . . *R.* × *harrisonii.* Tomlinson (1978) considers *R.* × *harrisonii* as a synonym for *R. brevistyla.*

3. Leaf apex initially with an indistinct but usually soon deciduous or recurved mucro. Peduncle 2.5–3 cm long but often longer, 2.6–3.2 mm wide, commonly branched beyond 2 orders. Flowers 2 to 9 per inflorescence. Bracteoles distinct, about 1 mm long. Mature flower buds white, neither sharply angular in cross section nor abruptly narrowed to a distinct shoulder at the base, 12–14 mm long. Apex of ovary extended into a distinct style 1–2 mm long. Plants sterile, lacking fruits and viviparous seedlings. (At present known from Fiji and New Caledonia; always in association with its putative parents.) (A probable F_1 hybrid, *R. stylosa* × *R. samoensis*) . . . *R.* × *selala.*

Bruguiera—key to the species (Ding Hou, 1958)

1. Flowers solitary. Bristle in the sinus between the petal lobes not exceeding the lobe tips or minute to absent.
   2. Tips of the petal lobes acute, each with 3 or 4 bristles 2–3 mm long, distinctly exceeding the lobe tips . . . 1. *B. gymnorrhiza.*

Distribution. Tropical South and East Africa, Madagascar, tropical Asia, Australia, Micronesia, Polynesia.

Synonyms. B. rheedii Bl.; *B. cylindrica* (non Bl.) Hance, etc.

   2. Tips of the petal lobes obtuse, without or each with 1 or 2, rarely 3 short bristles usually less than 1 and 0.25 mm, not or hardly exceeding the apex.
      3. Leaves usually elliptic-oblong, flat 8–13 by 3–6 cm, base acute to obtuse. Petals 13–15 mm long, the lobe tips with 1–3 bristles usually less than 1 and 0.25 mm; a distinct bristle in the sinus between the lobes . . . 2. *B. sexangula.*

Distribution. Tropical Asia, New Britain, eastern New Guinea (introduced in Hawaii).

Synonyms. B. sexangularis Spreng.; *B. eriopetala* W. and A.; *B. australis* A. Cunn. ex Arn.; *B. malabarica* (non Arn.) F.

      3. Leaves usually obovate (rarely elliptic), with revolute margins, 5–9 and 0.5 by 3–4 and 0.5 cm, base acuminate. Petals 9–11 mm long, exaristate, without (or very rarely with an obscure) bristle in the sinus of the lobes . . . 3. *B. exaristata.*

Distribution. Timor, Lesser Sunda Islands, northern Australia.

Synonyms. None known.

1. Peduncles 2–5 flowered (in some specimens of *B. hainesii* a few solitary flowers have been observed along with 2–3 flowered peduncles). Bristle in the sinus of the petal lobes far exceeding the lobe tips.

4. Calyx lobes a quarter to one-fifth the length of the calyx, erect of slightly spreading in fruits. Petals 1.5–2 mm long . . . 4. *B. parviflora.*

Distribution. Tropical Asia to Australia (North Queensland and Northern Territory), and Melanesia (introduced in Hawaii).

Synonyms. B. ritchiei, Rhizophora parviflora Roxb.

4. Calyx lobes half length of the calyx, reflexed in fruit. Petals 4–9 mm long.

5. Mature flowers 10–12 mm long. Petals 3–4 mm long. Calyx tube 2 mm diam., lobes 8, completely reflexed in fruit . . . 5. *B. cylindrica.*

Distribution. Almost like the former but not recorded from Melanesia.

Synonyms. B. caryophylloides Bl.; *B. malabarica* Arn.; *Rhizophora ceratophylloides* Gmel etc.

5. Mature flowers 18–22 mm long. Petals 7–9 mm long. Calyx tube 5 mm diam., lobes 10, patent at right angles in fruit . . . 6. *B. hainesii.*

Distribution. Burma, Thailand, Malay Peninsula, Papua New Guinea

Synonym. Rhizophora caryophylloides (non Burmf.) Griff.

Ceriops—key to the species

1. Petals with 3 clavate appendages at the apex. Anthers obtuse at the apex, about a quarter to one-sixth the length of the filament. Calyx lobes on the fruits reflexed or widely patent. Hypocotyl 15–25(–35) cm . . . 1. *C. tagal.*

Distribution. East Africa, Madagascar, Seychelles, tropical Asia, Australia, Micronesia.

Synonyms. C. candolliana Arn., *C. forstenianna* Bl., *C. boviniana* Tul. etc.

1. Petals fringe-like divided at the apex. Anthers apiculate, longer than the filament or rarely as long as it. Calyx lobes in fruit erect or ascending. Hypocotyl 9–15 cm long . . . 2. *C. decandra.*

Distribution. Almost everywhere in tropical Asia, including Kalimantan, Philippines and New Guinea but not reported from Sumatra.

Synonym. C. roxburghiana Arn.

Kandelia—key to the species

Monotypic genus (see key above), known by *K. candel* (L.) Druce.
Distribution. generally rare but found in many tropical countries of Asia (India, Burma, Thailand, Malay Peninsula, north-west Kalimantan) up to Hong Kong, Taiwan and southern Japan.
Synonyms. K. rheedei W. & A., *Rhizophora candel* L.

Sonneratiaceae

This family is represented in the mangroves by five species of *Sonneratia* only. As far as we know, the best, simplest and most reliable classification is that of Backer and van Steenis (1951), which is mainly based on flowering and fruiting material. The leaf shape usually is not a good indicator for specific differentiation.

KEY TO THE SPECIES OF SONNERATIA

1. *Petals absent*

2. Stigma large fungiform, 6 mm. Calyx usually 4 merous 1.5 to 2 cm long. Ovary 4–8 celled. Fruit 1.5 to 2 cm in diameter . . . *S. apetala* Buch-Ham.

Distribution. Mainly in India and Sri Lanka. Muddy flats in less saline localities.

2. Stigma capitale, 3 mm diam. Calyx usually 6–8 merous 2.5 to 4.5 cm long.
3. Calyx smooth throughout, not ribbed. Leaves obovate to suborbicular (subcircular in outline) 7–10 by 5–9 cm, rather thickish; nerves distinctly prominent on the upper surface . . . *S. griffithii* Kurz.

Distribution. Chittagong, Burma, Thailand (Phuket District), Andaman Islands. Its ecology is almost unknown.

3. Calyx finely verruculose; inner side of the segments strongly tinged red. Leaves broadly rounded at the top; 4–10 by 3–9 cm. Nerves not distinctly prominent on the upper surface . . . *S. ovata* Backer.

Distribution. Land side, in the less saline areas on muddy flats; never on coral reefs. Not common.

1. *Petals present*

4. Petals linear, 1.3 to 2 cm by 1.5 mm, white or tinged red in the lower part. Inner side of the petals red. Leaves obovate or oval 5–12 by 3–9 cm . . . *S. alba* J. Smith.

Distribution. Tropical East Africa, tropical Asia, northern Australia. Prefers salt-water, rocks as well as sand.

4. Petals linear lanceolate, dark red, 1.6–3.5 cm by 1–3 mm. Inner side greenish or yellowish-white. Leaves elliptic-oblong, base often cuneate; 5–13 × 2–5 cm . . . *S. caseolaris* L. Engl.

Distribution. Tropical Asia, northern Australia, Solomon Islands, New Hebrides. Never found on coral banks, frequent on muddy soils along slow moving waters, ascending as far as the high-tide line.

Several hybrids have been described recently by Muller and Hou Liu (1966), but contrary to what has been done for the *Rhizophora* group, they have not been given names (e.g. *Sonneratia alba* × *ovata* and *Sonneratia alba* × *Caseolaris*). On the whole this group of mangrove plants is well defined and well known. However, the lack of herbarium material and of field observations for *S. griffithii* is very conspicuous. It is hoped that this gap will be filled in the near future.

References

BACKER, C. A.; VAN STEENIS. 1951. Sonneratiaceae. *Flora Malesiana, Ser. I.*, Vol. 4, No. 3, pp. 280–9.

BRETELER, F. J. 1969. The Atlantic Species of *Rhizophora*. *Acta Bot. Neerl.*, Vol. 18, No. 3, pp. 434–41.

——. 1977. America's Pacific Species of *Rhizophora*. *Acta Bot. Neerl.*, Vol. 26, pp. 225–30.

DING HOU. 1958. Rhizophoraceae. *Flora Malesiana, Ser. I.*, Vol. 5, pp. 429–72.

——. 1960. A Review of the Genus *Rhizophora*, with Special Reference to the Pacific Species. *Blumea*, Vol. 10, No. 2, pp. 625–34.

MULLER, J.; HOU-LIU, S. Y. 1966. Hybrids and Chromosomes in the Genus *Sonneratia*. *Blumea*, Vol. 14, No. 2, pp. 337–43.

SAHAVACHARIN, O., et al. 1979. *Some Orchids Found in Mangrove Forests of Thailand*, pp. 51–8. (Biotrop Special Publication No. 10.)

TOMLINSON, P. B. 1978. *Rhizophora* in Australasia—Some Clarification of Taxonomy and Distribution. *J. Arnold Arboretum* (Harvard University), Vol. 59, pp. 156–69.

Bibliography

BAILLON, H. *Histoire des plantes*, Vol. VI. Paris, Hachette, 1877. 523 pp.

CHAPMAN, V. J. *The Algae*. London, Macmillan, 1961. 356 pp.

——. *The Marine Algae of Jamaica, Part 1; Mycophyceae and Chlorophyceae*. Kingston, Institute of Jamaica, 1961.

HUTCHINSON, J.; DALZIL, J. M. *Flora of West Tropical Africa*, Vol. I, Part 1. London. The Crown Agents for the Colonies, 1927. 246 pp.

IRVINE, F. R. *Woody Plants of Ghana*. Oxford University Press, 1961. 868 pp.

JOHNSON, A. A Survey of the Occurrence of Epiterranean Soil Algae in Singapore. *Gardens Bull*, Vol. XXVI, No. II, 1973, pp. 289–302.

——. *The Algae of Singapore Mangrove*, 1979. pp. 45–9. (Biotrop Special Publ. No. 10)

LEECHMAN, A. The Genus *Rhizophora* in British Guiana. *Royal Bot. Gardens, Kew, Bull. of Miscellaneous Information*, 1918, pp. 4–8.

LEVRING, T.; HOPPE, H. A.; SCHMID, O. J. *Marine Algae: A Survey of Research and Utilization*. 1969, Hamburg, De Gruyter.

LINNAEUS, C. *Species Plantarum* (Facsimile of the First Edition 1753). London, Bernard Quaritch, 1957. 560 pp.

SALVOZA, F. M. *Rhizophora*. *Nat. Appl Sc. Bul. Un. Philip.*, Vol. 5, 1936, pp. 179–237.

SAVORY, H. J. A Note on the Ecology of *Rhizophora* in Nigeria. *Kew Bull.*, 1953, No. 1, pp. 127–8.

6

Methods for studying mangrove structure

Gilberto Cintrón
Marine Resources Section
Department of Natural Resources
Government of Puerto Rico
Box 5887
Puerta de Tierra, Puerto Rico 00906

Yara Schaeffer Novelli
Instituto Oceanográfico
Universidad de São Paulo
Cidade Universitaria
Butanta, CEP–05508, Brazil

Introduction

In spite of numerous papers on mangrove floristics, systematics, phytogeography and related topics, there is little published information on mangrove forest structure. The comparative value of many of the available studies is not as great as it could be; authors' differing research goals have led to the adoption of non-uniform measurement techniques, yielding results that are often difficult to compare with others. This chapter reviews the development of structural measurements applicable in mangrove forest ecosystems.

The architecture of a forest is influenced by the magnitudes and periodicities of such forcing functions as tides, nutrients, hydroperiod, and stressors like hurricanes, drought, salt accumulation and frost. Because the action of these factors varies widely over geographic regions, mangrove stands exhibit wide regional and local variation in structural characteristics. Also, where species diversity is high, structural variation is even greater.

In 1974, Lugo and Snedaker developed a classification scheme for mangroves based on tidal and hydroperiod characteristics. Implicit in their scheme was the assumption that this environmental factor was the most important component of the energy signature of a mangrove forest. This classification system, further modified by Cintrón et al. (1980), serves to identify some common patterns of mangrove response to varying environmental conditions and it remains a useful framework for the first-approximation classification of mangrove stands. As amended, it recognized three general forest types: riverine, fringe and overwash, and basin. Dwarf, scrub and hammock mangroves are recognized as special sub-types responding to localized geologic or edaphic conditions. As a general rule, riverine forests exhibit the highest level of structural development, followed by basin, and finally by fringe and overwash types.

In 1973, researchers began searching for the most meaningful ecosystem parameters to use in rapid characterization of mangrove stands over wide geographic areas; the methods and parameters selected needed to be simple, time-cost effective and universally applicable. A survey of some twenty-five mangrove stands in Florida, Puerto Rico and Costa Rica was undertaken to test the methods chosen (Pool et al., 1977). Since then similar techniques have been used to describe Puerto Rican mangroves (Martínez et al., 1979), mangrove stands in Brazil (Schaeffer Novelli et al., 1980) and Costa Rica (Jiménez, 1981). The structural data available as of 1980 have been reviewed by Cintrón et al. (1980) and by Cintrón and Schaeffer Novelli (1983). In this chapter we shall explain some of the structural parameters measured by us and by others in the description of mangrove stands, and some commonly used measurement techniques. We shall also try to show how our structural measures relate to tree and forest architecture.

Structural attributes

DIAMETER

One of the simplest forms of forest stand characterization is the measurement of tree diameters. Diameter is closely related to stand development, and can easily be converted to basal area (the area occupied by the tree stems). Other stand characteristics such as height, crown diameter and biomass can often be predicted from stem diameter. Since a tree stem is a tapering solid, and its diameter decreases towards the crown, it is vital to standardize the point along the stem at which diameter is measured. By convention, diameter is always measured at 1.3 m above ground level and this measurement is referred to as diameter at breast-height (dbh). In mangrove stands the measurement is taken outside the bark and, in general, unless specified to the contrary, dbh is assumed to include the thickness of the bark.

Diameter measurements of trees are usually made with a special tape. This tape actually measures girth, not diameter, making use of the mathematical relationship between circumference of a circle and its diameter ($c = \pi \times d$). On a diameter tape the graduations are in intervals of 3.14 units, allowing a direct reading of diameter. If no diameter tape is available, the girth or circumference of the tree can be measured with any tape. The readings than can be converted to diameter, dividing by π. Diameter estimation based on measurements of circumference, as described above, will not be perfectly accurate unless the tree is circular in cross-section. When the stem has an eccentric form, use of a tape will lead to error. For these stems the desired measurement would be the diameter of a circle having a cross-sectional area equal to that of the tree being measured. Tape measurements of these elliptical or ovoid stems overestimate the diameter, since the circumference of an eccentric stem is proportionally greater than that of a circular stem of equal area. For these stems it is best to measure the diameter with a tree caliper, taking 'true' diameter as the average of the long and short axes. Fortunately, serious eccentricity of form is not common in most mangrove species or stands, so for most purposes the errors resulting from stem eccentricity can be ignored. When diameter is being measured, it is important to take care when wrapping the tape round the stem to ensure that it is level and stretched firmly against the trunk. Fibreglass tapes are preferable to steel tapes in the highly corrosive mangrove environment. Tapes need to be cleaned and lubricated often.

Because of the shape and growth forms of mangrove trees, it is sometimes difficult to decide where to measure the diameter. For this reason we recommend the following standard procedures:

When a stem forks below breast height, or sprouts from a single base close to the ground or above it, measure each branch as a separate stem.

When the stem forks at breast height or slightly above, measure the diameter at breast height or just below the swelling caused by the fork.

When the stem has prop roots or fluted lower trunk, measure the diameter above them.

When the stem has swellings, branches or abnormalities at the point of measurement, take the diameter slightly above or below the irregularity where it stops affecting normal form.

BASAL AREA

Basal area is the space covered by a tree stem. By convention, basal area is the cross section of a stem at the point where dbh is measured. The basal area of a stand is the sum of the individual basal areas of all trees greater than a certain diameter per unit ground area. Basal area is a valuable descriptive parameter. It is usually expressed as m^2 per hectare, for diameter limits of $\geqslant 2.5$ cm, $\geqslant 5$ cm or $\geqslant 10$ cm. Basal area ia a good measure of the overall stand development, and it can be related to wood volume and biomass.

The basal area (g) of a stem is computed using the formula, $g = r^2$, but since $r = \text{dbh}/2$, therefore, $g = \pi/4\ (\text{dbh}^2)$.

To express g in m^2 as a function of dbh measured in centimetres: $g(m^2) = \text{dbh}^2/4(10{,}000) = 0.00007854\ \text{dbh}^2$.

The basal area of a stand is estimated by measuring the diameters of all the trees larger than a specified size in a given area. Individual diameters are converted to basal area and added together. This method requires marking out a plot of known area. There are also techniques that permit estimation of basal directly, without the necessity of measuring and marking a plot. These techniques will be discussed later (see section on plotless methods below).

DENSITY

The density of a stand is the number of stems greater than a given diameter per unit area. In stands with many trees of poor form, it may become difficult to define what constitutes a single tree. In the section on diameter we suggested counting all branches formed below breast height as individual stems, or trees. When measuring species that coppice like *Avicennia* or *Laguncularia* it may be desirable to record the number of clusters and the mean number of stems per cluster as well as the total stem number. In order to measure stem density, it is necessary to select a plot size and shape in which to do the counting. These choices are influenced by the density and variability of the forest being studied. Plot shape should ideally yield a low ratio of boundary-to-plot area, thus reducing the chance of including trees not belonging to the plot in the count. The size of the plot should allow counting and other measurements to be completed in a reasonable time span.

Pool et al. (1977) used 0.1 hectare plots (subdivided into twenty 5 × 10 m subplots in which all steams greater than 2.5 cm in diameter were counted). A smaller plot size may be more practical for young stands with very high stem densities, since the accuracy of a count is not a function of the area sampled but rather of the number of enumerations. Where individuals are far apart, as in a mature stand, fewer trees would be counted than in a younger, denser stand. As a result a larger plot size is needed in more mature, open forests, but in a young stand the plot can be relatively small.

Another fixed-area sampling unit often used in ecological studies is the strip transect. A strip transect is essentially a very long rectangular plot. Because of its shape it has several disadvantages for mangrove work. First, it has a large amount of edge relative to its area, so that many decisions about whether or not to include individual stems in the count must be made relative to, say, a rectangular plot. Second, it is difficult to mark the boundaries as well as in a shorter, wider plot. The most serious drawback of the strip transect in the mangrove environment, however, is the high probability that it may cross soil and vegetational gradients, leading to sampling of heterogeneous plots. Properly laid-out strip transects in mangrove stands should run at right angles to all environmental gradients which lead to homogeneous sample units that include all structural variation associated with the gradient. A series of several such strips, parallel to each other and perpendicular to the environmental gradient at different points, can then be used to assess effect of the gradient on stand structure. A strip transect may be composed of ten 10 × 10 m plots placed end to end for a total plot area of 0.1 ha.

In mangroves, research experience is the best guide in selecting an appropriate plot size. We recommend field trials before the final selection of plot size and shape.

MEAN STAND DIAMETER

The mean diameter of a stand is defined simply as the diameter of the stem of mean basal area. The mean basal area per tree is g = basal area per hectare (BA)/number of trees per hectare (n). The diameter of the stem of mean basal area is given by $(\text{dbh})^2(0.00007854)$ so that $\text{dbh} = \sqrt{(\text{BA}/n)/(0.00007854)}$ or, $\text{dbh} = \sqrt{(\text{BA})(12732.39)/n}$, where dbh is the diameter of the tree of mean basal area, BA is the stand basal area and *n* is stand density. This is not the same as the true arithmetic mean of tree diameters in the stand; it is a useful descriptive measure and can be used for comparisons between stands and for correlation with other forest structural measures. Figure 6.1 shows density values for 114 mangrove stands and plotted against dbh. The relationship in inverse: density (also called stocking level by foresters) decreases as the stand matures and mean diameter increases.

TREE HEIGHT

Total height is the linear vertical distance between the ground and the tip of the tree crown. This measurement is often made with clinometers. A variety of models are available, including Haga®, Blume-Leiss® and Suunto®. Height measurements with a clinometer are based on the trigonometric relationships of right triangles. It is, therefore, also necessary to measure the distance from the observer to the base of the tree. The distance to the tree, multiplied by the tangent of the angle measured from the observer to the top of the tree, yields a height. Some clinometers are precalibrated to read height directly for some measured observer-to-tree distances. Others require computation. In any case, it is necessary to add the height of the observer's eye to the height measured by the clinometer to obtain total tree height. Often it is difficult to see the top of the tree in a closed stand and some moving around may be necessary to select a good viewing angle.

For rapid reconnaissance studies an optical rangefinger is often useful in height measurements. In using this device the observer stands directly below the crown and sights vertically at the topmost branches. Split images in the rangefinder are superimposed by turning a range knob, and height is read directly off a scale. Again, the observer's eye height must be added to the reading to obtain total tree height. With the rangefinder it is crucial that the observer be directly under the tree measured, since horizontal displacement of the observer will cause overestimation of tree height. It is sometimes difficult to distinguish the topmost crown branches from directly underneath them, but this disadvantage compared to the clinometer is offset by the convenience of the relatively small radius of horizontal movement needed around the trees to be measured, an advantage that any researcher who has tried to navigate on the ground in most mangrove environments can appreciate.

Both clinometer and rangefinder-based height measurements assume that the observer and tree are on level ground. The assumption is usually fulfilled in a mangrove stand, and slope corrections are hardly ever necessary. In young stands with many small trees, it is far easier to measure height using a graduated telescoping rod. This method is practical only in stands with a height less than 4–6 m. Although the method has been used to measure trees up to 7 m tall, the length of the extended rod makes it cumbersome to move and each measurement takes longer than similar measurements taken with a rangefinder or clinometer.

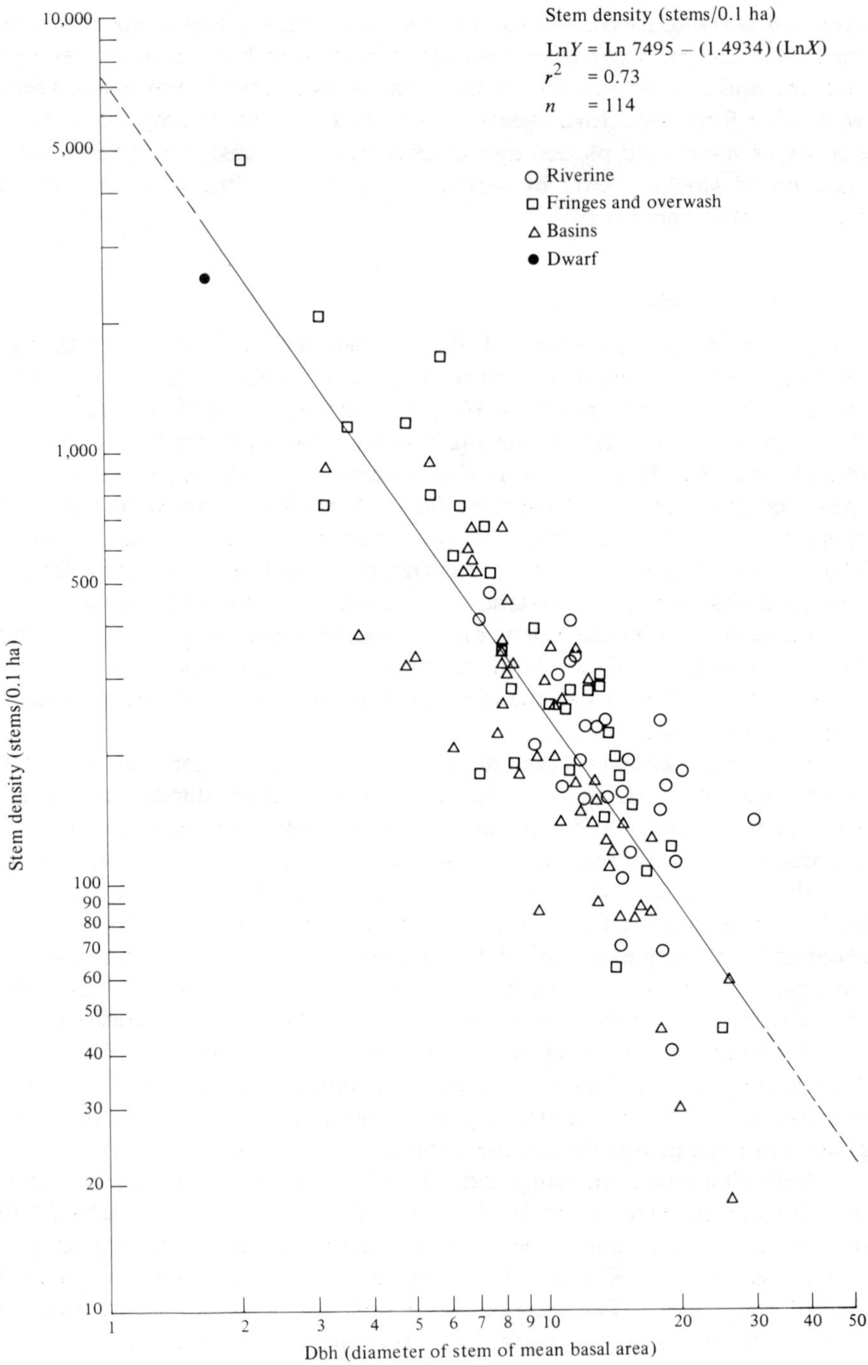

Figure 6.1
Density and dbh for 114 New World mangrove stands. Solid line is the least square fit for the data set.

Relative density, dominance and frequency

Fixed-area plots and distance methods yield three quantitative parameters which are useful in describing the plant community. Together, they can be used to interpret the importance of the contribution of each component species to the stand in terms of density, contribution to basal area (dominance) and the probability of occurrence throughout the plot (frequency). The three measures are computed as follows:

Relative density = (number of individuals of a species/total number of individuals) × 100.

Relative dominance = (total basal area of a species/basal area of all species) × 100.

Relative frequency = (frequency of a species/sum-frequency of all species) × 100.

Frequency is the percentage of plots in which a given species is present. Two problems arise with frequency measurements which considerably reduce its value. The most serious disadvantage is that the probability of a plant occurring in a given plot is a function of plot size, increasing as plot size increases. As a result large plots will yield larger frequencies of a given species than smaller plots. Therefore, comparison of frequencies can only be made between plots of a similar size. In addition, aggregation of plant species will influence frequency measurements, since clumping will lead to a lowered frequency value. Finally, frequency cannot be computed when using variable plot sampling techniques, since the presence of a given species in a plot is not a function of plot size but rather of individual stem diameter, and plot sizes are not fixed.

In quantitative vegetation studies, the three relative measures discussed above are commonly used in combination. This is the so-called importance value of Curtis (1959). The importance value is the sum of relative density, relative dominance and relative frequency. The importance value of a species reaches 300 in monospecific stands. These relative measurements should generally be used as a supplement to absolute values. Relative data alone are of limited value, since both densely and sparsely vegetated sites could have similar relative values if species composition is similar. Relative measurements are useful in assessing the contribution of a given species to the total density and basal area of a stand whose absolute attributes are known.

Plotless methods

BASAL AREA BY POINT SAMPLING

In 1948, Walter Bitterlich developed a remarkably effective system (angle-count cruising) to estimate stand basal area without requiring direct measurement of either plot area or tree diameter. In this method the observer occupies a sampling point and sights through a device designed to subtend a fixed and specified horizontal angle. The observer rotates around the central sampling point and tallies only the trees that are greater in size than the specified angle. Only those stems close enough or large enough to fill the sighting angle are counted. The number of trees counted is multiplied by a constant (depending on the angle used) to yield directly the basal area per unit ground area. Each tree counted, regardless of its dbh, contributes the same basal area per hectare to the estimate.

In order to understand how the technique works, imagine a cross-piece 50 cm long with a flat plate 1 cm across attached to the end (Fig. 6.2). This device subtends an angle of 1°10′. With this gauge, trees will count when they are not further away than fifty times their diameter. A tree 10 cm in diameter must be within 5 m of the sampling point to count. The plot area is therefore $\pi(5)^2$ or 78.54 m^2 and this particular stem's contribution to the stand basal area is 0.007854 m^2/78.54 m^2 or 0.0001 m^2/m^2 which is the same as 1 m^2/ha. If a 4 cm tree were sighted its plot radius would be 2 m, the plot area would be 12.57 m^2 and the tree's contribution to stand basal area would be 0.0012257 m^2/12.57 m^2 or 1 m^2/ha. Thus, regardless of the stem's diameter, if it falls within the gauge's angle and is tallied, it contributes 1 m^2/ha to the stand basal area. It is unnecessary to measure the diameter of each tree; due to the angle chosen each tree that falls within the count contributes a unit basal area to the total. Total basal area using this particular gauge is simply equal to the number of stems counted from the sampling point.

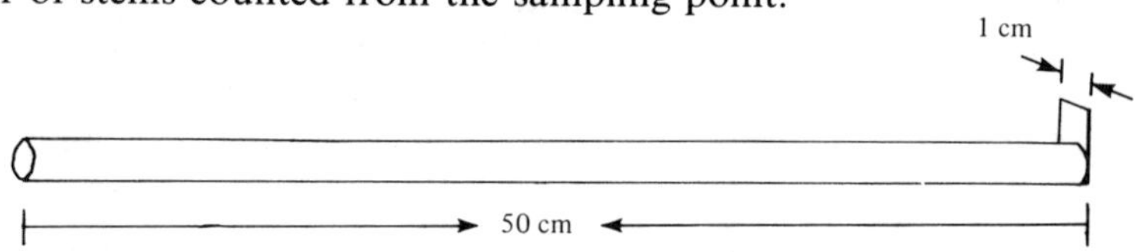

Figure 6.2
Bitterlich angle-gauge for measuring basal area by counting trees.

The gauge constant at the borderline condition is given by the formula $k = \text{dbh}/100R = 2 \sin \alpha/2$, where dbh is the diameter in centimetres of the tree and R is plot radius in metres. The constant 100 converts the denominator units into metres. The symbol α is the gauge angle. Substituting in this formula, we see that for a 10 cm tree with a plot radius of 5 m, $k = 0.02$. The basal area factor (BAF) per hectare is: $\text{BAF} = \pi(\text{dbh})^2/4(10{,}000) \times 10{,}000/\pi R^2 = (\text{dbh})^2/4R^2 = 2{,}500\ k^2$. In our example, BAF = 1 m^2 per hectare.

DENSITY BY POINT SAMPLING

It is possible to calculate the density contribution of a given stem by measuring its diameter. Since each enumerated tree in the example given contributed 1 m^2 per ha to the stand's basal area, this number divided by the basal area of the counted stem (in m^2) must correspond to the number of stems represented by the counted stem. For example, a 10 cm dbh stem (basal area 0.007854 m^2)

represents 1/0.007854 or 127.3 stems per hectare. In general, stem density per hectare is equal to the sum of the densities represented by individual counted trees: stems/ha = $BAF/(0.00007854)(dbh)^2$.

For convenience a table of values for this function has been prepared (Table 6.1). In this table the diameter of the tallied stem is entered to obtain its contribution (in stems/ha) to total stand density.

TYPES OF ANGLE GAUGES AND THEIR USE

We earlier described a simple, easily constructed device for projecting an angle of 1°10′ corresponding to a BAF of 1 m^2/ha. This device must have a peep-sight at one end although if a hollow tube is used, a peep-hole or notch becomes unnecessary. In using the angle gauge the observer's eye becomes the vertex of the sighting angle and therefore the gauge must be revolved about this point. When properly constructed and calibrated this type of device is as accurate as any other, more expensive point-sampling device.

The most commonly used angle gauge for horizontal sampling is the wedge prism. These prisms are made of optical glass and are square, rectangular or circular. Because of the tapered glass surfaces light rays are refracted at a given angle. When a stem is viewed through the prism the bole appears displaced as shown in Figure 6.3. The amount of displacement is a function of the prism's strength. When using a wedge prism, it is important to hold it precisely over the sample point at all times, rotating the observer's body around the point and counting only those stems in which image separation is not complete. Borderline trees should be checked as described in the following section to determine if they are 'in'. If 'in', these trees contribute half the normal basal area and also contribute half their computed stem density. Each tree is sighted at breast height. The prism may be held at any convenient distance from the eye provided a clear image is obtained and that the prism is always positioned exactly over the sample point. It is important to hold the prism vertically, at right angles to the line of sight (Fig. 6.4). If the prism is held carelessly, large errors in the tally may occur. If a tree is leaning, the prism or angle gauge must be vertically rotated so that the vertical axis of the gauge matches that of the leaning tree. If a prism is used then the sides of the prism should be parallel to the sides of the tree. When the tree leans towards or away from the observer it should be checked by measuring its distance to see if it should be included in the count. In dense stands some trees may be hidden behind others. In this case the observer can step to the side of the sampling point but maintain the original distance from the observer to the tree in question. This will also be necessary where two or more stems overlap.

CHECKING BORDERLINE TREES

When an apparent borderline tree is encountered, its status is resolved by checking the plot radius corresponding to the tree's diameter. The plot radius

TABLE 6.1. Stems per hectare represented by a stem of a given diameter. For angle gauges of BAF 1 m^2/ha

Diameter (cm)	Tenth of a centimetre 0.0	0.1	0.2	0.3	0.4	0.5	0.6	0.7	0.8	0.9
0	0	1 273 239.5	318 309.9	141 471.1	79 577.5	50 929.6	35 367.8	25 984.5	19 894.4	15 718.0
1	12 732.4	10 522.6	8 841.9	7 534.0	6 496.1	5 658.8	4 973.6	4 405.7	3 929.8	3 527.0
2	3 183.1	2 887.2	2 630.7	2 406.9	2 210.5	2 037.2	1 883.5	1 746.6	1 624.0	1 514.0
3	1 414.7	1 324.9	1 243.4	1 169.2	1 101.4	1 039.4	982.4	930.1	881.7	837.1
4	795.8	757.4	721.8	688.6	657.7	628.8	601.7	576.4	552.6	530.3
5	509.3	489.5	470.9	453.3	436.6	420.9	406.0	391.9	378.5	365.8
6	353.7	342.2	331.2	320.8	310.8	301.4	292.3	283.6	275.4	267.4
7	259.8	252.6	245.6	238.9	232.5	226.4	220.4	214.7	209.3	204.0
8	198.9	194.1	189.4	184.8	180.4	176.2	172.2	168.2	164.4	160.7
9	157.2	153.8	150.4	147.2	144.1	141.1	138.2	135.3	132.6	129.9
10	127.3	124.8	122.4	120.6	117.7	115.5	113.3	111.2	109.2	107.2
11	105.2	103.3	101.5	99.7	98.0	96.3	94.6	93.0	91.4	89.9
12	88.4	87.0	85.5	84.2	82.8	81.5	80.2	78.9	77.7	76.5
13	75.3	74.2	73.1	72.0	70.9	69.9	68.8	67.8	66.9	65.9
14	65.0	64.0	63.1	62.3	61.4	60.6	59.7	58.9	58.1	57.4
15	56.6	55.8	55.1	54.4	53.7	53.0	52.3	51.7	51.0	50.4
16	49.7	49.1	48.5	47.9	47.3	46.8	46.2	45.7	45.1	44.6
17	44.1	43.5	43.0	42.5	42.1	41.6	41.1	40.6	40.2	39.7
18	39.3	38.9	38.4	38.0	37.6	37.2	36.8	36.4	36.0	35.6
19	35.3	34.9	34.5	34.2	33.8	33.5	33.1	32.8	32.5	32.2
20	31.8	31.5	31.2	30.9	30.6	30.3	30.0	29.7	29.4	29.1
21	28.8	28.6	28.3	28.1	27.8	27.5	27.3	27.0	26.8	26.5
22	26.3	26.1	25.8	25.6	25.4	25.2	24.9	24.7	24.5	24.3
23	24.1	23.9	23.7	23.5	23.3	23.1	22.9	22.7	22.5	22.3
24	22.1	21.9	21.7	21.6	21.4	21.2	21.0	20.9	20.7	20.5
25	20.4	20.2	20.0	19.9	19.7	19.6	19.4	19.3	19.1	19.0
26	18.8	18.7	18.5	18.4	18.3	18.1	18.0	17.9	17.7	17.6
27	17.5	17.3	17.2	17.1	17.0	16.8	16.7	16.6	16.5	16.4
28	16.2	16.1	16.0	15.9	15.8	15.7	15.6	15.5	15.4	15.2
29	15.1	15.0	14.9	14.8	14.7	14.6	14.5	14.4	14.3	14.2
30	14.1	14.1	14.0	13.9	13.8	13.7	13.6	13.5	13.4	13.3
31	13.2	13.2	13.1	13.0	12.9	12.8	12.8	12.7	12.6	12.5
32	12.4	12.4	12.3	12.2	12.1	12.1	12.0	11.9	11.8	11.8
33	11.7	11.6	11.6	11.5	11.4	11.3	11.3	11.2	11.1	11.1
34	11.0	10.9	10.9	10.8	10.8	10.7	10.6	10.6	10.5	10.5
35	10.4	10.3	10.3	10.2	10.2	10.1	10.0	10.0	9.9	9.9
36	9.8	9.8	9.7	9.7	9.6	9.6	9.5	9.5	9.4	9.4
37	9.3	9.3	9.2	9.2	9.1	9.1	9.0	9.0	8.9	8.9
38	8.8	8.8	8.7	8.7	8.6	8.6	8.5	8.5	8.5	8.4
39	8.4	8.3	8.3	8.2	8.2	8.2	8.1	8.1	8.0	8.0
40	8.0	7.9	7.9	7.8	7.8	7.8	7.7	7.7	7.6	7.6
41	7.6	7.5	7.5	7.5	7.4	7.4	7.4	7.3	7.3	7.3
42	7.2	7.2	7.1	7.1	7.1	7.0	7.0	7.0	7.0	6.9
43	6.9	6.9	6.8	6.8	6.8	6.7	6.7	6.7	6.6	6.6
44	6.6	6.5	6.5	6.5	6.5	6.4	6.4	6.4	6.3	6.3
45	6.3	6.3	6.2	6.2	6.2	6.2	6.1	6.1	6.1	6.0
46	6.0	6.0	6.0	5.9	5.9	5.9	5.9	5.8	5.8	5.8
47	5.8	5.7	5.7	5.7	5.7	5.6	5.6	5.6	5.6	5.5
48	5.5	5.5	5.5	5.5	5.4	5.4	5.4	5.4	5.3	5.3
49	5.3	5.3	5.3	5.2	5.2	5.2	5.2	5.2	5.1	5.1
50	5.1	5.1	5.1	5.0	5.0	5.0	5.0	5.0	4.9	4.9

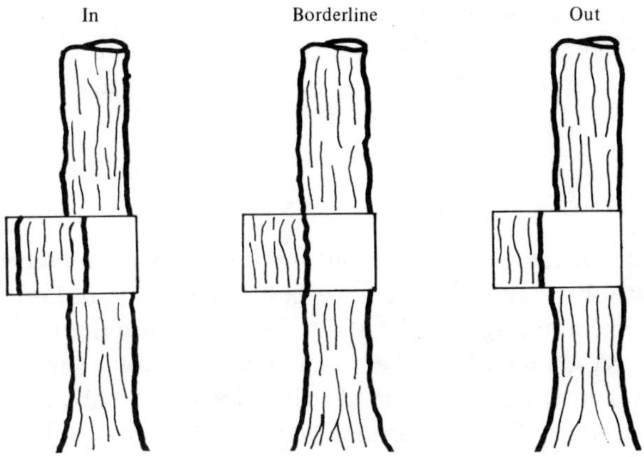

Figure 6.3
Image displacement by a prism showing trees which are recorded as count, borderline and stems outside of the 'plot'.

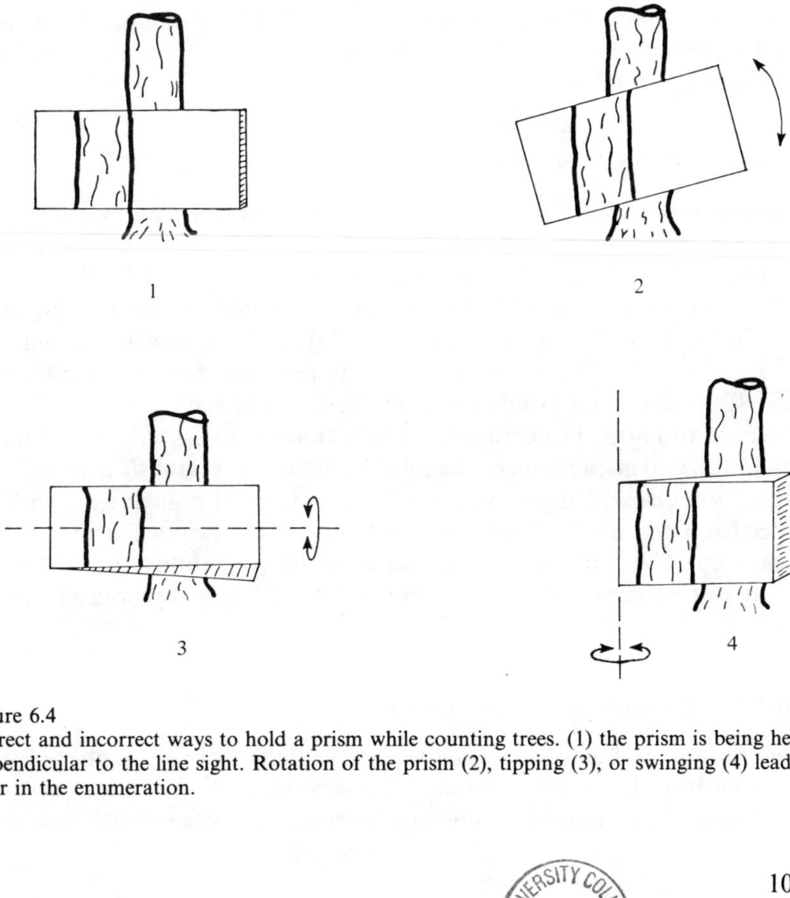

Figure 6.4
Correct and incorrect ways to hold a prism while counting trees. (1) the prism is being held perpendicular to the line sight. Rotation of the prism (2), tipping (3), or swinging (4) leads to error in the enumeration.

is given by the relationship: plot radius (metres) $= \text{dbh (cm)}/\sqrt{2\ \text{BAF}\ (\text{m}^2/\text{ha})}$. A tree is 'in' when the measured horizontal distance to its centre is equal to or less than the computed plot radius. If it is measured as 'out' it is omitted from the count.

SELECTING A GAUGE CONSTANT

The selection of the best gauge constant, or basal area factor (BAF) is determined mostly by stand density. The best accuracy is obtained with counts of 7–10 trees per sampling point. Large counts increase the probability of finding borderline trees, increasing the number of decisions (on whether or not to include stems in the count), and slowing down the tally, as well as increasing chances for error. Furthermore, a high-count density makes viewing more difficult and the tallying process more tedious. On the other hand, low counts are subject to greater variation, since inclusion or deletion of one individual makes a proportionately greater difference in the total. Small counts lead to undersampling and greater variation whereas large counts ('overlapping') are uneconomical.

Knowing that 10 is a convenient number of stems to count, we can use the following formula to find a suitable BAF: BAF = estimated basal area $(\text{m}^2/\text{ha})/10$, for a forest with an estimated basal area of 20 m^2/ha, a suitable prism would have a BAF of 2.

NUMBER OF POINTS REQUIRED

Point sampling is considered to be sampling with replacement and it can be assumed that the sampling population is infinite. To estimate the number of points needed the following formula may be used: $n = t^2(\text{CV})^2/(\text{SE})^2$, where n = number of points required for the accuracy selected or the sampling error and t (= Student's t) for the probability level defined for the stated sampling error (SE); CV = coefficient of variation in per cent for the forest to be sampled; SE = size of allowable sampling error in per cent.

The need to know the coefficient of variation for the specific forest under study requires that a preliminary sample be taken to establish a reasonable approximation. Grosenbaugh (1952) published a table of numbers of sampling points needed based on the desired sampling error and the coefficient of variation measured. The table is prepared so that the probability that the given error limit will be exceeded is one in three. This table is reproduced here as Table 6.2.

DISTANCE MEASURING TECHNIQUES

There are various techniques that allow the calculation of the number of trees per unit area from the average distance between trees. These techniques have the advantage of not requiring plot boundaries and are generally fast, since

TABLE 6.2. The number of samples needed (from an infinite population with a given coefficient of variation) so that the standard error of the sample will tend to be near the given limit of error

Coefficient of variation (%)	Specified limit for standard error ±1–0.5%	±5%	±10%	±20%
	N			
5	12	1	1	1
10	45	4	1	1
15	100	9	3	1
20	178	16	4	1
25	278	25	7	2
30	400	36	9	3
35	545	49	13	4
40	712	64	16	4
45	900	81	21	6
50	1 112	100	25	7
55	1 345	121	31	8
60	1 600	144	36	9
65	1 878	169	43	11
70	2 178	196	49	13
75	2 500	225	57	15
80	2 845	256	64	16
85	3 212	289	73	19
90	3 600	324	81	21
95	4 012	361	91	23
100	4 445	400	100	25
125	6 945	625	157	40
150	10 000	900	225	57
175	13 612	1 225	307	77
200	17 778	1 600	400	100

Source: Grosenbaugh (1952).

inter-tree distances tend to be low and therefore rapidly measured. In this section we will describe the so-called Point-Centred Quarter Method (PCQM). This method is fast, simple and is the most efficient of the distance methods. For these reasons it has had a wide acceptance.

THE POINT-CENTRED QUARTER METHOD

In this sampling method, points to be sampled are located randomly along a transect line. At the sampling point four quarters are established by crossing the compass direction of the transect line with a perpendicular line (Fig. 6.5). The distance from the sampling point to the midpoint of the nearest tree in each quadrant is measured. The four distances are averaged. The mean of the four distances measured has been found empirically to be the square root of the mean area per tree (Cottam and Curtis, 1956). The total stem density is

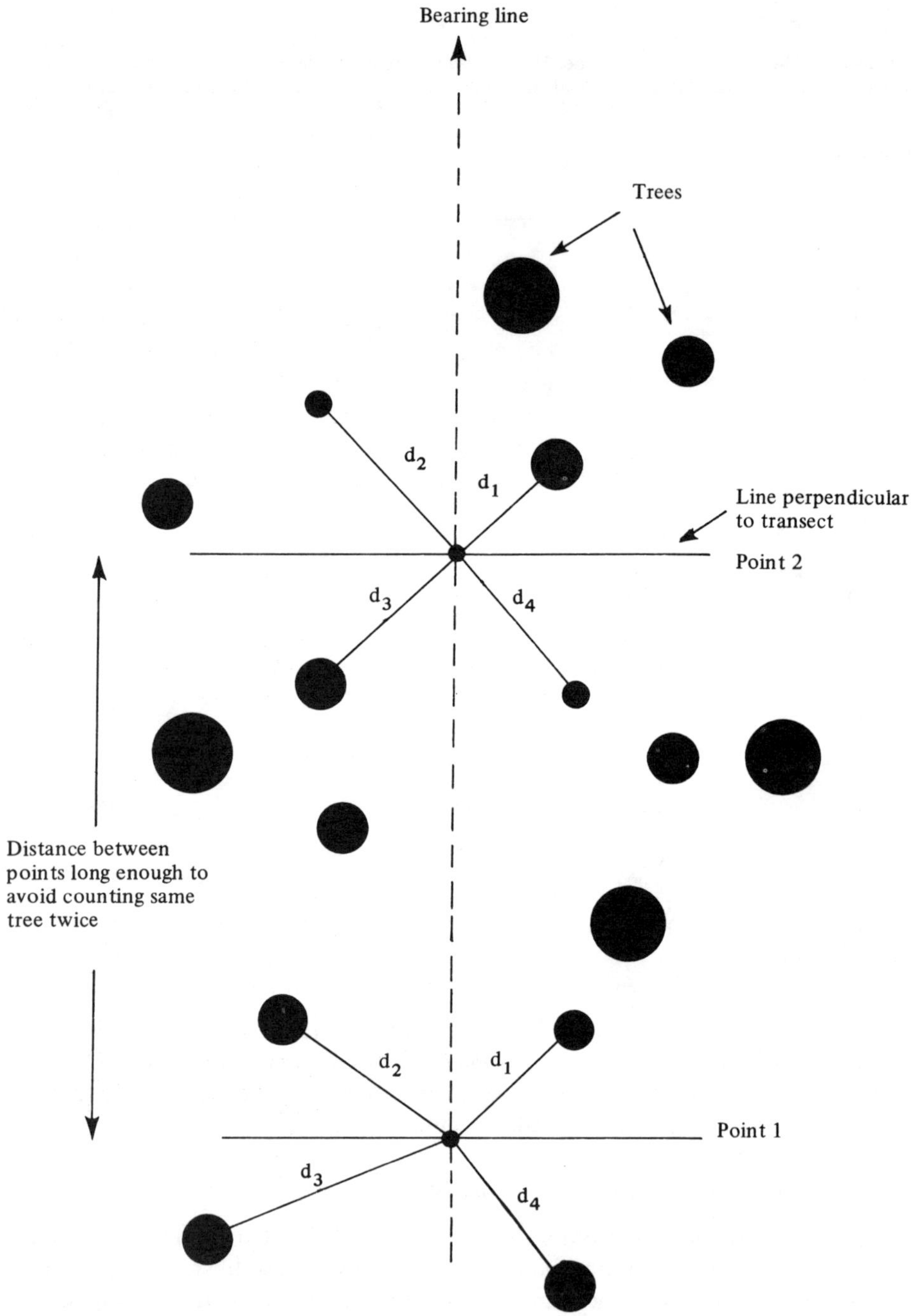

Figure 6.5
The Point-Centred Quarter Method. Four distances are measured from each point to the nearest trees in each of the quarters. The area around the point is divided into four quarters by a line running perpendicular to the bearing line. The points are separated by a distance long enough so that the same trees are not enumerated twice.

obtained by dividing the mean area per individual into the unit area on which density is to be expressed. Mathematical proof of the workability of the method was given by Morisita (1954). In contrast to other distance methods, no correction factor is needed.

The accuracy of the method increases with the number of sampling points and a minimum of twenty points is recommended (Cottam and Curtis, 1956). Although this system requires more time per point than other distance methods, this disadvantage is compensated for by the fact that it is intensive, requiring fewer sampling points. A biased result will be obtained, however, if tree measurements are made to the bark surface rather than to the centre of the trunk. The method has two limitations that may cause problems in some circumstances. First, an individual tree must be located in each quarter. Second, an individual tree must not be measured twice. Problems are most likely to be encountered in widely space stands. This method is useful for measuring species occurrence, density, basal area and frequency.

Example of PCQM analysis

Table 6.3 shows the anslysis procedure for the PCQM data. Only five points are shown to save space, but at least twenty points need to be sampled per stand. All trees 2.5 cm in diameter or greater were tallied. The data is tabulated as shown for convenience in the analysis. The mean distance for the twenty

TABLE 6.3. Point-Centred Quarter Method (PCQM) data analysis for five sampling points

Sampling point	Quarter number	Distance (m)	Species	dbh (cm)
1	1	2.5	*L. racemosa*	10.0
1	2	0.7	*A. germinans*	5.0
1	3	3.5	*R. mangle*	15.5
1	4	1.8	*A. germinans*	8.0
2	1	1.1	*A. germinans*	5.0
2	2	1.7	*A. germinans*	6.0
2	3	1.9	*A. germinans*	10.0
2	4	2.0	*R. mangle*	8.0
3	1	0.7	*A. germinans*	6.0
3	2	2.2	*L. racemosa*	15.0
3	3	1.5	*R. mangle*	8.0
3	4	2.8	*R. mangle*	12.0
4	1	1.3	*L. racemosa*	15.0
4	2	0.8	*A. germinans*	5.0
4	3	1.1	*A. germinans*	8.5
4	4	2.0	*R. mangle*	16.0
5	1	3.1	*L. racemosa*	12.0
5	2	1.6	*L. racemosa*	15.0
5	3	1.4	*A. germinans*	12.5
5	4	1.9	*L. racemosa*	13.0
TOTAL		35.6		
Mean distance		1.78		

measurements (five sample points with four distances each) was 1.78 m which corresponds to a mean density of $1/d^2 - 0.3156$ stems/m^2.

Density calculations are shown in Table 6.4(a). The species *L. racemosa*, for instance, occurred in six out of twenty plots or 0.3 times per quarter. This figure is multiplied by the total stem density per hectare (previously computed as the reciprocal of the square of the mean distance times 1,000) to compute density of this particular species. Other species are treated similarly.

Basal area (Table 6.5) is calculated by converting diameter data to basal area and then calculating mean basal area, individually for each species. Mean basal area for a species multiplied by absolute species density yields basal area contribution by species (Table 6.4(b)). Total stand basal area is the sum of individual species' contributions.

TABLE 6.4. Computation of density, basal area and absolute frequency using PCQM data

(a) Density computation

Species	Number in quarters	Number of stems per 0.1 ha
L. racemosa	6/20 = 0.30	(0.30) (315.6) = 95
R. mangle	5/20 = 0.25	(0.25) (315.6) = 79
A. germinans	9/20 = 0.45	(0.45) (315.6) = 142
Total per 0.1 ha		316

(b) Basal area computation

Species	Basal area (m^2)	Rank
L. racemosa	(95) (0.0142) = 1.349	2
R. mangle	(79) (0.0604) = 4.772	1
A. germinans	(142) (0.0047) = 0.667	3
	Total basal area = 6.788 m^2/0.01 ha	

(c) Absolute frequency $= \frac{\text{number of points with species}}{\text{total number of points}} \times 100$

Species	Absolute frequency
L. racemosa	4/5 × 100 = 80%
R. mangle	4/5 × 100 = 80%
A. germinans	5/5 × 100 = 100%
TOTAL	260%

Absolute frequency is calculated next by dividing the number of points at which a species occurs by the total number of sample points and multiplying by 100 (Table 6.4(c)). In Table 6.6, example computations for relative density, dominance and frequency, as well as importance values for each species, are shown.

TABLE 6.5. Computation of mean basal area from PCQM data

L. racemosa		*A. germinans*		*R. mangle*	
Dbh (cm)	Basal area (m^2)	Dbh (cm)	Basal area (m^2)	Dbh (cm)	Basal area (m^2)
10.0	0.0078	5.0	0.0020	15.5	0.0189
15.0	0.0177	8.0	0.0050	8.0	0.0050
15.0	0.0177	5.0	0.0020	8.0	0.0050
12.0	0.0113	6.0	0.0028	12.0	0.0113
15.0	0.0177	10.0	0.0078	16.0	0.0202
13.0	0.0133	6.0	0.0028	—	—
—	—	5.0	0.0020	—	—
—	—	8.5	0.0057	—	—
—	—	12.5	0.0123	—	—
Mean basal area	0.0142		0.0047		0.0604

TABLE 6.6. Computation of relative density, dominance, frequency and importance value (IV) from PCQM data

$$\text{Relative density} = \frac{\text{Number of individuals of species}}{\text{Total number of individuals}} \times 100$$

Species	Relative density
L. racemosa	95/316 × 100 = 30%
R. mangle	79/316 × 100 = 25%
A. germinans	142/316 × 100 = 45%
TOTAL	100%

$$\text{Relative dominance} = \frac{\text{Dominance of a species}}{\text{Dominance for all species}} \times 100$$

Species	Relative dominance
L. racemosa	1.349/6.788 × 100 = 20%
R. mangle	4.772/6.788 × 100 = 70%
A. germinans	0.667/6.788 × 100 = 10%
TOTAL	100%

$$\text{Relative frequency} = \frac{\text{Frequency of a species}}{\text{Sum frequency of all species}} \times 100$$

Species	Relative frequency
L. racemosa	80/260 × 100 = 31%
R. mangle	80/260 × 100 = 31%
A. germinans	100/260 × 100 = 38%
TOTAL	100%

TABLE 6.6.—*continued*

Species	Relative density	Relative dominance	Relative frequency	IV	IV rank
L. racemosa	30	20	31	81	3
R. mangle	25	70	31	126	1
A. germinans	45	10	38	93	2

Crown diameter

Crown diameter is important as a measure of the crowding or stocking of stems in an area. Unfortunately tree crowns are difficult to measure because of their irregularity, intermixing and the fact that their outline must be projected to the ground. Although there are instruments that can project a vertical line of sight, they are often difficult to use. Crown diameter may be measured with the aid of a fibreglass rod in lower trees. Tall and leaning trees present serious problems. Since the canopy outline is often irregular, it is desirable to measure at least the major and minor axes, taken at right angles to each other. The mean of these two measurements is taken to be the canopy diameter. Canopy diameter is often compared with stem diameter and this relation yields the crown-diameter/bole-diameter relationship. If we assume that canopy outlines are circular, the forest is even aged and therefore canopy size is uniform, and that there is no canopy overlap, then canopy diameter can be related to spacing. Figure 6.1 shows the 'normal' stocking levels determined from a best-fit analysis. From these data it is possible to estimate canopy diameters for stands of different mean diameters.

At full stocking the mean tree spacing is also the crown diameter, since canopies do not overlap. The mean tree spacing is also the square root of the average area per stem. Therefore the crown diameter (CD) or spacing may be calculated for different 'normally stocked' stands using the relation: $\mathrm{CD} = \sqrt{1000/n} = 31.6228/\sqrt{n}$, where n is the number of stems per 0.1 ha.

The crown diameter at a dbh of 1 cm (the y intercept of the derived CD/dbh relationship) can be easily calculated as:

$$b = 31.6228/\sqrt{7496}.$$

The slope parameter (m) is similarly derived from Figure 6.1 as:

$$m = \mathrm{Ln}\ Nb - \mathrm{Ln}\ Na/2(\mathrm{Ln\ dbh}_a - m\mathrm{dbh}_b) = 0.7476.$$

This relationship suggests that at normal stocking and throughout the development of a stand, crown diameter is related to dbh as a function of the type:

$$\mathrm{CD} = b(\mathrm{dbh})^m.$$

In Figure 6.6 we have plotted the canopy diameter for forty-five trees of *R. mangle*. The solid line is the best fit for the data points used. The broken line is that derived from the above computed parameters.

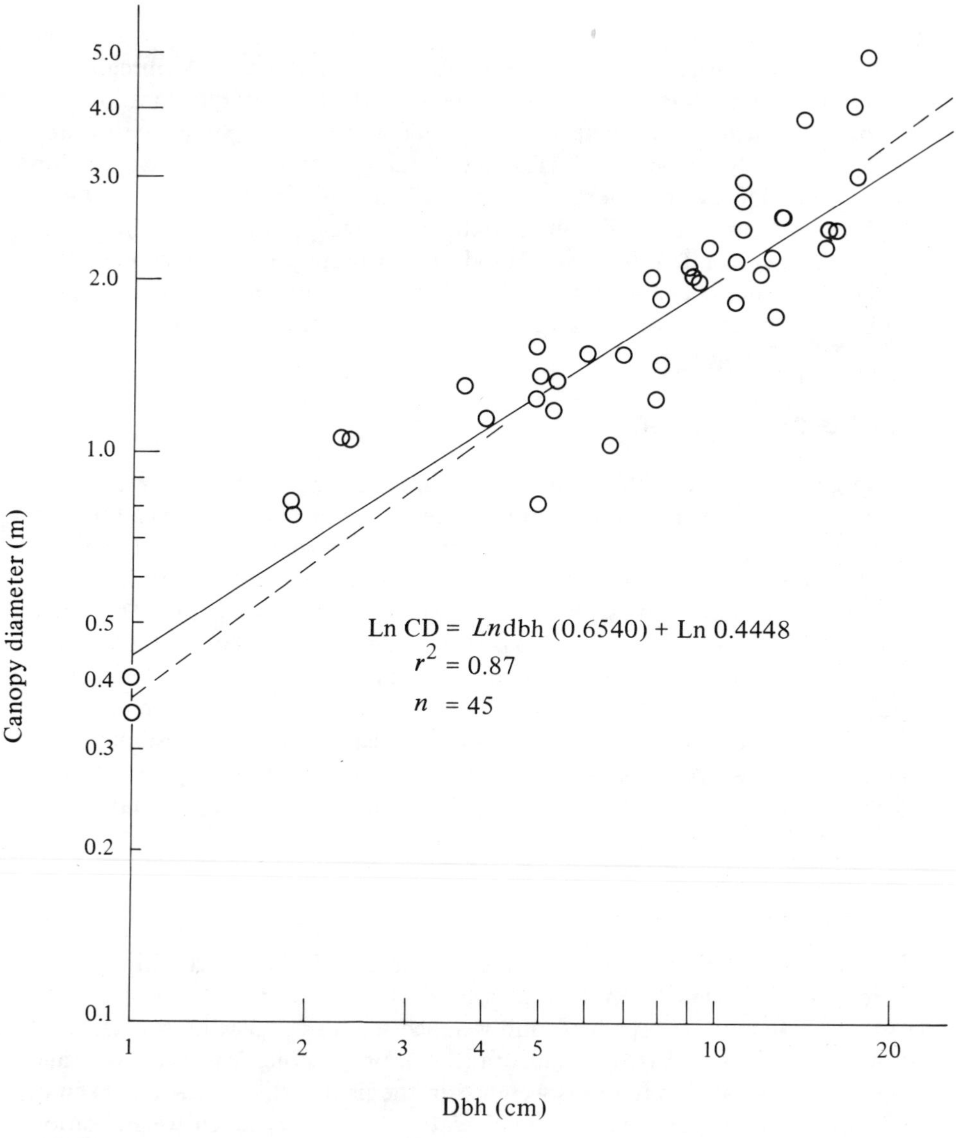

Figure 6.6
Canopy diameter in *R. mangle* stands of different mean diameter (dbh).

Leaf area index

Leaf area index (LAI) is defined as the total area of (one-sided) leaf surface per unit ground-surface area. In tall forests, LAI is most easily determined using plumb-bob methods. A heavily weighted line is lowered through the canopy from above and the number of leaves touching the line are counted.

The average number of contacts per 'drop' is the LAI. A fishing-pole device can be built to measure LAI in forests up to ten metres tall. A fibreglass or aluminium pole with extensions can be used to drop the line from above the canopy, but measures become more difficult as forest height increases, and can be made only on calm days. The line should be strong, heavyweight nylon fishing line. Fluorescent coloration helps to make the line more visible.

Measurements are made on randomly selected points with five or more replications per point. With this method fifty or more points can be measured per day. The results can also be used to determine canopy cover. Cover is the number of non-zero drops as a percentage of total measurements.

Above-ground biomass

Above-ground biomass is the amount of standing organic matter per unit area at a given time. The amount of standing biomass stored in a forest is a function of the systems's productivity, age and organic matter allocation and exportation strategies. Biomass data for mangrove forests are scarce; the data available up to 1982 were summarized by Cintrón and Schaeffer Novelli (1983). There are two main approaches to biomass determination, clear-cut and allometric techniques. Clear-cutting is only recommended in young or scrub stands. Allometric measurement is preferred for tall forests. The development of regressions of biomass on structural measurements of harvested trees allows estimation of standing biomass from easily measured parameters, such as dbh and height. Allometric biomass estimation is non-destructive, once samples have been taken and the regression equations calculated.

CLEAR-CUT HARVESTS

The clear-cut technique is easy but it is recommended only for scrubby growth where stem diameters cannot be measured. The trees are cut off at ground level, separated into components and weighed separately. The most often-used components are wood (trunk and branches), prop roots, leaves, flowers and fruits. The harvested material is weighed in the field and samples are taken of each compartment for drying and calculation of the dry/green weight ratios. This material should be dried to constant weight in an oven at 70 °C. Wet weights are multiplied by the above ratio and results are expressed as dry weight.

ALLOMETRIC TECHNIQUES

This method requires that sets of trees of a given species be felled, cut and separated into compartments and weighed. The sizes of sample trees selected must represent a spread of the sizes present in the community. Diameter and height are measured carefully on each selected tree. Height can be measured after the tree has been felled, adding the stump height to the total length measured. The tree is divided into trunk, branches, prop roots (when present), and leaves and the wet weight of each compartment is taken. Samples are collected of each of these compartments to determine the dry-to-wet-weight ratio, as discussed above (trunk and branch discs are cut for this purpose). Dry weights may then be tabulated as shown in Table 6.7. Least-squares regression analysis is used to find the curve of best fit, using diameter and height as independent variables. Some fifteen to thirty trees per species are needed for this type of analysis.

The dry weight of the tree and its components are usually best estimated as a function of the square of the diameter multiplied by tree height. The relationship is a power function:

$$\text{biomass} = b[(\text{dbh})^2\ (\text{ht})]^m.$$

The regression parameters for the tree samples, shown in Table 6.7 taken from a mangrove swamp in Puerto Rico, are shown in Table 6.8.

TABLE 6.7. Dry weights of prop roots, trunk, branches, leaves and total weight for 26 harvested *Rhizophora mangle* trees

Dbh (cm)	Height (m)	Dbh².Height (cm)² (m)	Oven-dry weight (g)				
			Prop roots	Trunk	Branches	Leaves	Total
1.1	1.90	2.299	20	167	70	48	305
1.2	2.60	3.744	35	322	188	118	663
1.3	2.90	4.901	0	270	59	35	364
1.6	3.10	7.936	35	425	113	76	649
1.8	3.35	10.854	61	573	263	143	1 040
2.0	3.45	13.800	122	683	209	73	1 087
2.5	4.00	25.000	152	1 063	242	105	1 562
2.6	4.90	33.124	192	1 700	398	225	2 515
2.9	4.90	41.209	238	1 630	1 021	553	3 442
3.3	4.50	49.005	152	1 417	1 612	750	3 931
3.5	7.22	88.445	947	3 525	924	159	5 555
3.8	7.00	101.080	1 175	3 903	1 032	222	6 332
4.2	7.70	135.828	577	4 844	1 134	191	6 746
4.4	8.07	156.235	1 327	6 454	1 005	248	9 034
4.6	8.35	176.686	821	6 905	1 187	311	9 224
5.2	10.10	273.104	2 619	8 277	1 907	375	13 178
5.6	8.50	266.560	3 505	9 803	3 406	816	17 530
6.0	9.05	325.800	2 482	13 591	2 321	854	19 248
6.1	9.30	346.053	1 879	10 512	2 026	575	14 992
6.7	10.95	491.545	2 801	18 363	3 020	832	25 016
7.5	10.70	601.875	6 534	19 194	2 670	765	29 163
8.3	10.40	716.456	9 877	14 582	3 278	664	28 401
9.0	11.50	931.500	6 726	25 100	3 530	1 350	36 706
10.9	10.00	1 188.100	15 539	50 594	10 875	3 395	80 403
11.1	10.45	1 287.545	11 270	42 569	8 581	2 827	65 247
15.4	11.10	2 632.476	28 602	88 480	17 833	5 494	140 409

TABLE 6.8. Allometric relations for *R. mangle* based on data contained in Table 6.7

X	*Y*		*b*	*m*	r^2
dbh²·height	Total above ground biomass		125.9571	0.8557	0.99
dbh²·height	Trunk and branches	⩾ 2.5 cm	70.6782	0.8773	0.99
dbh²·height	Branches	< 2.5 cm	35.2197	0.7358	0.93
dbh²·height	Prop root		5.4406	1.0694	0.97
dbh²·height	Leaf biomass		23.6398	0.5902	0.80
dbh	Leaf biomass		27.551	1.7914	0.81

BIOMASS PARTITIONING

The allometric relation developed for a given species may be used to develop a table showing the allocation of biomass into the different anatomical compartments in trees of different diameter and height. This table can be made to show the relation between structural (nonphotosynthetic) components and the foliar component (Table 6.9).

TABLE 6.9. Biomass partitioning in *R. mangle*

Dbh²·height (cm)² (m)	Compartment contribution (%)				Structural components (%)
	Prop roots	Trunk	Branches	Leaves	
1	3.7	52.6	25.9	17.8	82.2
10	7.3	60.5	21.8	10.4	89.6
100	12.1	65.1	16.9	5.8	94.2
200	14.0	65.7	15.5	4.8	95.2
300	15.2	65.9	14.6	4.3	95.7
400	16.0	65.9	14.1	3.9	96.1
500	16.7	65.9	13.6	3.7	96.3
1 000	19.0	65.6	12.3	3.0	97.0
2 000	21.5	65.0	11.0	2.4	97.6

STAND BIOMASS

The biomass of an entire stand can be estimated by multiplying the biomass of the stem of mean basal area by stand density. This estimation technique is reported to give acceptable results (Loetsch et al., 1973). Another technique is to determine the frequency distribution of the diameters in a stand and estimate the biomass of each diameter class separately. The average biomass of a class is multiplied by the number of stems in that class and the results for all the classes are summed to obtain total stand biomass. For plots containing a reasonably small number of trees (30–40), it is practical to calculate the biomass of each tree from the allometric relation and sum to obtain the plot's total biomass.

It is worth noting that the LAI of a stand can be calculated if the foliar biomass is known. Leaf weight must be converted to leaf area by using a leaf-weight/leaf-area relationship (remembering, that for purposes of LAI, one-sided area is measured), determined by planimetering a sample of leaves. Leaf area is divided by stand area (in m²) to obtain forest LAI.

Conclusion

The methods discussed in this chapter are suggested standard field research methods for mangrove ecosystems. Comparative standardized data from diverse locations and settings are needed to interpret the complex relationships between structure and the major forcing functions. A more complete understanding of the structural responses of mangrove ecosystems to natural variations in the energy signature of the environment will undoubtedly lead to a better understanding of the responses of these systems to stressors and human manipulations.

Acknowledgements

Research in Puerto Rico was made possible by the Puerto Rico Department of Natural Resources through funding provided by the United States Department of Interior, Bureau of Sport Fisheries and Wildlife. In Brazil, our research effort has been made possible through the support of National Research Council (CNPq), FAPESP (Fundacão de Amparo a Pesquisa do Estado de São Paulo) and CAPES (Coordenacão do Aperfeicoamento de Pessoal de Nivel Superior) and the Oceanographic Institute of the University of São Paulo. Bárbara Cintrón reviewed the manuscript and made valuable comments.

References

BITTERLICH, W. 1948. Die Winkelzahprobe. *Allg. Forst- u. Holzwirtsch. Ztg.*, Vol. 59, pp. 4–5.

CINTRÓN, G.; SCHAEFFER NOVELLI, Y. 1983. *Introducción a la ecologia del manglar*. Montevideo, Unesco.

CINTRÓN, G.; LUGO, A. E.; MARTÍNEZ, R. 1980. Structural and Functional Properties of Mangrove Forests. *Symposium Signalling the Completion of the 'Flora of Panama'*. Panama City, University of Panama.

COTTAM, G.; CURTIS, J. T. 1956. The Use of Distance Measures in Phytosociological Sampling. *Ecology*, Vol. 37, pp. 451–60.

CURTIS, J. T. 1959. *The Vegetation of Wisconsin. An Ordination of Plant Communities.* Madison, University of Wisconsin Press. 657 pp.

GROSENBAUGH, L. R. 1952. *Shortcuts for Cruisers and Scalers.* U.S. Forest Service, Southern Forest Expert. Station. 24 pp. (Occasional Paper, 126.)

JIMÉNEZ, J. A. 1981. The Mangroves of Costa Rica: A Physiognomic Characterization. Coral Gables, Fla., University of Miami. 130 pp. (MS thesis.)

LOETSCH, F.; ZOHRER, F.; HALLER, K. E. 1973. *Forest Inventory*, Vol. II. Munich, BLV Verlagsgesellschaft.

MARTÍNEZ, R.; CINTRÓN, G.; ENCARNACIÓN, L. A. 1979. *Mangroves in Puerto Rico: A Structural Inventory*. San Juan, Department of Natural Resources. 149 pp. (Final Report, Office of Coastal Zone Management, NOAA.)

MORISITA, M. 1954. Estimation of Population Density by Spacing Method. *Kyushu Univ. Mem. Fac. Sci.*, Vol. I(E), pp. 187–97.

POOL, D. J.; SNEDAKER, S. C.; LUGO, A. E. 1977. Structure of Mangrove Forests in Florida, Puerto Rico, Mexico and Costa Rica. *Biotropica*, Vol. 9, No. 3, pp. 195–212.

SCHAEFFER NOVELLI, Y.; CINTRÓN, G.; ADAIME, R. R. 1980. *Algumas caracteristicas dos manguezais de ragião de Cananeia, São Paulo, Brasil.* Seminarios de Biologia Marinha, Academia Brasileira de Ciencias, Instituto do Biologia Marinha da U.S.P.

7

Waterlogged saline soils

Kevin G. Boto
Australian Institute of Marine Science
PMB No. 3
Townsville, M.S.O. 4810
Queensland, Australia

There are many environmental factors that can affect primary production and plant-species distribution in intertidal ecosystems such as mangroves and salt marshes. Of these, the substrate characteristics are possibly the most important in terms of direct influence on plant growth. However, it must be noted that the type of soil and its chemical state are in turn affected by factors such as topography, tidal or riverine sedimentation patterns, climate, tidal range and long-term sea-level changes. Therefore, studies of substrate conditions should not only include characterization of the present physico-chemical status but also include information derived from hydrological and geological studies. Information on long-term sedimentation patterns and rates and origins of 'recent' sediments can be invaluable for the chemist attempting to describe nutrient cycling in these systems. Hence, while this chapter will concentrate on the relatively short, time-scale, chemical and physical studies of waterlogged saline soils, it is important to remember that longer time-scale effects are also operative and that a complete picture can only emerge by a combination of both sets of information.

Characteristics of waterlogged saline soils

It is beyond the scope of this chapter to review basic aspects of soil chemistry. The purpose of this brief discussion is to acquaint the reader with some important aspects of intertidal marsh soils with an emphasis on potential plant-soil interactions.

When a soil is flooded, the rate of oxygen diffusion is greatly reduced. Depending on the depth of flooding and the amount of time the soil remains flooded, the oxygen supply to the soil can be almost completely shut off. Under these conditions, bacterial populations in the soil are drastically altered. In normal, well-aerated soils, aerobic bacteria are dominant. These species require oxygen for respiration. When starved of oxygen, they are reduced in population and other types of bacteria which do not depend on oxygen become dominant. Facultative anaerobes, while not requiring oxygen, do need an alternative oxidant source (e.g. nitrate ion, iron (III), manganese (IV)) for respiration. These chemical entities act as electron acceptors for various types of facultative anaerobes and hence, as the soil becomes increasingly anaerobic, are responsible for drastic and very significant chemical changes. Another group of anaerobes that become dominant when the soil becomes anaerobic are species which are responsible for reactions such as the reduction of sulphate to sulphide and the production of methane utilizing carbon and 'excess' hydrogen. A comprehensive discussion of soil microbiology is given by Alexander (1961).

Figure 7.1 is a schematic summary of some of the more important reactions occurring in flooded soils. The redox potential (Eh) referred to in this diagram is a convenient measure of the extent to which the soil is in a reducing or anaerobic state. Such measurements are easily made in field or laboratory studies by simple insertion of an inert platinum electrode which 'senses' the redox state of the soil. These measurements will be discussed in more detail later. Note that a large positive Eh indicates a well-oxidized or aerated state, while a large negative Eh indicates a highly reduced state. Various bacterially

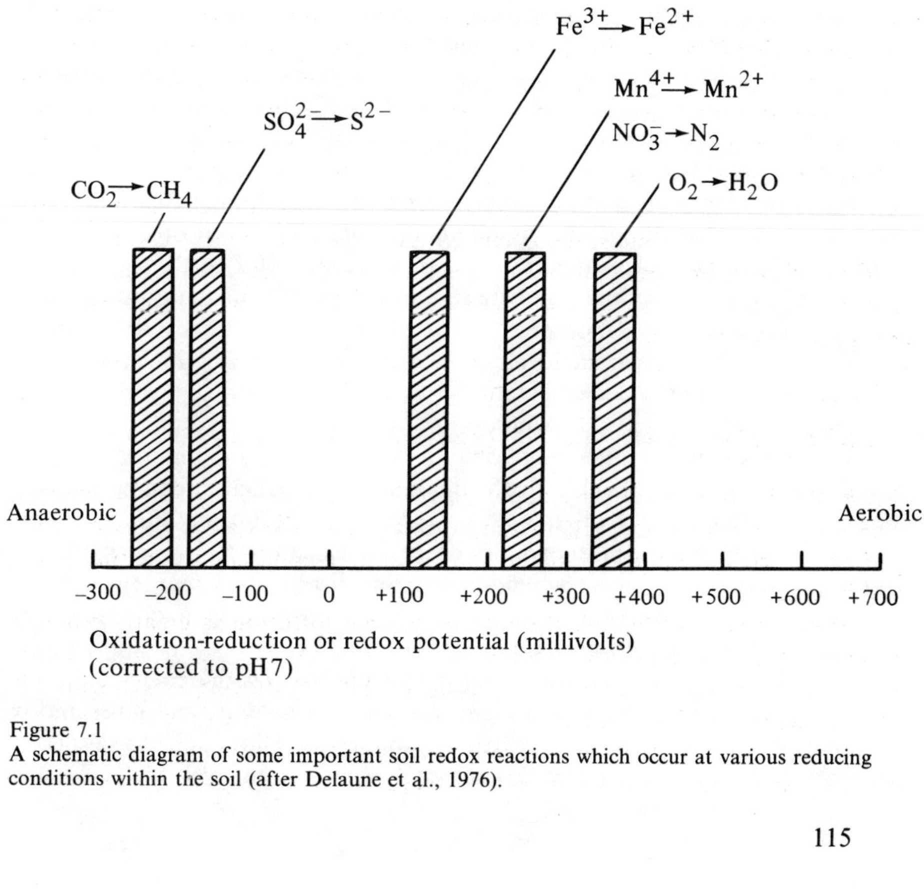

Figure 7.1
A schematic diagram of some important soil redox reactions which occur at various reducing conditions within the soil (after Delaune et al., 1976).

mediated redox reactions can be associated with intermediate Eh values, as shown in Figure 7.1. Referring to this diagram, it can be seen that as the oxygen supply becomes limited, it is rapidly consumed by bacterial respiration. This process has been shown to take place over the Eh range of +350 to +380 mV. When all the oxygen is consumed, and only when it is consumed, Mn^{4+} and NO_3^- are converted to Mn^{2+} and N_2 (gaseous) respectively. In turn, when Mn^{4+} and NO_3^- are completely consumed, then Fe^{3+} is reduced to Fe^{2+} and so on, until the soil eventually reaches a highly anaerobic state where the further reduction of dioxide to methane (marsh gas) occurs. The rate at which all these processes occur depends on such factors as depth and time of flooding (i.e. to what extent oxygen is limited) and the amount of organic carbon present in the soil.

Bacteria (except chemotrophs) require carbon compounds as energy sources and here the type of organic carbon is also critical. For example, if a sample of well-aerated soil is flooded and glucose is added to the soil, then the various reactions proceed at a very rapid rate and the soil could become highly reducing in a matter of days, or even hours. If other, more refractory carbon compounds (such as ground straw) are used instead, then the reactions will be less rapid.

The changes in the redox states of NO_3^-, Fe and Mn can have profound effects on plant growth. Iron in the divalent state, for example, is, at any given pH, more soluble and thus becomes more available for plant uptake. In very iron-rich soils, the iron uptake can be extreme and can result in iron toxicity. Similarly, manganese becomes more available to the plant. Another very important but less direct effect is that inorganic phosphorus which may have been previously 'trapped' or sequestered in the insoluble Fe^{III} and Mn^{IV} oxides/hydroxides is at least partially released. The resulting increase in phosphate availability is important, as plants can only utilize phosphorus in the inorganic phosphate forms (e.g. PO_4^{3-}, HPO_4^{2-} and $H_2PO_4^-$).

Nitrogen is an essential and often limiting nutrient for plant growth and can only be utilized in inorganic forms (NO_3^-, NH_4^+). Severe losses of nitrogen can occur in flooded anaerobic soils via denitrification. As shown in Figure 7.2, many flooded soils have a thin (1–5 mm) aerobic layer overlying the more highly reduced soil. Ammonium in the anaerobic zone diffuses into the aerobic layer where nitrifying bacteria can convert ammonium to nitrate. The nitrate may diffuse back into the anaerobic zone where it can be reduced to gaseous nitrogen and nitrous oxide (N_2O) and lost from the soil. This reaction scheme has been shown to be responsible for losing a significant fraction of nitrogen fertilizer amendments from rice paddyfields. However, in natural intertidal systems where inorganic nitrogen content is much lower, it is probable that less severe losses occur.

Soil pH is also an important parameter controlling the chemical status and mobility of many important elements. For example, at any given redox potential, many metals exhibit greater solubilities at a low pH. As discussed above, this can have an effect on phosphate binding in the soil. If a soil has a pH greater than about 7, it is possible that ammonia volatilization, with a resulting loss of nitrogen, can occur. In some instances, however, flooding may have the effect of increasing the pH of acid soils and decreasing the pH of alkaline soils so that most flooded soils converge to about pH 7. The flooded soils are buffered by iron, manganese and aluminium oxides/hydroxides and carbonates. Organic matter, such as humic acids and organic acids from root exudates, are also thought to play an important role in the buffering capacities of these soils. Also, soils rich in 'labile' materials (i.e. reactive organic matter) exhibit increased microbiological activity and thus intensified reduction. This in turn may contribute to an increased buffering capacity near pH 7.

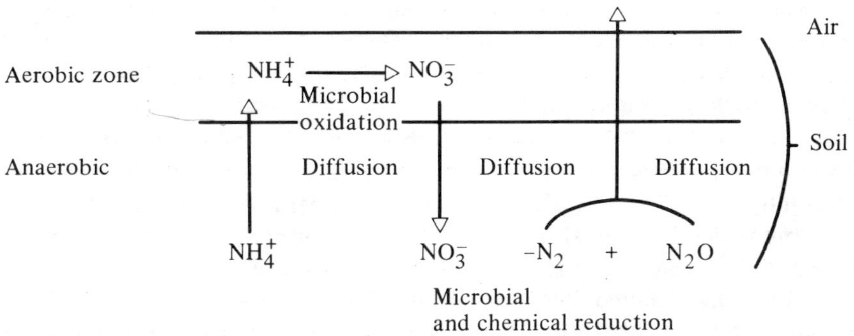

Figure 7.2
A schematic diagram of the denitrification process in anaerobic soil.

In saline flooded soils, the chemistry of sulphur is of great importance. For example, sulphate in sea-water is reduced to sulphide in highly anaerobic soils. Sulphide ion is highly toxic to plants and many intertidal mangrove and marsh plants have developed strategies to overcome sulphide toxicity (see next section). Most metals have highly insoluble sulphides; hence the mobility of metals in highly anaerobic soils is limited if sulphide is in excess. Over long time-periods and given the appropriate conditions in terms of iron concentration and redox potentials etc., iron pyrite is formed according to the (simplified) equation:

$$Fe^{2+} + S^{2-} + S^{o} \longrightarrow FeS_2$$

where S^o is elemental sulphur formed by re-oxidation of some of the sulphide in the soil. If the soil containing sulphide and pyrite numerals becomes aerated again, e.g. by sea-level changes or man-induced effects, formation of jarosite $KFe_3^{III}(SO_4)_2(OH)_6$ occurs. This hydrolyses to form sulphuric acid, thus creating very acidic soils. Such soils, known as acid sulphate soils, are almost useless for agricultural purposes, as has been found in areas of South-East Asia and Africa.

Overall, it can be seen that a plant growing in a saline, flooded soil is in a completely different chemical environment compared with that of a well-aerated terrestrial soil. Not only do the plants have to contend with high salinity, but also with the strongly reducing conditions present. The presence of sulphide and the higher solubility of many metals can give rise to toxicity effects if the metal or sulphide is in excess. Nutrient availability can be decreased markedly by denitrification or volatilization losses of inorganic nitrogen. Alternatively, phosphate mobility may be greater. Decomposition processes tend to be slow in anaerobic soils and this could lead to decreased recycling of key nutrients. Over long time-periods, nitrogen and phosphorus may be gradually 'locked up' in organic compounds which are not decomposed or mineralized as rapidly as in well-aerated soils. Soil salinity is also a very important factor. Various halophytes have quite different salinity tolerances; and changes in salinity, either in the long term (sea-level changes, sediment accretion) or short term (damming of rivers, etc.), can have drastic effects on species distribution and growth.

Therefore, in order to gain a complete picture of the effects of flooded soils on plant productivity, relatively long-term studies of nutrient status, redox potential, pH, soil mineralogy and salinity changes must be made. Longer term predictions of marsh viability require information on sea-level changes, sedimentation rates, topography and hydrological regime.

For more detailed information on these various aspects of waterlogged soil characteristics, the reader is referred to the reviews and papers listed in the literature cited (De Laune et al., 1976; Brummer et al., 1971; Connell and Patrick, 1968; Gotoh and Patrick, 1972; Isirimah and Keeney, 1973; Patrick et al., 1973; Patrick and De Laune, 1972; Ponnaperuma, 1972; Shapiro, 1958*a*, 1958*b*; Turner and Patrick, 1968; Holford and Patrick, 1979).

Problems of research

As can be seen from the above discussion, intertidal soils represent a complex biological and chemical system. The way in which the soils and plants interact also presents many problems of a conceptual nature. Even apart from these factors, there are a number of more immediate practical and logistical difficulties in dealing with research into saline, waterlogged soils. Most wetlands or mangrove forests are located in relatively inaccessible areas. Vehicular traffic is usually impossible and boat access is usually difficult. Field sampling normally involves walking, and transporting equipment, through very inhospitable terrain. For all these reasons, it is hardly surprising that this type of research is not widespread, and many important questions remain unanswered because of the general reluctance or logistical inability to investigate these systems. This is in spite of the generally accepted importance of marshes and mangroves in inshore ecosystems and, hence, their relationship to important socio-economic factors such as fisheries production, coastline stability, 'pollution sinks', etc. (Greeson et al., 1978).

Most studies of salt marsh soils and, to a lesser extent, mangrove soils, have focused on the following aspects:

Chemical and physical characterization of field samples.

Variation of nutrient status and other parameters with time, topography, hydrological regime, etc.

Correlation of soil characteristics and variation thereof, with plant productivity and/or species distribution.

Laboratory studies of chemical and microbiological reactions in flooded soil and soil-plant interaction.

Some of the basic principles and problems associated with these studies and the extent to which these problems can be overcome will now be examined.

FIELD SAMPLING AND SOIL CHARACTERIZATION

The only major problems encountered here are logistical. To obtain truly representative samples of the soils of any given marsh, it is often necessary to obtain a large number of samples from many different sites within the intertidal complex, along with estimates of the relative areal extent of different types of sites. The number of sites chosen will depend on the observed structural heterogeneity of the system in terms of topography and/or plant species distribution. Many subtropical marsh grasslands, for example, are quite uniform in structure. Obvious levee bank formation at streamsides sloping slowly downwards to lower inland areas, along with associated differences in observed plant viability at these sites, allows for a fairly simple and obvious sampling strategy. That is, a straight line transect perpendicular to a stream with sampling at, for example, ten intervals will give representative samples of the area. On the other hand, many tropical mangrove forests are much less uniform with high species diversity and marked differences in plant growth over apparently small distances. Here the choice must be made between, say, random sampling or transects or use of some fixed pattern that attempts to incorporate as many different site types as possible. In practice, the latter is often the only feasible strategy because of accessibility, even though it is less likely to give good representation of the system.

Having chosen the sites, it is then most desirable to sample the soil at all depths throughout the whole rooting depth of the plants. This is usually not simple as many mangrove species have rooting depths of one metre or more. If a large number of sites are to be sampled, it is usually only possible to take samples at, for example, two different depths and again this reduces the chances of obtaining representative samples. Subsequent field and laboratory analyses of the soil present no major conceptual problems and these techniques are more appropriately discussed in the methodology section of this chapter.

VARIATION OF SOIL PROPERTIES WITH TIME

Apart from the problems in sampling strategies, as discussed previously, the major problems here involve the time-scale of sampling. It is usually desirable to examine changes in redox conditions, salinity and nutrient availability in terms of seasonal variation throughout one year. The cycles may not necessarily be annual, however. For example, major changes may perhaps occur only during severe storms or exceptional spring tides. These events may occur, for example, at ten- to twenty-year intervals and thus it is difficult to assess such effects. Often attempts to predict such effects can only be made on the basis of results obtained under normal seasonal conditions. Again this emphasizes the need for a multidisciplinary approach. Past records of, for example, sediment accretion rates are often extremely valuable in estimating the effects of infrequent events. Long-term meteorological and oceanographic records are also necessary.

Short-term seasonal studies are necessary to assess the average nutrient status of a marsh soil in 'normal' years. Measurement of nitrogen status will often show a marked variation in the amount of inorganic nitrogen (the only form of immediately available N) during periods of low or high plant productivity. Figure 7.3 shows the variation of ammonium-N with the seasons for a Louisiana salt-marsh soil (De Laune et al., 1976). During the spring and summer months, when plant production is very high, the plant uptake of ammonium exceeds the rate at which it is replenished through anaerobic de-

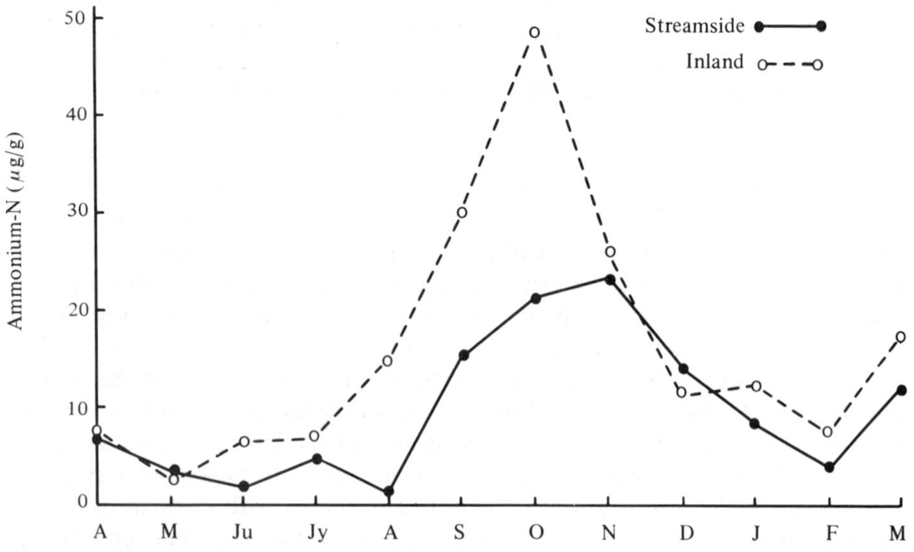

Figure 7.3
Variation of soil ammonium with time for a Louisiana salt marsh (after De Laune et al., 1976).

composition and remineralization and the concentration in the soil drops to a very low level. Conversely, during the winter, the ammonia level increases again due to decreased plant uptake.

In principle, there are relatively few conceptual problems associated with field sampling and characterization of intertidal soils. Variation of soil properties and nutrient status is amenable to seasonal studies but longer-term affects can normally only be guessed. How good the guess is depends upon the availability of data from many other disciplines.

CORRELATIONS WITH PLANT PRODUCTIVITY

Having measured the present status of an intertidal marsh substrate and having some idea of at least the seasonal variation of the soil, it is then necessary to determine whether any of these factors may be growth limiting. To do this, reliable estimates of plant standing crop and the rate of turnover of the standing biomass (productivity) are needed. Also, in order to attempt sensible correlations, the measurement of productivity must be sensitive enough to detect changes on the same time-scale as the measurement of the variation of soil parameters.

It is not appropriate to examine this topic in detail here and the reader is referred to reviews on this topic (Turner, 1976; Clough and Attiwill, 1982). It is sufficient to say that methods of estimating productivity still leave much to be desired, especially in mangrove forests. More success has been achieved in studies of marsh grasslands where harvesting techniques are possible. This topic is raised only to point out that it often represents a limiting factor in research on intertidal soils. One can overcome the logistical and sampling problems and obtain good measurements of soil properties, often to be confronted with the problem of correlation of this data with questionable estimates of productivity.

LABORATORY STUDIES OF SOIL TRANSFORMATIONS

A great deal of information on short- and long-term soil transformations can be derived from controlled laboratory studies. Most notable in this field are Dr W. H. Patrick, Jr, and his co-workers at Louisiana State University. These workers have successfully applied techniques of controlling the redox potential and pH in soil slurries to investigate the kinetics of many microbial and chemical reactions in rice soils and marsh soils.[1] Very valuable information on denitrification mechanisms, reactions of metals, sulphur chemistry and nitrogen transformations, for example, has become available through these studies. With a knowledge of how various reactions are altered with changes in redox status and pH of the soil, and using information derived from other disciplines, such as sea-level changes, sedimentation rates, etc., long-term changes in soil

1. The reader is referred to the references mentioned in the first section for details of these experiments.

conditions and in turn how these changes affect the important chemical reaction in the soil can be predicted.

Major problems arise, however, when soil–plant interactions are considered. Here it is questionable as to whether controlled experiments using well-stirred soil slurries in any way simulates the all-important micro-environment at the root–soil interface. In fact, very little is known about this chemical environment in the immediate vicinity of the root surface. Redox and pH conditions in the bulk soil are thought to be drastically altered near the root surface due to exudation of organic acids (often in mucilage form) and oxygen 'pumping' from roots. Certain types of marsh plants, including rice, translocate oxygen to the roots for respiration and presumably to aid in removal of toxic chemicals (e.g. sulphide and/or iron). In the usual controlled experiments the stirring would have the effect of making the bulk soil conditions more closely approximate the conditions at the root surface but then the root-surface conditions are quite different from what they would be normally. On the other hand, in an unstirred experiment, the conditions at the root surface would be closer to those found in the natural intertidal environment, but then the conditions are unknown and not easily measured or monitored. Studies of plant–soil interactions await the availability of methods to monitor conditions in the root-surface zone. This problem in fact relates also to field measurements of Eh and pH in the soils. The techniques used (see next section) really only measure the bulk soil conditions or perhaps some average of the root/soil and bulk conditions. This therefore makes these measurements less suitable if it is hoped to correlate plant growth with soil parameters.

BASIC METHODS

Having discussed some of the important aspects of the chemistry of anaerobic soils and the problems associated with interpretation of plant–soil interactions, some of the basic methodology of soil sampling and analysis will now briefly be discussed.

Sampling and field measurements

Because of the difficulty associated with transportation in most intertidal habitats, it is crucial to keep field equipment simple, light and robust. The usual requirement in sampling is that core samples be taken with as little compaction as possible. Here, the depth of the core is critical. For example, if the plants are shallow-rooted, sampling may only be necessary to 15–20 cm. This means that a large-diameter corer can be used without taking a sample which would be too heavy to lift. The larger the diameter, the easier it is to prevent compaction during coring. Thus, in many cases, a 20–30 cm diameter by 20 cm long, thin-walled, aluminium cylinder is ideal. A small shovel can usually be used to dig out the corer and sample. It is then quite simple to push the core sample intact from the sampler for further sampling and measurement.

In our studies of mangrove soils where roots usually extend down to about one metre, it was obviously not possible to use a large-diameter corer because of the weight of the resulting sample. Figure 7.4 shows a diagram of the compromise solution. The dimensions (1.2 m × 5 cm) allow for relatively easy transportation, and minimal compaction is encountered in most waterlogged soils. Stainless steel is used for strength and to minimize corrosion problems. A major advantage of this design is that the holes at 5-cm intervals allow for easy removal of soil samples corresponding to a certain depth and/or insertion of electrodes of Eh and pH measurements. When taking the core, the holes are covered by PVC sleeving which can be slid back to expose the holes after withdrawal of the corer. The major drawback of this otherwise simple but effective corer is that it is 'home-made' and thus requires reasonable workshop facilities. Unfortunately, no commercially available samplers of this type were found to be suitable.

Once a suitable core sample has been taken, smaller subsamples at various depths are taken for laboratory analysis. These should be taken as quickly as possible and stored in airtight containers on ice. This is to keep the soils anaerobic and to inhibit further microbiological activity. The samples should never be frozen as this may lead to bursting of living cells with release of contaminating nutrients. If, albeit minimal under most circumstances, samples are kept airtight and cold, we have found that there are no apparent ill-effects on storage for up to two weeks, although it is normally possible and desirable to extract and analyse (see below) within one or two days after collection.

Electrode measurements must be made immediately in the field, and before the soil core absorbs too much oxygen. Eh measurements are, of course, especially susceptible to oxygen contamination and these measurements should be done first. Commercially available platinum-disc or wire electrodes, in combination with a suitable reference electrode, are quickly (but carefully) inserted into the soil at the appropriate depth along the core sample. Entering the core sample from the side, it is desirable to insert the electrode as far as possible into the soil in order to minimize oxygen interference. Such interference is, unfortunately, inevitable and field Eh measurements can only be considered an approximate indication of the soil-redox status. Nevertheless, the measurements are a very useful guide and we have generally found quite good correlations between field Eh readings and the chemical status of the soil as would be predicted from Figure 7.1, for example. Note that redox potential measurements made, using calomel electrodes as reference, must be corrected by +240 mV to convert to Eh which is simply the redox potential with respect to the standard hydrogen electrode. Measurements of pH are made using a robust thick-walled glass electrode. The thick membranes give slower responses than usual but are far less susceptible to breakage. Small battery-operated millivoltmeters with waterproof housings are commercially available.

In saline soils, measurement of the pore water or interstitial water salinity is important. For most purposes, an average salinity throughout the sampling depth is adequate. Such a sample is usually easily obtained in waterlogged

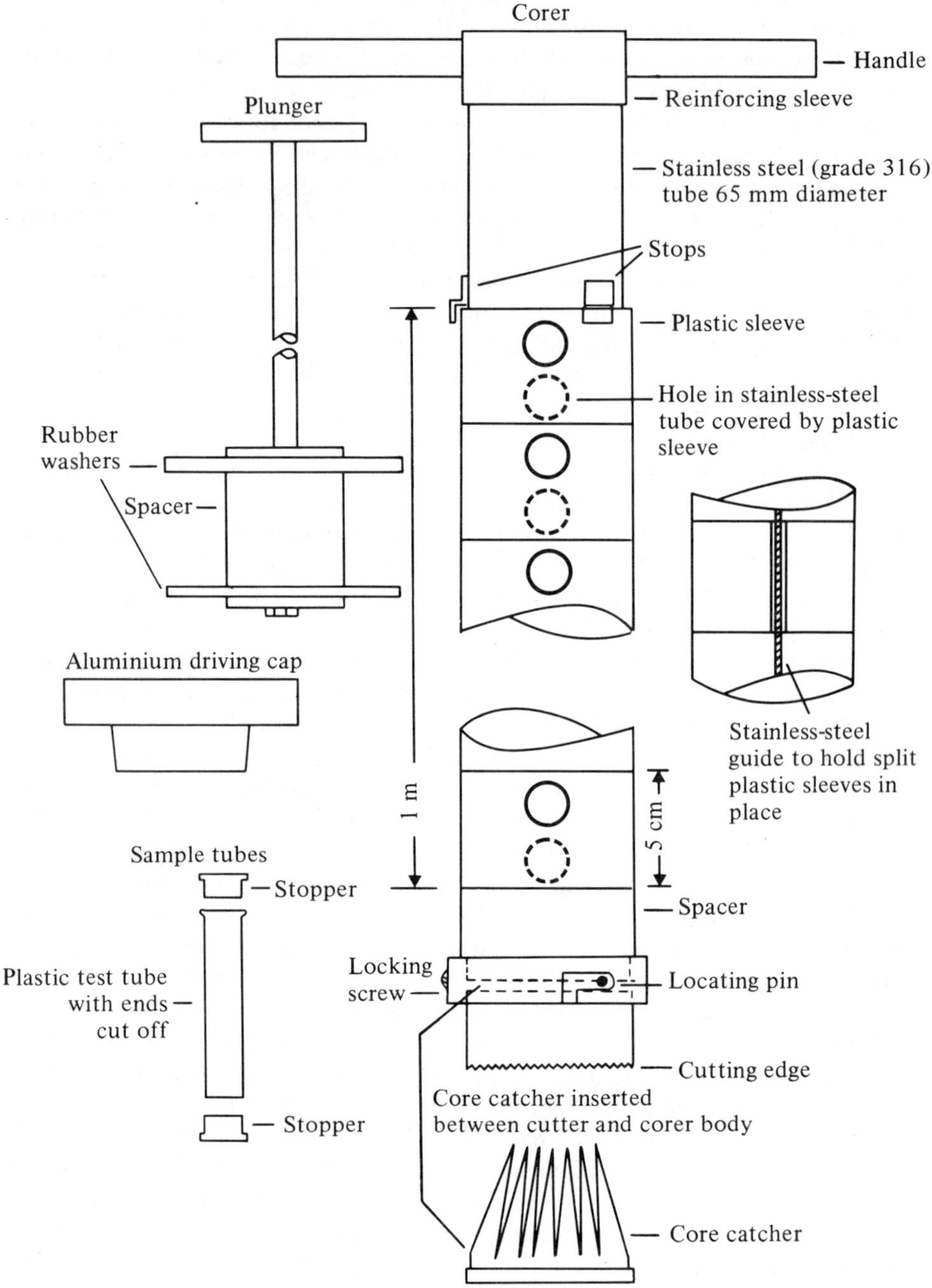

Figure 7.4
Core-sampling apparatus used for sampling through deep rooting zones in mangrove forests.

soils by withdrawing a portion of the water that fills the hole in the soil after the sampler is withdrawn. For impermeable thick clays or where the soil is less waterlogged, it may be necessary to use a pore water squeezer, or centrifuge the soil to extract a water sample. The salinity measurements are then made either by classical titration techniques with silver nitrate or by using a conductivity cell (Strickland and Parsons, 1972). Both methods are very accurate. The titration technique requires no special equipment but is quite tedious, whereas the conductivity measurements can involve expensive equipment. Normally, however, cheap conductivity meters give sufficient accuracy. Hand-held refractometers are inexpensive and extremely simple to operate. Accuracy is normally limited to about ± 2 parts per thousand but this is usually sufficient. Salinity-depth profiles are sometimes used by soil hydrologists as a sensitive tracer of below-ground water movement. For these purposes, very accurate measurements may be required.

Basic laboratory methods

Most of these analytical methods are standard practice in soils laboratories. Allen et al. (1974) give an excellent account of these techniques along with problems associated with any given method. Sampling techniques and statistical analysis are also described in detail. Here I shall attempt a brief synopsis of the methods most relevant to waterlogged soils and outline the more important elements to be measured. The techniques described are essentially those that we have applied successfully in our studies.

Physical measurements

The most commonly measured physical properties are bulk density and particle-size fractionation. Bulk density, defined as the ratio of dry mass of soil to the total wet volume, is important both in terms of physical description and characterization of the soil as well as for correction of dry weight analyses to volume expression. Many workers consider that, while it is easier to analyse for any component in terms of the dry weight of the sample, it is more realistic to convert this to a volume basis (e.g. μg NH_4-N/ml wet soil). This, more realistically reflects the actual concentration of the element in the environment of the plant roots. The measurement simply involves the removal of a soil core of known volume, with no compaction, drying and weighing. For some soils it is easier to cut out blocks of known dimensions.

Particle-size analysis is a widely accepted method of soil classification. Allen et al. (1974) give details of these classifications. A commonly used method is to disperse 50 g of air-dried soil into a 1-litre measuring cylinder, insert a Bouyocos soil hydrometer and take density measurements at time intervals of: 4 min 48 s—International silt and clay (<20 μm); and, 5 h—International clay (<2 μm). The measurement is based on Stoke's law which relates the settling rate of a particle to its size. It assumes that all the particles behave as perfect spheres and all have the same density (see Allen et al., 1974, for the equations used in the calculations). We have found that, for mangrove soils, which usually contain large amounts of fine fibrous root material, a sieving technique is useful. A wet-soil sample equivalent to about 50 g dry weight is carefully and thoroughly washed through a nest of British Standard sieves ranging from a mesh of 2 mm to 50 μm. Most of the organic matter is retained on 150-μm mesh or greater. The fine silts and clays that are washed through into a large collection bucket are then dispersed thoroughly, samples being taken for analysis using a Coulter counter. Alternatively, the silt and clay fraction can be analysed by the standard hydrometer method. The advantage is that estimation of soil macro-organic matter can be done simultaneously.

Basic chemical analyses

The amount of organic carbon in the soil is important in soil classification and chemical characterization. Many soil bacteria require organic carbon as an energy source. Hence, organic soils (organic C $\geqslant$ 6 per cent) are likely to exhibit greater microbiological activity than mineral soils. The simplest method to determine organic matter is by loss of weight after ignition at 550 °C. The soil is initially dried at 100 °C to constant weight and then ignited to constant weight. Division of the weight loss by a factor of about 1.8 gives an approximate measure of organic carbon. However, this method is really only useful as a guide, or for comparative studies of similar soil types. Other weight losses, including volatilization of structural water, ammonia and volatile heavy metals, can cause significant overestimates in some soils. 'Dry' or 'wet' combustion methods (Allen et al., 1974) give more accurate results but are very tedious when large numbers of samples are involved. Commercially available carbon, hydrogen and nitrogen analysers give excellent results with the added advantage of simultaneous total nitrogen determination (compare with other methods discussed below). These instruments are, however, quite expensive and often require a full-time operator.

Some intertidal systems are considered to be growth-limited by deficiencies in nitrogen or phosphorus and hence only these elements will be discussed here. Other elements such as Fe, Mn, K, Al, Co or Cu may be of local importance. Analyses of these elements and their speciation (chemical state) are treated in detail by Allen et al. (1974) and references therein.

Nitrogen exists in a variety of forms in soils. These can be subdivided into organic and inorganic nitrogen. Organic nitrogen exists in a large variety of compounds such as proteins, amino acids, etc., in both living and non-living material. By far the major form of total nitrogen in soils is bound up in organic form (99 per cent or more). The inorganic forms, i.e. ammonium, nitrate and nitrite ions, are the only forms that can be directly utilized by plants. In most waterlogged soils, nitrate and nitrite are rapidly reduced to nitrogen gas and are therefore usually very low in concentration. Ammonium is therefore often the only measurable component of inorganic nitrogen. Organic nitrogen is made available through the breakdown of organic matter by heterotrophic organisms. This is termed mineralization. This process is often very slow in anaerobic soils and so the total nitrogen level stays relatively constant with time. Measurement of total N represents a theoretical 'pool' but does not give any indication of nitrogen availability.

Bremner (1965*a*, 1965*b*) details methods for measurement of total N and inorganic N, and these methods are almost universally accepted. A brief outline of these methods along with some minor adaptations, will be given here. For total N the soil is dried, finely ground and a small sample (<1 g) is weighed into a digestion tube. The sample is then digested using the classical Kjeldahl reagents (acid and catalyst). This converts all organic N to ammonium-N which is then estimated following steam distillation into excess boric acid, then titrated with standard sulphuric acid. The method is simple and requires no expensive instrumentation. If a colorimeter is available, the ammonium-N in the digest solution can be estimated using the classical Nessler colour reaction. Autoanalysers are well suited to ammonium analyses and provide a rapid, accurate method for large numbers of samples. The cost of these is becoming less prohibitive with a number of manufacturers now producing similar instruments. Information on autoanalyser techniques can be obtained from any of the manufacturers.

Analysis of inorganic nitrogen must be carried out on fresh, wet samples. Small subsamples of these are set aside to obtain wet to dry weight conversion factors. The amount of wet sample used for analysis depends on the method used for determination of ammonium in the extract solution. If the classical titration method is used, for example, then sample weights of up to 50 g may be required. This is no great disadvantage and, in fact, it is easier to obtain representative samples if large quantities are analysed. Use of autoanalyser methods, on the other hand, usually require only small samples and it is then easier to run replicates to ensure representative sampling. To extract ammonium (or nitrate) from the soil, a high ionic-strength solution of sodium chloride or sodium-acetate/acetic-acid buffer (pH 4) is used to remove the ions from ion-adsorption sites on the clay minerals. The extractant-solution/soil ratio is usually greater than 10 : 1 and the soil is shaken in the solution for thirty minutes, then filtered or centrifuged to give a clear extract. The extract is then analysed for ammonium (and nitrate if desired) by the distillation or colorimetric methods outlined above. Nitrate is analysed by distillation over Devada's alloy which reduces nitrate to ammonium. Alternatively, reaction with phenoldisulphonic acid gives a colour reaction for colorimetric analysis. Nitrite ion is very reactive and exists in very low concentrations and is, therefore, rarely analysed in soils.

A single measurement of inorganic nitrogen does not give a reliable estimate of available nitrogen. This is because of the marked seasonal variation. Regular monitoring of ammonium-N over a time period of at least one year is necessary to evaluate the average nitrogen availability. There are laboratory methods that have been designed to measure mineralization in more short-term studies, but these methods are very artificial and do not give a realistic estimate of nitrogen availability in the natural system.

Phosphorus also exists in a variety of chemical forms which again can be broadly subdivided into organic and inorganic phosphorus. However, the situation is even more complex here in that many of the inorganic phosphorus compounds are very insoluble and not immediately available. Various extraction media have been designed to attempt an estimation of plant-utilizable P. All of these give somewhat artificial results and it is best to simply choose a convenient method and use the results on a comparative basis. For example, many workers use a sodium acetate/acetic acid, buffered to pH 4. Use of this

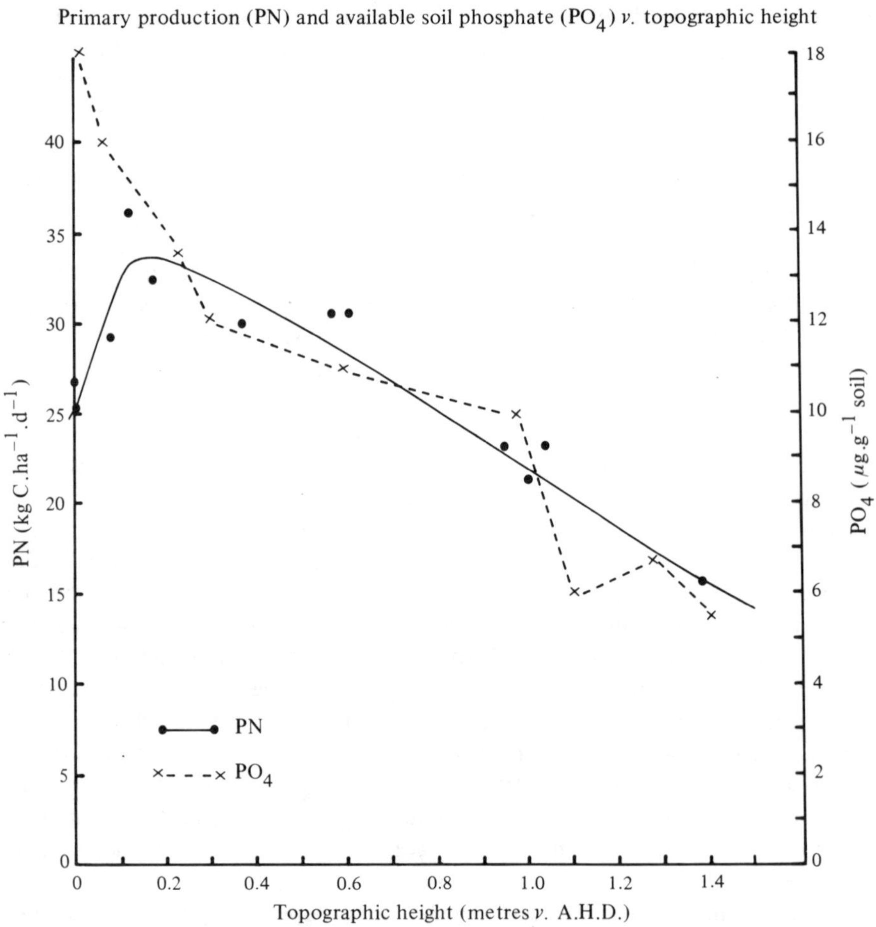

Figure 7.5
Variation of mangrove productivity and soil-extractable phosphate with topographic height for the coral-creek mangrove forest system in northern Australia.

medium gives a measure of extractable phosphorus which, while not giving an absolute measure, at least can be compared with results from other laboratories. We are at present using this extraction medium and the results obtained appear to correlate with plant productivity estimates (Fig. 7.5). This is all that can be expected of any such artificial simulation of a biological system.

The fresh, wet soil is shaken in the extraction solution (10:1 solution/soil ratio) for thirty minutes and cleared by filtration or centrifugation. The extract is then analysed for phosphate ion species (PO_4^{3-}, HPO_4^{2-} and $H_2PO_4^{-}$) by the standard molybdenum-blue method. Again, autoanalysers are highly suited to this method of analysis. Total P in soil is easily measured by digesting a small sample of dried, finely ground, soil in perchloric-nitric-sulphuric-acid mixture, followed by colorimetric analysis of the digest. As for nitrogen, a measurement of total P indicates a 'pool' size only, and gives no indication of availability.

It is worth while noting that acetate/acetic acid is suitable for extraction of both N and P. This is particularly convenient for autoanalyser methods, as a single extract can be used on a multichannel system, thus decreasing analysis time considerably.

To summarize, there are many well-tried and reliable methods available for basic soil analyses. In almost all cases, as exemplified by the methods outlined above, these analyses can be carried out with the most basic laboratory apparatus. More sophisticated equipment, if available, can be easily and effectively adapted to most of the classical methods. This will not normally give any greater accuracy but does allow a greater number of replicate determinations to be carried out with minimized risk of operator error.

References

ALEXANDER, M. 1961. *Introduction to Soil Microbiology*. New York, John Wiley.

ALLEN, S. E. (ed.); GRIMSHAW, H. M.; PARKINSON, J. A.; QUARMBY, C. 1974. *Chemical Analysis of Ecological Materials*. New York, John Wiley.

BREMNER, J. M. 1965*a*. Total Nitrogen. In: C. A. Black (ed.), Methods of Soil Analysis, Part 2. *Agronomy*, Vol. 9, pp. 1149–78. Madison, Wis., American Society of Agronomy.

——. 1965*b*. Inorganic Forms of Nitrogen. In: C. A. Black (ed.) *Methods of Soil Analysis*, Part 2: Agronomy, pp. 1179–237. Madison, Wis., American Society of Agronomy.

BRUMMER, G.; GRUNWALT, H. S.; SCHROEDER, D. 1971. Contribution to the Genesis and Classification of Marsh Soils. II: Sulphur Metabolism of Muds and Salt Marshes. *Z. Pfl. Ernabr. Bodenk.*, Vol. 128, pp. 208–20.

CLOUGH, B.; ATTIWILL, P. 1982. Primary Production in Mangroves. In: B. F. Clough (ed.), *Mangrove Ecosystems in Australia; Structure, Function and Management*, pp. 213–22. Canberra, ANU Press.

CONNELL, W. E.; PATRICK, W. H. Jr. 1968. Sulphate Reduction in Soil: Effects of Redox Potential and pH. *Science*, Vol. 159, pp. 86–7.

DE LAUNE, R. D.; PATRICK, W. H. Jr; BRANNON, J. M. 1976. *Nutrient Transformations in Louisiana Salt Marsh Soils*. Baton Rouge, La., Center for Wetland Resources, Louisiana State University. (Sea Grant Publication, LSU–T–76–009.)

GOTOH, S.; PATRICK, W. H. Jr. 1972. Transformation of Manganese in a Waterlogged Soil as Affected by Redox Potential and pH. *Proc. Soil Sci. Soc. Amer.*, Vol. 36, No. 5, pp. 738–42.

GREESON, P. E.; CLARK, J. R.; CLARK, J. E. (eds.). 1978. Wetlands Functions and Valves. The State of our Understanding. *Proceedings of the National Symposium on Wetlands*. Minneapolis, Minn., American Water Resources Association. (Publication No. TPS 79–2.)

HOLFORD, I. C. R.; PATRICK, W. H. Jr. 1979. Effects of Reduction and pH Changes on Phosphate Sorption and Mobility in an Acid Soil. *Soil Sci. Soc. Amer. J.*, Vol. 43, pp. 292–7.

ISIRIMAH, N. O.; KEENEY, D. R. 1973. Nitrogren Transformations in Aerobic and Waterlogged Histosols. *Soil Sci.*, Vol. 115, No. 2, pp. 123–9.

PATRICK, W. H., Jr.; DE LAUNE, R. D. 1972. Characterization of the Oxidized and Reduced Zones in Flooded Soil. *Proc. Soil. Sci. Soc. Amer.*, Vol. 36, No. 4, pp. 573–6.

PATRICK, W. H., Jr.; GOTOH, S.; WILLIAMS, B. G. 1973. Strengite Dissolution in Flooded Soils and Sediments. *Science*, Vol. 179, pp. 564–5.

PONNAPERUMA, F. N. 1972. The Chemistry of Submerged Soils. *Advan. Agron.*, Vol. 24, pp. 29–88. New York, Academic Press.

SHAPIRO, R. E. 1958*a*. Effect of Flooding on the Availability of Phosphorus and Nitrogen. *Soil Sci.*, Vol. 85, No. 4, pp. 190–7.

——. 1958*b*. Effect of Flooding on the Availability of Phosphorus and Nitrogen. *Soil Sci.*, Vol. 85, No. 5, pp. 267–72.

STRICKLAND, J. D. H.; PARSONS, T. R. 1972. *A Practical Handbook of Seawater Analysis*. (Bull. Fish. Res. Board Can., No. 167, 2nd ed.)

TURNER, F. T.; PATRICK, W. H., Jr. 1968. Chemical Changes in Waterlogged Soils as a Result of Oxygen Depletion. *Trans. 9th Intern. Congr. Soil Sci.* (Adelaide, Australia), Vol. 4, pp. 53–65.

TURNER, R. E. 1976. Geographic Variations in Salt Marsh Productivity: A Review. *Contribution in Marine Sci.*, Vol. 20, pp. 47–68.

8

Analyses of soil, plant and water components

Federico Pannier
Universidad Central de Venezuela
Facultad de Ciencias
Escuela de Biología, Departamento de Botánica
Apartado 80.390
Caracas, Venezuela

This chapter is intended to facilitate the selection of methods suitable for the acquisition of chemical data required for the treatment of the diverse problems of mangrove ecology, such as nutrient cycling, plant nutrition, soil dynamics and water quality. However, as is the case with most complex ecosystems and their associated research, the investigator is well advised to use care both in selecting the methods to be used and the criteria for data acceptance. It is not the intention in this chapter to describe all applicable chemical methods thoroughly; this would be a near-impossible task. However, we do hope to show the sequence of the most common procedures necessary to obtain a quantitative picture of the most important plant nutrients present in the system.

Detailed descriptions of methods have been avoided; instead, reference has been made to original papers and general analytical handbooks containing detailed descriptions.

Soil analysis [1]

SAMPLING AND SAMPLE PROCESSING

Taking samples

The number of soil samples to be taken depends on the precision required for minimizing variability of the data, and of the analytical facilities available.

It is recommended that samples be taken in three plots of 3 × 3 m each, inside the selected stand, with a minimum of 3 samples per plot. The samples are taken, preferably with a soil-borer for waterlogged soils, to a depth at which water infiltration appears. The separation of samples of each core, according to the depth, should include the little layer (if this exists) followed by sections of 10–15 cm each.

1. Allen (1974), Black (1965), Hesse (1971), Piper (1950).

Sample processing

The samples, after measurement *in situ* of actual (Bates, 1964) and potential pH (Ellenberg, 1958), are put into closed plastic bags and brought to the laboratory for immediate processing.

It is recommended that a sample of infiltrated water be taken from each bored hole, and analysed according to the methods indicated in the section 'Water analysis' (page 138 below).

The processing of the samples taken in the field varies according to the preferred type of analysis, as indicated in Table 8.1.

TABLE 8.1. Processing of samples

Measurement	Recommended treatment
pH Redox potential	From freshly collected samples
Organic nitrogen and phosphrous fractions Extractable nitrate, nitrite and ammonium nitrogen Peat extractions	In samples maintained only for a short time in freezer below −5 °C
Specific ions such as K, Na, Ca, Ng, Fe, Mn, Zn, Cu, B, etc. Available nitrogen Cation exchange capacity Organic analysis	In samples air-dried at temperatures not exceeding 40 °C
Total mineral constituents (B, S, C, N) Ignition loss	In oven-dried samples at 105 °C

Source: Allen (1974).

The soil samples must be thoroughly mixed before being crushed and sieved. Air-dried soils should be sieved to 2 mm mesh-size; oven-dried soils must be ground mechanically by mortar or mill.

In every case, and independently of the previous processing, a measurement of water content (Cope and Trickett, 1965; Nutting, 1943) in subsamples has to be made in order to correct the analytical values obtained for each specific element to the corresponding dry weight.

Soils of high organic content require an estimation of their organic carbon content by weighing the samples after ignition (Ball, 1964).

For a more precise measurement of organic carbon content of the soil sample, the method of wet oxidation (Walkley, 1947; Steubing, 1965) is advisable.

SAMPLE ANALYSIS

According to the kind of research intended, the methods can be grouped under:

Procedures for the determination of the global content of plant-important nutrients.
Procedures for the determination of plant-available nutrients.
Procedures to measure nutrient dynamics in waterlogged soils.

PROCEDURES FOR THE DETERMINATION OF THE GLOBAL CONTENT OF PLANT-IMPORTANT NUTRIENTS

Sample digestion

In general, the method of digestion of the previously air-dried soil is performed with a mixture of perchloric-nitric-sulphuric acid (Piper, 1950) which gives satisfactory results, guaranteeing the breakdown of silicates, except in the case where a precise measurement of aluminium and potassium content is needed. In these cases a method by sample fusion (Piper, 1950) or digestion by a mixture of HF-H_2O_4 (Allen, 1974) should be employed.

Analysis for reductible macronutrients (nitrogen, phosphorus, sulphur, iron and manganese)

Nitrogen, total nitrogen. Method by Kjeldahl-distillation (Bradstreet, 1965): Nesslerization (Allen, 1974).

Ammonium. Method by Conway-distillation (Conway 1950); ion selective electrode (Merks, 1975; Gilbert and Clay, 1973).

Nitrate. Method by ion-selective electrode (Milham et al., 1970); colorimetric method by 2,4-xylenol (Balks and Reekers, 1955); colorimetric method by phenyl-disulphonic acid.

Nitrite. Method by diazotization with N-(1-naphthyl-ethylendiamine-HCl) (Barnes and Folkhard, 1951).

Phosphorus. Colorimetric method by stannous chloride (Jackson, 1958; Allen, 1974); method by Amidol (2,4,diaminopheno-dihydrochloride) (Allen, 1940).

Sulphur, sulphate. Method by turbidimetry (Butters and Chenery, 1959); ion-selective electrode (Hulanicki et al., 1976; Goertzen and Oster, 1972).

Sulphide. Method by iodometry (Allen, 1974).

*Iron.**[1] Method using sulphonated bathophenantroline (Riley and Williams, 1959).

Manganese.* Colorimetric method by formaldoxime (Bradfield, 1957).

Analysis for non-reductible macronutrients

Potassium. Method by emission-photometry (Allen, 1974); ion-selective electrode (Orion Research, n.d.).

1. Asterisk (*) indicates atomic absorption spectrometry determination possible.

Sodium. Method by emission-photometry (Allen, 1974); ion-selective electrode (Secor et al., 1976).
*Calcium.** Method by emission-photometry (Allen, 1974); ion-selective electrode (Woolson et al., 1970; Hulanicki and Trojanowicz, 1973).

Analysis for micronutrients

*Boron.** Method by formation of tetrafluoroborate-complex (Pasztor et al., 1960).
Chloride. Method by titration (Allen, 1974; Lieth, 1964); ion-selective electrode (Hipp and Langdale, 1971; La Croix et al., 1970; Cantliffe et al., 1970).
*Copper.** Colorimetric method by carbamate (Allen, 1974).
*Magnesium.** Method by titration with EDTA in presence of Eriochrome black as indicator (Allen, 1974); colorimetric method by Titanium yellow (Cornfield and Pollard, 1950).
*Molybdenum.** Colorimetric method by thiocyanate (Allen, 1974).
*Zinc.** Method by dithizone (Cowling and Miller, 1941).
The use of an atomic absorption spectrophotometer gives highly accurate determinations and should be employed whenever possible. Specific analytical procedures can be studied in the literature.

PROCEDURES FOR THE DETERMINATION OF PLANT-AVAILABLE NUTRIENTS

Direct sample extraction for exchangeable ions

To extract acid organic soils, a procedure using ammonium acetate at pH 7 is recommended. Calcareous soils, due to the solubility of calcium carbonate, are to be extracted with alkaline ammonium-acetate solution at pH 9 (Bower et al., 1952).

The conditions for extraction should maintain a constant ratio of soil-weight to extractant volume. That is, 25 g of organic soil for 250 ml extractant solution. It is important to keep in mind the need to determinate the water content in corresponding subsamples, to relate the final values obtained to dry weight. The manipulations for extraction should be done at ambient temperature, taking care that the soil during the filtration process is constantly saturated with the extractant. A convenient laboratory technique for continuous extraction with (H_4EDTA) has been described (Marks and Osden, 1979).

Specific ions

Nitrogen (*ammonium and nitrate*). Leaching with distilled water and decoloration with aluminium hydroxide.
Phosphorus. In acid or neutral organic soils with 1 per cent v/v citric acid, or

2.5 per cent v/v acetic acid, or 0.02 M citric acid +0.02 M KCl. In calcareous soils with 0.5 M $NaHCO_3$ at pH 8.5.
Iron. With 3 per cent oxalic acid.
Sulphur. With distilled water.
Sodium. With distilled water.
Determination of cationic exchange capacity (Rendig, 1947; Mehlich, 1945).
Determination of total bases exchange capacity (Allen, 1974; Bray and Willhite, 1929).

Sample analysis

Determination of specific ions (as under analysis for reductible macronutrients, non-reductible macronutrients and micronutrients).

Biological sample incubation for the determination of mineralization of organic nitrogen (Bremmer, 1960; Zottl, 1958); organic phosphorus (Rodriguez, 1979; Ozbek and Aksoy, 1975; Fardeau and Triboi, 1974).

Radiochemical sample incubation: isotopic-dilution-method for the determination of plant-available soil phosphorus (Larsen, 1952; Larsen and Gennary, 1964; Andersen et al., 1961).

PROCEDURES TO MEASURE NUTRIENT DYNAMICS IN WATER-LOGGED SOILS

Determination of the oxido-reduction status: method for measuring redox potential along soil profiles (Patrick and Delaune, 1972; Urquart, 1966; Urquart and Gore, 1973); method for maintaining plants under controlled pH and redox potential (Reddy et al., 1976).

Determination of the global biological activity of the soil: method for the determination of dehydrogenase activity (Stevenson, 1959; Lenhard, 1956); of ATP (Cunningham and Wetzel, 1978; Conklin and Macgregor, 1972; Christian et al., 1976; Ferguson and Murdoch, 1975).

Methods to measure dynamics of nitrogen

Measurement of nitrogen input: measurement of nitrogen fixation by the acethylene-ethylene conversion (Zuberer, 1976; Capone et al., 1977); measurement of ammonification capacity by the colorimetric determination of urease activity in incubated samples (Hoffmann and Teicher, 1961; Gerlach, 1980); measurement of nitrification rate in soil columns by the continuous percolation technique (Macura, 1966; Reddy and Patrick, 1975; Patrick and Reddy,1976; Skjins et al., 1978).

Measurement of nitrogen output (as denitrification processes)

Indirect methods. By measuring respiratory microbial CO_2 evolution (Paul et al., 1970; De Jong et al., 1978; Schwartzkopf, 1978; Brown, 1979); by measuring water-soluble carbohydrates (Burford and Bremner, 1975); by measuring redox potential *in situ* (Bohn, 1971; Whitfield, 1969; Kokholm Petersen, 1966); by measuring activity of denitrifying bacteria (Valera and Alexander, 1961; Focht and Joseph, 1973).

Direct methods. By nitrate removal from water overlaying soils and sediment, measured by colorimetry (Engler and Patrick, 1974); by continuous measurement of O_2 depletion and redox-potential of flooded samples (Engler et al., 1976; Rickman, 1968); by sample incubation with N^{15} (Bremner, 1965; Focht, 1978).

Measurement of nitrogen output (as volatilization processes)

By ammonium loss in soil columns incubated under anaerobic conditions (Delaune and Patrick, 1970); by gas chromatography (Payne, 1973; Bailey and Beauchamp, 1973); by infra-red analysis (Arnold, 1954; Hauck and Melsted, 1956; Schwartzbeck et al., 1961).

Methods for measurement of organic phosphorus mineralization rate

Measurement of labile organic phosphorus by the method of sequential alkali-acid extraction (Anderson, 1960; Ruiz and Lopez Hernandez, 1978); measurement of phosphate release (Patrick and Khalid, 1974; Lopez Hernandez and Flores-Aguilar, 1979; Patrick and Mahapatra, 1968).

Methods to measure dynamics of sulphur

Method to measure elementary sulphur by extraction with acetone followed by attack with hydrochloric acid and iodometry (Hart, 1961).

Method to measure H_2S and soluble sulphurs by attack with normal hydrochloric acid and iodometry (Thornton and Giglioli, 1965; Goni and Parent, 1966).

Method to measure soluble sulphates in the soil solution (Chesnin and Yien, 1950; Jouis and Lecacheux, 1962).

Method for determination of sulphite (Wainwright and Johnson, 1978).

Method to measure total sulphur (Chaudhry and Cornfield, 1966).

Method to measure iron sulphurs (Smittenberg et al., 1952).

Method to measure sulphate produced by dryness of the sample (Vieillefon, 1971).

Method to measure biological sulphate reduction (Connell and Patrick, 1968, 1969; Jaaq, 1973).

Method to measure biological sulphur reduction (Baldensperger, 1973).

Plant analysis[1]

COLLECTING SAMPLES OF PLANT BIOMASS FOR CHEMICAL ANALYSIS

Trees

Sample branches at different levels of the canopy; separate leaves and stems. Pool the components of each tree in apical middle and basal level of extension of the crown. Take wood-discs at different heights of the tree trunk.

Herbaceous Vegetation

Collect a predetermined number of individuals of the same species in microplots inside the sampling area. According to the kind of research intended, separate the plant parts to be studied (leaves, stems, roots, hypocotyls, etc.) and obtain homogenous subsamples of each part using scissors or cork-borer.

Litter

Collect cores of the litter layer inside different microplots of the sampling area. Separate leaves and woody material. Pool and process.

SAMPLE PROCESSING

The processing of fresh plant samples has to be done as soon as possible. The transportation of the plant material, collected in plastic bags in the field, to the laboratory should not take more than several hours.

The material has to be oven-dried at 105 °C to constant weight. The dried material is then ground in a mortar or mechnical mill to a fine powder.

SAMPLE ASHING

The analytical procedure has to be preceded by ashing the material in order to remove any traces of contamination by organic material.

The precautions to be employed, such as temperature adjustment, additives, specific mixtures for fusion or digestion, in order to prevent losses of elements like phosphorus, chlorine and boron, are described in Piper (1950).

Before starting the analytical procedure, a method of concentrating the elements, usually with 8-hydroxiquinoline (Mitchell, 1948) should be employed.

SAMPLE ANALYSIS

The analytical methods for the determination of specific elements in plant

1. Allen (1974); Piper (1950).

materials are generally the same as those described in the 'Procedures for the Determination of the Global Content of Plant-important Nutrients' section (page 133 above) for soil analysis.

The method to be chosen depends upon the instruments available. Atomic absorption spectrophotometry and ion-specific electrodes are to be preferred.

Valuable descriptions concerning atomic absorption spectrophotometry are: in relation to procedures for wet ashing and ignition (Castellano-Giron, 1973), determination of contamination (David, 1962) and specific applications in agricultural crops (Slavin, 1962) and marine research (Burrell, 1968).

Details on the application of ion-specific electrodes can be found in *Analytical Methods Guide* (Orion Research, 1978).

Water analysis [1]

SAMPLING

Rainwater

Rainwater is collected using a system consisting of polyethylene or borosilicate glass funnels of 25 cm diameter adapted to hold amounts up to 5 litres. These devices should be distributed over the sampling area and fixed to supports to prevent their displacement by tidal currents.

To minimize contamination problems due to microbial activity, the samples should be collected daily, filtered to eliminate solid particles and processed as soon as possible. It is advisable to wash the walls of the collecting funnels with chloroform and to analyse the collected washings since they contain some of the nutrients.

Soil (infiltration) water

After taking a soil sample in an organic mangrove soil, an accumulation of infiltrated water can usually be observed at the bottom of the hole. Once the column reaches equilibrium, the water in the hole is rapidly displaced with a suction bomb. From the newly infiltrated water, a sample of 500 to 1,000 ml is transferred to a bottle, a few drops of preservative (usually mercuric chloride, 20 mg/l, or toluene) added, and processed immediately after arrival at the laboratory.

Flowing or stagnant water

Sampling techniques vary according to whether one is studying stagnant to low-flowing waters or those with a strong flow. In the first case sampling

1. Allen (1974); Burton and Liss (1976); Good et al. (1978); Unesco/WHO (1978); Lee (1980); APHA (1965); Rainwater and Thatcher (1960); Strickland (1960).

devices normally used in limnology can be employed to obtain nearly undisturbed samples. A simple method of taking water samples at precise depths is described by Mackereth (1963). In the case of flowing water in channels and creeks, the samples should be taken at half-average depth. In both cases, it is convenient to take transects of channels and creeks of strategic importance inside the communicating hydrological system of the experimental area, selected by maximum water-volume displacement per unit of time.

It is important to measure rate of water flow in the sampling site, and in the case of areas exposed to tidal fluctuation, to indicate the direction of the current at the moment of taking the samples. A careful sampling periodicity has also to be considered.

Sample analysis

Solids. Method to measure total solids in suspension (APHA, 1965); method to measure total soluble solids (Howard, 1933); method to measure particulate organic carbon (Mulholland, 1981); method to measure total organic matter (Strickland, 1960); pH (Strickland, 1960); alkalinity (Strickland, 1960); electrical conductivity (Allen, 1974; Sjors, 1950).

Dissolved gases. O_2—method by portable oximeter (Clark et al., 1953; Boto and Bunt, 1981; Beechey and Ribbons, 1972; Fork, 1972) (Clark electrode); Winkler (Ohle, 1953); CO_2—method of ion-selective electrode (Ladenson et al., 1975); H_2S—method of ion-selective electrode (Barica, 1973; Bauman, 1974; Golterman, 1969); spectrophotometry (Cline, 1969); SO_2—method by precipitation with barium chloride (Picou and Waterlow, 1963; Walker, 1972; Massoumi and Cornfield, 1963).

Cation determination

Direct measurement without previous concentration of the sample by ion-selective electrodes (Duist, 1971; Pungor and Toth, 1970; Moody et al., 1969; Andelmann, 1968). This method is very adequate for waters in mangrove communities where the colour does not allow direct colorimetry.

Method by atomic absorption spectrophotometry, following extraction from the water sample using organic solvents. The most convenient extractant is ammonium pirolodine dithiocarbamate, which, in an acid medium, forms complexes with most elements. The spectrophotometrical analysis follows the indications for the specific ions (Christian and Feldman, 1970; Watson, 1968).

Method by ion exchange (Dow Chemical Company, 1974). (See also specific applications of resins in the technical booklets of the manufacturers.)

Method by spectrophotometry, following digestion with oxidant acids (Edisbury, 1966).

Anion determination

Method for measurement of soluble constituents by ion-exchange columns (Allen, 1974; Mackereth, 1963).

References

ALLEN, R. 1940. The Estimation of Phosphorus. *Biochem. J.*, Vol. 34, p. 858.

ALLEN, S. E. (ed.). 1974. *Chemical Analysis of Ecological Materials.* Oxford, Blackwell Scientific Publications.

APHA (American Public Health Association). 1965. *Standard Method for the Chemistry of Water and Waste-waters*, 12th ed. New York, APHA.

ANDELMANN, J. B. 1968. *J. Wat. Pollut. Contr. Fed.*, Vol. 40, p. 1844.

ANDERSEN, A.; GREGERS-HANSEN, B.; NIELSEN, G. 1961. Determination of the Phosphate Condition of Soils by Means of Radioactive Phosphorus in Pot Experiments. *Acta Agricult., Scandinavia*, Vol. 11, pp. 270–90.

ANDERSON, G. 1960. *J. Sci. Fd. Agric.*, Vol. 11, p. 497.

ARNOLD, P. W. 1954. *J. Soil. Sci.*, Vol. 5, p. 116.

BAILEY, L. D.; BEAUCHAMP, E. G. 1973. *Can. J. Soil Sci.*, Vol. 53, p. 122.

BALDENSPERGER, J. 1973. The Use of a Respirometric Method for the Evaluation of Sulfur Oxidation in Soils. *Acid Sulphate Soils. Proc. Int. Symp., 13–20 August 1972*, pp. 141–58. Wageningen, International Institute for Land Reclamation and Improvement. (Publication No. 18, Vol. 2.)

BALKS, R.; REEKERS, I. 1955. Bestimmung des Nitrat- und Ammoniakstickstoffs im Boden. *Landw. Forschung*, Vol. 8, p. 7.

BALL, D. F. 1964. *J. Soil. Sci.*, Vol. 15, p. 84.

BARICA, J. 1973. Use of a Silver-Sulfide Electrode for Standardizing Aqueous Sulfide Solution in Determining Sulfide in Water. *J. Fisheries Res. Bd. Canada*, Vol. 30, No. 10, p. 1589.

BARNES, H.; FOLKHARD, A. R. 1951. The Determination of Nitrites. *Analyst* (London), Vol. 76, p. 599.

BATES, R. G. 1964. *Determination of pH: Theory and Practice.* New York, John Wiley.

BAUMAN, E. W. 1974. Determination of Parts per Billion Sulfide in Water with the Sulfide-Electrode. *Anal. Chem.*, Vol. 46, No. 9, p. 1345.

BEECHEY, R. B.; RIBBONS, D. W. 1972. Oxygen Electrode Measurements. *Methods Microbiol.*, Vol. 6B, pp. 25–53.

BLACK, C. A. (ed.). 1965. *Methods of Soil Analysis*, Part 2. Madison, Wis., American Society of Agronomy.

BOHN, H. L. 1971. *Soil Sci.*, Vol. 112, p. 39.

BOTO, K. G.; BUNT, J. S. 1981. Dissolved Oxygen and pH Relationships in Northern Australian Mangrove Waterways. *Limnol. Oceanogr.*, Vol. 26, No. 6, pp. 1176–8.

BOWER, C. A.; REITEMEIER, R. F.; FIREMAN, M. 1952. *Soil Sci.*, Vol. 73, p. 251.

BRADFIELD, E. G. 1957. An Improved Formaldoxime Method for the Determination of Manganese in Plant Material. *Analyst* (London), Vol. 82, p. 254.

BRADSTREET, R. B. 1965. *The Kjeldahl Method for Organic Nitrogen.* New York. Academic Press.

BRAY, R. H.; WILLHITE, F. M. 1929. Determination of Total Replaceable Bases in Soils. *Ind. Engng. Chem. Analyt. Edn.*, Vol. 1, p. 144.

BREMMER, J. M. 1960. Forms of Nitrogen in Soils and Plants. *Rothamsted Exp. Stnt. Rept. for 1959*, p. 59.

——. 1965. In: C. A. Black (ed.) *Methods of Soil Analysis* Part 2, p. 1256. Madison, Wis., American Society of Agronomy.

BROWN, K. A. 1979. A Modified Procedure for Determining $^{14}C/CO_2$ produced by Respiration in Water or Sediment Samples. *Limnol. Oceanogr.*, Vol. 24, No. 6, pp. 1141–5.

BURFORD, J. R.; BREMNER, J. M. 1975. *Soil Biol. Biochem.*, Vol. 7, p. 389.

BURRELL, D. C. 1968. Atomic Absorption Spectrophotometry in the Field of Marine Research. *Atomic Absorption Newsletter*, Vol. 7, No. 4, pp. 65–8.

BURTON, J. D.; LISS, P. S. (eds.). 1976. *Estuarine Chemistry.* London, Academic Press. 220 pp.

BUTTERS, B.; CHENERY, E. M. 1959. A Rapid Method for the Determination of Total Sulphur in Soils and Plants. *Analyst* (London), Vol. 84, p. 239.

CANTLIFFE, D. J.; MACDONALD, G. E.; PECK, N. H. 1970. The Potentiometric Determination of Nitrate and Chloride in Plant Tissue. *Food and Life Sci. Bull.*, No. 3: *Plant Sciences*.

CAPONE, D. G.; OREMLAND, R. S.; TAYLOR, B. F. 1977. Significance of N_2 Fixation to the Production of *Thalassia testudinum* Communities. *Symp. on Progress in Marine Research in the Caribbean and Adjacent Regions*, pp. 71–85. (FAO Fisheries Report No. 2000.)

CASTELLANO-GIRON, H. 1973. Comparison between Dry Ashing and Wet Digestion in the Preparation of Plant Materials for Atomic Absorption Analysis. *Atomic Absorption Newsletter*, Vol. 12, No. 1, pp. 28–9.

CHAUDHRY, I. M.; CORNFIELD, A. H. 1966. The Determination of Total Sulphur in Soil and Plant Material. *Analyst* (London), Vol. 91, pp. 358–530.

CHESNIN, L.; YIEN, C. H. 1950. *Proc. Soil Sci. Soc. Am.*, Vol. 15, p. 149.

CHRISTIAN, G. D.; FELDMAN, F. J. 1970. *Atomic Absorption Spectroscopy. Applications in Agriculture, Biology and Medicine*. New York, Wiley-Interscience.

CHRISTIAN, R. R.; BANCROFT, K.; WIEBE, W. J. 1976. Distribution of Microbial Adenosine Triphosphate in Salt Marsh Sediments at Sapelo Island, Georgia. *Soil Science*, Vol. 119, No. 1, pp. 89–97.

CLARK, L. C., Jr.; WOLD, R.; GRANGER, D.; TAYLOR, Z. 1953. Continuous Recording of Blood Oxygen Tensions by Polarography. *J. Appl. Physiol.*, Vol. 6, pp. 189–93.

CLINE, J. D. 1969. Spectrophotometric Determination of Hydrogen Sulfide in Natural Waters. *Limnol. Oceangr.*, Vol. 14, pp. 454–8.

CONKLIN, A. R.; MACGREGOR, A. N. 1972. Soil Adenosine Triphosphate: Extraction, Recovery and Half-life. *Bull. Environm. Contamination and Toxicol.*, Vol. 7, pp. 296–300.

CONNELL, W. E.; PATRICK, W. H., Jr. 1968. Sulfate Reduction in Soil: Effects of Redox Potential and pH. *Science*, Vol. 159, No. 3,810, pp. 86–7.

——. 1969. Reduction of Sulfate in Waterlogged Soil. *Soil Sci. Soc. Amer. Proc.*, Vol. 33, pp. 711–14.

CONWAY, E. 1950. *Microdiffusion Analysis and Volumetric Error*. London, Crosby Lockwood.

COPE, F. and TRICKETT, E. S. 1965. *Soils Fertil.*, Vol. 28.

CORNFIELD, A. H.; POLLARD, A. G. 1950. *J. Sci. Fed. Agric.*, Vol. 1, p. 357.

COWLING, H.; MILLER, E. H. 1941. Determination of Small Amounts of Zinc in Plant Materials. A Photometric Dithizone Method. *Ind. Engng. Chem. Analyt. Edn.*, Vol. 13, p. 145.

CUNNINGHAM, H. W.; WETZEL, R. G. 1978. Fulvic Acid Interferences on ATP Determinations in Sediments. *Limnol. Oceanogr.*, Vol. 23, pp. 166–73.

DAVID, D. J. 1962. Atomic Absorption Spectrochemical Analysis of Plant Materials with Particular Reference to Manganese and Iron. *Rev. Univ. Ind. Santander* (Colombia), Vol. 4, No. 3/4, p. 207.

DE JONG, E.; REDMAN, R. E.; RIPLEY, E. A. 1978. A Comparison of Method to Measure Soil Respiration. *Soil Sci.*, Vol. 127, No. 5, pp. 300–6.

DELAUNE, R. D.; PATRICK, W. H., Jr. 1970. Urea Conversion to Ammonia in Waterlogged Soils. *Soil Sci. Soc. Am. Proc.*, Vol. 34, No. 4, pp. 603–7.

DOW CHEMICAL COMPANY. 1974. *Dowex: Ions Exchange. Principles of Ion Exchange. Applications of Ion Exchange. Test Procedures for Dowex Resins. Laboratory Guide*. Midland, Mich., Dow Chemical Co. 80 pp.

DUIST, R. A. 1971. *Amer. Sci.*, Vol. 59, p. 353.

EDISBURY, J. R. 1966. *Practical Hints on Absorption Spectrometry*. Hilger & Watts.

ELLENBERG, H. 1958. Bodenreaktion (einschliesslich Kalkfrage). *Handb. Pflanzenphysiol.*, Vol. IV, p. 638. Berlin, Springer Verlag.

ENGLER, R. M.; PATRICK, W. H., Jr. 1974. Nitrate Removal from Floodwater Overlaying Flooded Soils and Sediments. *J. Environm. Quality*, Vol. 3, No. 4, pp. 409–13.

ENGLER, R. M.; ANTIE, D. A.; PATRICK, W. H., Jr. 1976. Effect of Dissolved Oxygen on Redox Potential and Nitrate Removal in Flooded Swamp and Marsh Soils. *J. Environm. Quality*, Vol. 5, No. 3, pp. 230–5.

FARDEAU, J. C.; TRIBOI, E. 1974. Mesure au champ du phosphore disponible 'L'. *Plant and Soil*, Vol. 41, pp. 293–302.

FERGUSON, R. L.; MURDOCH, M. B. 1975. Microbial ATP and Organic Carbon in Sediments of the Newport River Estuary, North Carolina. In: L. E. Cronin (ed.), *Estuarine Research*, Vol. 1, pp. 229–49.

FOCHT, D. D. 1978. Methods for Analysis of Denitrification in Soils. In: D. R. Nielsen and J. G.

MacDonald (eds.), *Nitrogen in the Environment*, Vol. 2, pp. 453–64. New York, Academic Press.

FOCHT, D. D.; JOSEPH, H. 1973. *Soil Sci. Soc. Amer. Proc.*, Vol. 37, pp. 698.

FORK, D. C. 1972. Oxygen Electrode. *Methods Enzymol.*, Vol. 24, pp. 113–22.

GERLACH, A. 1980. Ein Vergleich von Methoden zur Bestimmung von Ammonium- und Nitratstickstoff in Boden. *Acta Oecologica, Oecol. Plant. No. 2*, Vol. 1, No. 15, pp. 185–200.

GILBERT, T. R.; CLAY, A. M. 1973. Determination of Ammonia in Aquaria and Sea Water Using the Ammonia Electrode. *Anal. Chem.*, Vol. 45, No. 9, p. 1757.

GOERTZEN, J. O.; OSTER, J. D. 1972. Potentiometric Titration of Sulfate in Water and Soil Extracts using a Lead Electrode. *Soil Sci. Soc. Am. Proc.*, Vol. 36, No. 4, p. 691.

GOLTERMAN, H. L. 1969. *Methods for Chemical Analysis of Freshwaters*. Oxford, Blackwell Scientific Publications. (IBP Handbook, No. 8.)

GONI, J.; PARENT, C. O. 1966. Étude géochimique d'une tranche de sediments fins actuels. *Bull. BRGM*, No. 5, pp. 19–33.

GOOD, R. E.; WHIGHAM, D. E.; SIMPSON, R. L. 1978. *Freshwater Wetlands. Ecological Processes and Management Potential.* New York, Academic Press. 378 pp.

HART, M. G. R. 1961. A Turbidimetric Method for the Determination of Elementary Sulphur. *Analyst* (London), Vol. 36, pp. 472–5.

HAUCK, R. D.; MELSTED, S. W. 1956. *Soil Sci. Soc. Amer. Proc.*, Vol. 20, p. 361.

HESSE, P. R. 1971. *A Textbook of Soil Chemical Analysis.*

HIPP, B. W.; LANGDALE, G. W. 1971. Use of a Solid State Chloride Electrode for Chloride Determinations in Soil Extracts. *Soil Sci. and Plant Anal.*, Vol. 2, No. 4, p. 237.

HOFFMANN, G.; TEICHER, K. 1961. Ein kolorimetrisches Verfahren zur Bestimmung der Ureaseaktivitat in Boden. *Z. f. Pflanzenern., Dungung, Bodenk.*, Vol. 95, p. 55. (Cited in L. Steubing, *Pflanzenokologisches Prakticum*, p. 240. Berlin/Hamburg, Paul Parey, 1965.)

HOWARD, C. S. 1933. Determination of Total Dissolved Solids in Water Analysis. *Ind. and Eng. Chem. Anal. Ed.*, Vol. 5, pp. 4–6.

HULANICKI, A.; TROJANOWICZ, M. 1973. Potentiometric Determination of Calcium and Magnesium in Water. *Chem. Anal.*, Vol. 18, p. 235.

HULANICKI, A.; LEWANDOWSKI, R.; LEWENSTAM, A. 1976. Elimination of Ionic Interferences in the Determination of Sulphates in Water using the Lead-sensitive Electrode. *Analyst*, Vol. 101, p. 939.

JAAQ, V. A. 1973. Biological Sulphate-reduction in the Spermosphere and the Rhizosphere of Rice in Some Acid Sulphate Soils of Senegal. *Acid Sulphate Soils. Proc. Int. Symp. 13–20 August 1972.* pp. 82–98. Wageningen, International Institute for Land Reclamation and Improvement. (Publication 18, Vol. 2.)

JACKSON, M. L. 1958. *Soil Chemical Analysis.* London, Prentice Hall.

JOUIS, E.; LECACHEUX, M. T. 1962. Une méthode nephelométrique rapide pour le dosage du soufre dans les sols. *Ann. Agron.*, Vol. 5, pp. 483–9.

KOKHOLM PETERSEN, G. 1966. *Redox Measurements. Their Theory and Technique Radiometer.* Copenhagen. 30 pp.

LA CROIX, R. L.; KENNEY, D. R.; WALSH, L. M. 1970. Potentiometric Titration of Chloride in Plant Tissue Extracts Using the Chloride Ion Electrode. *Soil Sci. and Plant Anal.*, Vol. 1, No. 1, p. 1.

LADENSON, J. H.; HUEBNER, M.; MARR, J. J. 1975. Continuous Real-time Monitoring of Metabolic Parameters in Growing Bacterial Cultures. *Anal. Biochem.*, Vol. 63, p. 56.

LARSEN, S. 1952. The Use of ^{32}P in Studies of the Uptake of Phosphorus by Plants. *Plant Soil*, Vol. 4, pp. 1–10.

LARSEN, S.; GENNARY, D. 1964. The Determination of Labile Soil Phosphate as Influenced by the Time of Application of Labelled Phosphate. *Plant and Soil*, Vol. 20, pp. 135–42.

LEE, R. 1980. *Forest Hydrology.* New York, Columbia University Press. 349 pp.

LENHARD, G. 1956. Dehydrogenaseaktivitat des Bodens al Mass fur die Mikroorganismentatigkeit im Boden. *Z. f. Pflanzenern, Dungung, Bodenk.*, Vol. 73, p. 1.

LIETH, H. 1964. Beziehungen zwischen der Bodensalinitat und dem Vorkommen von Pflanzenarten in den Salzmarschen der Gaspe-Halbinsel. *Studien über die Vegetation der Gaspe-Halbinsel*, Vol. III. Stuttgart. (Beitr. z. Phytologie Eugen Ulmer, 7.)

LOPEZ HERNANDEZ, D.; FLORES-AGUILAR, D. 1979. La desorcion de fosfatos en suelos. Implicaciones fisioecologicas en el proceso. *Acta Cientif. Venezol.*, Vol. 30, pp. 25–35.

MACKERETH, E. J. H. 1963. *Water Analysis for Limnologists.* Fresh Water Biology Association (Scientific Publication No. 21.)

MACURA, J. 1966. Article in: I. Matek and Z. Fenci (eds.), *Theoretical and Methodological Basis of Continuous Culture*, pp. 462–92. New York, Academic Press.

MARKS, K.; OSDEN, R. A. A Convenient Laboratory Technique for Continuous Extraction of Ions from Soils. *Soil Science*, Vol. 128, No. 3, pp. 163–70.

MASSOUMI, A.; CORNFIELD. 1963. A Rapid Method for Determining Sulfate in Water Extracts of Soils. *Analyst*, Vol. 88, pp. 321–2.

MEHLICH, A. 1945. *Soil Science*, Vol. 66, p. 429.

MERKS, A. G. A. 1975. Determination of Ammonia in Sea-water with an Ion-selective Electrode. *Noth. U. Sea Res.*, Vol. 9, No. 3/4, p. 371.

MILHAM, P. J., et al. 1970. Analysis of Plants, Soils and Waters for Nitrate by Using an Ion-selective Electrode. *Analyst*, Vol. 95, pp. 751–7.

MITCHELL, R. L. 1948. *Tech. Commun.*, Vol. 44, pp. 473–4. Harpenden, Commonwealth Bureau of Science. (Quoted in: E. C. Humphries, *Mod. Meth. Pl. Anal.*, Vol. 1.)

MOODY, G. J.; OKE, R. B.; THOMAS, J. D. R. 1969. *Lab. Pract.*, Vol. 18, p. 941.

MULHOLLAND, P. J. 1981. Formation of Particulate Organic Carbon in Water from a Southeastern Swamp-stream. *Limnol. Oceanogr.*, Vol. 25, No. 4, pp. 790–5.

NUTTING, P. G. 1943. *Some Standard Thermal Dehydration Curves of Minerals.* U.S. Geological Survey. (Professional Paper, 197-E.)

OHLE, W. 1953. *Die chemische und elektrochemische Bestimmung de molekular gelosten Sauerstoffs der Binnengewasser.* Int. Ver. Limnolog. (Mitteilung, 3.)

ORION RESEARCH. 1978. *Analytical Methods Guide*, 9th ed. Cambridge, Mass., Orion Research, Inc.

——. *Orion Instruction Manual, Model 93-Electrode.* Cambridge, Mass., Orion Research, Inc.

OZBEK, N.; AKSOY, T. 1975. 'A', 'L' and 'E' Values of Three Soils Differing in Their Properties Including their Cation Exchange and Phosphorus Fixation Capacities. *Ziraat Fakultesi Yilligi* (Ankara University), Vol. 25, pp. 76–94.

PASZTOR, L.; BODE, J. D. FERNANDO, Q. 1960. Determination of Micro Quantities of Boron in Steel by a Solvent Extraction Method. *Analyt. Chem.*, Vol. 32, p. 277.

PATRICK, W. H. Jr; DELAUNE R. D. 1972. Characterization of the Oxidized and Reduced Zones in Flooded Soil. *Soil Sci. Soc. Am. Proc.*, Vol. 36, pp. 573–6.

PATRICK, W. H., Jr.; KHALID, R. A. 1974. Phosphate Release and Sorption by Soils and Sediments: Effect of Aerobic and Anaerobic Conditions. *Science*, Vol. 186, pp. 53–5.

PATRICK, W. H., Jr.; MAHAPATRA, I. C. 1968. Transformation and Availability to Rice of Nitrogen and Phosphorus in Waterlogged Soils. *Advances in Agronomy*, Vol. 20, pp. 339–53.

PATRICK, W. J., Jr.; REDDY, K. R. 1976. Nitrification-Denitrification Reactions in Flooded Soils and Water Bottoms: Dependence on Oxygen Supply and Ammonium Diffusion. *J. Environm. Quality*, Vol. 5, pp. 469–72.

PAUL, E. A.; BIEDEREBECK, V. O.; ROSHA, N. S. 1970. Investigations of the Metabolism of Soil Micro-organisms by the Use of C^{14}. *Methods of Study in Soil Ecology. Proceed. Paris Symp. Ecology and Conservation*, Vol. 2. pp. 111–16. Paris, Unesco.

PAYNE, W. J. 1973. *Bull. Ecol. Res. Comm.* (Stockholm), Vol. 17, p. 263.

PICOU, D.; WATERLOW, J. C. 1963. Estimation of Small Amounts of Sulfate by Precipitation With Radioactive Barium. *Nature*, Vol. 197, pp. 1103–4.

PIPER, C. S. 1950. *Soil and Plant Analysis.* New York, Interscience Publishers.

PUNGOR, E.; TOTH, K. 1970. Ion-selective Membrane Electrodes. *Analyst* (London), Vol. 95, pp. 625–48.

RAINWATER, F. H.; THATCHER, L. L. 1960. *Methods for Collection and Analysis of Water Samples.* Washington D.C., U.S. Government Printing Office. 301 pp. (Geological Survey Water-Supply Paper, 1454.)

REDDY, K. R.; PATRICK, W. H., Jr. 1975. Effect of Alternate Aerobic and Anaerobic Conditions on Redox Potential, Organic Matter Decomposition and Nitrogen Loss in a Flooded Soil. *Soil Biol. Biochem.*, Vol. 7, pp. 87–94.

REDDY, C. N.; JUGSUJINDA, A.; PATRICK, W. H., Jr. 1976. System for Growing Plants under Controlled Redox Potential-pH Conditions. *Agronomy J.*, Vol. 68, pp. 987–9.

RENDIG, V. V. 1947. *Proc. Soil Sci. Am.*, Vol. 11, p. 449.

RICKMAN, R. W. 1968. Effect of Salts on Oxygen Diffusion Rate Measurements in Unsaturated Soils. *Soil Sci. Soc. Amer. Proc.*, Vol. 32, No. 5, pp. 618–22.

RILEY, J. P.; WILLIAMS, H. P. 1959. The Microanalysis of Silicate and Carbonate Materials. 1: Determination of Ferrous Iron. 2: Determination of Water and Carbon Dioxide.

RODRIGUEZ, D. 1979. Determinación de la incorporación de fósforo en *Rhizophora mangle* como índice de la disponibilidad de este nutrinte en suelos de manglar, pp. 9–19. Caracas, Facultad de Ciencias, U.C.V. (Biology thesis.)

RUIZ, J.; LOPEZ HERNANDEZ, D. 1978. Tasa de mineralización de fósforo orgánico en suelos tropicales in relación al contenido de humedad y la temperatura. (IAEAA–SM211/68.)

SCHOLLENBERGER, C. J. 1927. Exchangeable Hydrogen and Soil Reaction. *Science*, Vol. 65, p. 552.

SCHWARTZBECK, R. A.; MACGREGOR, J. M.; SCHMIDT, E. L. 1961. *Soil Sci. Soc. Amer. Proc.*, Vol. 25, p. 186.

SCHWARTZKOPF, S. H. 1978. An Open Chamber Technique for the Measurement of Carbon Dioxide Evolution from Soils. *Ecology*, Vol. 59, No. 5, pp. 1062–8.

SECOR, G. F.; MCDONALD, G. M.; MCCREADY, R. M. 1976. Determination of Sodium in Alkali-treated Cereal Straws, Ley-peel Wastes and Brine-frozen foods with an Ion Selective Electrode Compared with Atomic Absorption Spectrometry. *J. AOAC*, Vol. 59, No. 4, p. 761.

SJORS, H. 1950. On the Relation between Vegetation and Electrolytes in North Swedish Mire Waters. *Oikos*, Vol. 2, pp. 241–58.

SKJINS, J.; TRUJILLO, A. FULGHAM, P. 1978. Nitrification in Great Desert Soils. In: N. E. West and J. J. Skjins (eds.), *Nitrogen in Desert Ecosystems*, pp. 60–74. (US/IBP Synthesis Series, No. 9.) Stroudsburg, Pa., Dowden, Hutchinson & Ross.

SLAVIN, W. 1962. Agricultural Applications of Atomic Absorption Spectrophotometry. *Atomic Absorption Newsletter*, Vol. 1/2, No. 4, pp. 13–19.

SMITTENBERG, J.; HARMSEN, C. W.; OTZEN, D. 1952. Rapid Methods for Determining Different Types of Sulphur Compounds in Soil. *Plant and Soil*, Vol. 3, pp. 353–60.

STEUBING, L. 1965. *Pflanzenokologisches Praktikum*, pp. 209–12. Berlin/Hamburg, Paul Parey.

STEVENSON, I. 1959. Dehydrogenase Activity in Soils. *Canadian J. Microbiol.*, Vol. 5, p. 229.

STRICKLAND, J. D. H. 1960. *A Practical Handbook of Seawater Analysis*. (Bull. Fish. Res. Bd Canada, No. 125.)

THORNTON, I.; GIGLIOLI, M. E. C. 1965. The Mangrove Swamps of Keneba, Lower Gambia River Basin. II: Sulphur and pH in the Profiles of Swamp Soil. *J. Appl. Ecol.*, Vol. 2, pp. 257–69.

UNESCO/WHO. 1978. *Water Quality Surveys*. Paris/Poole, Unesco/WHO. 350 pp. (Studies and Reports in Hydrolosy, 23.)

URQUART, C. 1966. An Improved Method of Demonstrating the Distribution of Sulphides in Peat Soils. *Nature* (London), Vol. 211, p. 550.

URQUART, C.; GORE, A. V. P. 1973. *Soil Biol. Biochem.*, Vol. 5.

VALERA, C. L.; ALEXANDER, M. 1961. *Plant and Soil*, Vol. 15, p. 268.

VIEILLEFON, J. 1971. Contribution a l'étude du cycle du soufre dans les sols de mangrove. Ses rapports avec l'acidification naturelle ou provoquée. *Cah. ORSTOM, ser. Pedol.*, Vol. 9, No. 3, pp. 241–70.

WAINWRIGHT, M.; JOHNSON, J. 1978. Determination of Sulphite in Mineral Soils.

WALKER, D. R. 1972. Soil Sulfate. I: Extraction and Measurement. *Can. J. Soil Sci.*

WALKLEY, A. 1947. A Rapid Method for Determining Organic Carbon in Soils. *Soil Sci.*, Vol. 63, pp. 251–64.

WATSON, C. A. 1968. *Ammonium Pyrollidine Dithiocarbamate*. New York, Hopkins & William. (Monograph, 74.)

WHITFIELD, M. 1969. Eh as an Operational Parameter in Estuarine Studies. *Limnol. Oceanography*, Vol. 14, pp. 547–58.

WOOLSON, E. A.; AXLEY, J. H.; KEARNEY, P. C. 1970. Soil Calcium Determination using a Calcium-specific Ion Electrode. *Soil Sci.*, Vol. 109, No. 5, p. 279.

ZOTTL, H. 1958. Die Bestimmung der Stickstoffmineralisation im Wald jumus durch den Brutversuch. *Z. Pflanzenernahrung, Dung. Bodenk.*, Vol. 81, p. 35.

ZUBERER, D. 1976. Biological Nitrogen Fixation: A Factor in the Establishment of Mangrove Vegetation. In: R. R. Lewis (ed.) *Third Annual Conference on Restoration of Coastline Vegetation in Florida*, pp. 37–57. Tampa Port Authority.

9

Methods for the study of mangrove fauna

A. Sasekumar
Department of Zoology
University of Malaya
Kuala Lumpur 22–11
Malaysia

The mangrove shore

FAUNA: DISTRIBUTION, BIOMASS AND PRODUCTION

Macrofauna

Many descriptive studies have been carried out on the distribution and zonation of macrofauna in mangrove swamps (Verwey, 1930; Berry, 1964, 1972; Gearlach, 1958; Sasekumar, 1974; Frith et al., 1976; Frith, 1977; Warner, 1969). Several of these studies give values on the density or abundance of fauna. The fauna living in the mangrove swamps are comprised of a few dominant groups represented mainly by polychaetes, gastropods, brachyurans and a sipunculid.

In a recent study carried out in Kuala Selangor, Malaysia, Macintosh (1979) estimated the annual production of fiddler crabs to be in the region of 15–20 kcal m^{-2} in open habitats and about 3 kcal m^{-2} in forested areas. These values are well below the estimates obtained for salt-marsh fiddler crabs. Teal (1962) obtained values of 35 kcal m^{-2} yr^{-1} for the salt marsh as a whole.

Quantitative sampling of macrofauna

To begin with, a series of stations are marked along a transect from low shore to the high shore. These stations may be located in the various forest zones or at specific intervals depending on the breadth of the shore and the objectives of the investigation.

The mangrove shore, being three-dimensional, provides a variety of habitats in which various animals live. Animals may be found living on different substrates such as soil, roots, stems, leaves and dead wood. Apart from the resident fauna, there are terrestrial species such as birds, monkeys, and marine fish and crustaceans which encroach on to the shore at regular intervals.

Soil epifauna and infauna

Collection of quantitative samples on a mangrove shore is necessary for productivity studies and for estimating the biomass per unit area. The sampling techniques given below are modified to suit muddy substrate.

Samples may be taken by driving a square, sheet-aluminium frame of 0.25 m^{-2} or 0.5 m^{-2} into the substratum to a depth of 20 cm. Samples below this depth will reveal few animals except for large sesarmid crabs and *Thalassina* mud-lobsters whose burrows may reach depths of 0.5 m or more. The edge of the sampling frame should be sharpened to form cutting edges on all four sides so as to cut through roots and wood.

Samples should normally be taken at low tide during a spring tidal period when the animals are most active. Once the quadrat is laid, the debris in the area is removed and all surface-dwelling animals are collected. Crabs are carefully picked out from their burrows by pressing one finger into the side of small burrows to prevent the crabs from retreating deeper. Large sesarmid crabs are trapped with the aid of a long knife or trowel which is used to block the burrow beneath the animal. The crab is then caught as it runs out.

The substratum is then dug out in small blocks using a spade, and placed on a large polyethylene sheet. This is then broken up by hand and examined carefully. All visible animals are picked out. Hand-picking in this manner is advantageous in muddy substrates such as those of the Selangor coast of Malaysia. Sieving is laborious as it is impractical to reach the sea or streams to sieve every sample. Furthermore, the sticky mud does not facilitate easy sieving. However, sieving of macro-infauna samples may be feasible in sandy substrates.

Macrofauna may also be collected using can cores (Frey et al., 1973; Cammen, 1979). The cans are dug out and later sieved in the laboratory.

Tree fauna

An important component of the mangrove fauna are the bivalves (e.g. oysters), barnacles and gastropods that dwell on the prop roots and lower trunks of trees. The density and vertical distribution of these fauna may be estimated by sampling adequate numbers of trees in each station. The trees may be zoned vertically above ground level into 25-cm zones. The number and species of epifauna in each vertical zone is recorded to estimate their density per unit length at different heights. On the seaward edges, encrusting epifauna and gastropods may be up to heights of two metres above the mud surface.

Biomass and productivity of mangrove macrofauna

Biomass is defined as the amount of living substance constituting the organisms and is normally expressed as the biomass per unit area of habitat. Crisp (1971) gives details for estimating biomass for various faunal groups.

Methods used to estimate the production of animal populations are given in Ricker (1968), Edmondson and Winberg (1971), Crisp (1971) and Grodzinski et al. (1975). Ansell, et al. (1978) discuss some of the problems encountered in the estimation of faunal production on tropical beaches.

Identification of macrofauna

For ecological studies, precise identification of the macrofauna is essential. Identification can be confirmed by dispatching specimens to a specialist in a museum, preferably in the same region (e.g. Queensland Museum, Australia, for South-East Asian countries). It would be advisable to write in advance to the curator of the faunal group before sending specimens for confirmation. Adult male specimens are preferred for brachyurans since taxonomic keys are normally based on male characters.

Crab specimens collected alive should be kept separately in polyethylene bags or vials to prevent their appendages breaking as a result of struggling or of fighting between individuals. Crabs should be killed instantly by placing them in ice or a deep freeze and then transferring them to formalin (4 per cent). Neutralized formalin is essential because ordinary formalin will soften the carapace after some time in storage. Russel (1963) gives useful notes on methods for preservation of marine animals.

Polychaetes

Polychaetes may be identified with the aid of Fauvel (1953), Day (1967) and Fauchald (1977).

Mollusca

Few keys are available for mangrove gastropods. Specimens should be sent to specialists in museums for identification. Some keys and descriptions of mangrove gastropods are included in keys of non-marine mollusca, e.g. description of Potamididae (*Telescopium*, *Cerithidea*) and the Amphibolidae are found in van Benthem-Jutting (1977). Keys to the Indo-Pacific Littorinidae are given in Rosewater (1970). Small gastropods belonging to the families Stenothyridae and Fairbankiidae may be identified with the aid of Brandt (1968).

Crustacea

Keys for the Penaeidae are given in Hall (1962), and for Alpheidae, in Banner and Banner (1966). Keys for the brachyuran fauna are available in the papers of Tweedie (1935, 1936, 1937, 1938, 1940, 1950, 1954),[1] Campbell and Griffin (1966), Campbell (1967) and Serene and Lundoer (1974). Recent taxonomic papers include those of Serene and Soh (1970), Soh (1978), and Crane (1975).

1. Published in the *Bulletin of the Raffles Museum* (now known as the *Bulletin of the National Museum, State of Singapore*).

Identification of marine fishes

The common marine fish species can be easily identified using local keys available in the various countries of South-East Asia. Munro (1955) and Smith (1950) and FAO Species Identification Sheets are useful aids for the uncommon species.

Meiofauna

The mangrove meiofauna consists predominantly of free-living nematodes, harpacticoid copepods, oligochaetes and a kinorhynch species (Sasekumar, personal observation). Mean densities vary from 1,130 individuals per 10 cm^2 in the lower shore *Avicennia* forest to 400 individuals per 10 cm^2 in the high shore *Bruguiera* forest in Kuala Selangor, Malaysia.

Various size categories of small animals have been considered previously as meiofauna. The most recent definition of meiofauna considers it as that group of animals in the size range of 40 μm to 1,000 μm (Ankar and Elmgren, 1978). In this size category fall the metazoans: nematodes, harpacticoid copepods, oligochaetes, kinorhynchs, a few other groups, the temporary meiofauna and the protozoan foraminifera.

Sampling and sorting poses difficult problems in meiofaunal investigations. The investigator should consult Hulings and Gray (1971), Hulings (1971) and Elmgren (1973) for details on sampling and sorting techniques. The mangrove forest floor has an uneven topography due to the presence of crab burrows, small mounds, aerial roots, decaying leaves and twigs and puddles of water. Sampling in such a substrate presents many problems because substrate relief influences the abundance and distribution of meiofauna. A multi-subsampled grid plot is desirable for each station. Random, 2-cm-diameter core samples are taken from a grid of 0.25 m^2. If stations on the transect are to be compared, then the variation in density of the meiofauna within a given station must be known. If too few samples are taken, the size of variation within a given station will, in all probability, be greater than any differences between stations (Gray, 1971).

Sampling technique

Sampling may be done using a simple brass tube (2 cm diameter), sharpened at one end to form a cutting edge to facilitate slicing through rootlets. The tube is gently pushed down to a depth of 10 cm and withdrawn slowly. A cork piston or plunger is introduced into the lower end of the tube and the core extruded. If the sample is to be used to provide information on the vertical zonation of the fauna, it should be cut into sections of suitable length immediately to avoid errors due to migration of the fauna or subsequent mixing of the sample.

The method of removal of the core is important. Pushing out the core with a plunger will cause compression of the sample. A method developed by

Jansson (1967) avoids compression of the core during removal and provides a series of vertical subsamples. The coring tube is composed of a series of rings of uniform height which fit closely within an outer tube. The subsampler is pushed into the substratum and a core is obtained. The inner, sectional tube is pushed out by a plunger which bears on the rings, and not on the sediment, thus avoiding compression. The rings are removed separately as they emerge.

All core samples are preserved whole by adding hexamine-buffered formalin in sea-water to a strength of 4–5 per cent.

Mangrove substrates contain high densities of meiofauna and densities of 1,200 individuals per 10 cm^2 are common. Sorting of the entire sample in a 2-cm-diameter core is time-consuming and tedious. Therefore, it is desirable to resort to subsampling. This may be performed with a cylindrical subsampler which has its bottom divided into eight equal chambers (Fig. 9.1). A preserved sample is poured into the sample divider, a little detergent added (to prevent copepods adhering to surfaces), the volume made up to one litre and a tightly fitting lid applied. The sample divider is then inverted and vigorously shaken for a short while so as to make the water swirl within it. It is then reversed and placed on a stand, its bottom horizontal. The sample is allowed about an hour for sedimentation to the bottom, during which time a few twists of the sample divider make material sedimenting on the dividing walls fall down into the chambers. The water is then slowly drained off through a tap, until it reaches the level of the dividing walls. Then the rubber stoppers in the bottoms of the chambers are removed and the subsamples drained into eight small containers. Remaining sediment is washed out by a gentle jet of water. After use, the sample divider is rinsed out, and the water from the rinse, as well as the water drained off all other spill, is carefully collected and sieved through a 40-μm sieve. The meiofauna is counted and the subsample figures subsequently corrected by adding one-eighth of the numbers found to each of them. The chambers in the sample divider are numbered from 1 to 8 and each subsample numbered. Subsamples to be examined are selected by means of a table of random numbers (Elmgren, 1973).

Taxonomic keys

Nematoda. The free-living nematodes comprise the most abundant metazoan group in the mangrove benthos. This is a difficult group for the non-specialist. Reference to Nicholas (1975) is recommended for tips on the preparation of specimens for identification. The most extensive keys are those of de Coninck (1965) and Wieser (1953, 1954, 1956). Tarjan's (1980) introductory manual classifies nematodes down to subfamily level.

Harpacticoidea. Harpacticoid copepods form the second most abundant group in the mangrove meiofauna. The most useful recent key for the group is that of Wells (1976), but reference should also be made to earlier works of Lang

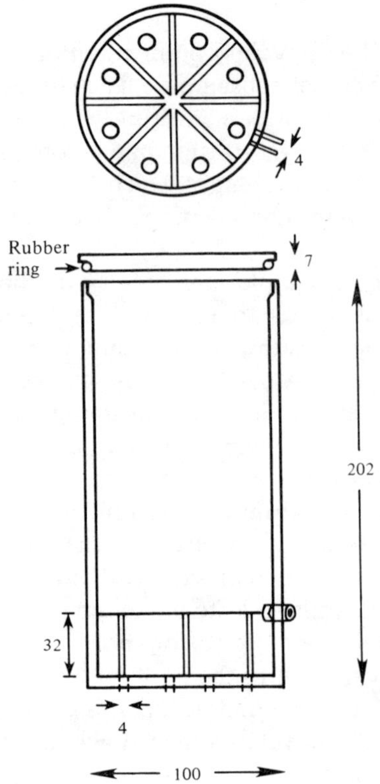

Figure 9.1
Sample divider. Measurements in millimetres (Elmgren, 1973).

(1948, 1965). Techniques for the preparation of harpacticoids for identification are given in Hulings and Gray (1971) and Hamond (1969).

For aids in the identification of other meiofaunal groups, the investigator should refer to Hulings (1971) and Hulings and Gray (1971). Recent references on meiofauna are available in *Psammonalia*, a quarterly newsletter of the International Association of Meiobenthologists.[1]

Meiofauna biomass

The entire numbers in random subsamples from a sample divider should be used for obtaining meiofauna biomass estimates. Dry weights are obtained by conversion from wet weights.

The most reliable wet weights are those for nematodes based on calculations of their volumes. Assuming that nematodes are cylindrical, measure maximum width and use the formula (length $\times$ width^2)/1.7 (Andrassy, 1956). The wet weight is calculated from volume multiplied by specific gravity for which values of 1.1 may be assumed (Ankar and Elmgren, 1978). For nem-

1. For details contact Dr Carlo Heip, Marine Biology Section, Zoology Institute, State University of Gent, Ledeganckstraat 35, B-9000 Gent, Belgium.

atodes, dry weight is calculated as 23 per cent of wet weight (Ankar and Elmgren, 1978). Table 9.1 gives conversion factors used for estimating shell-free dry weight and energy content for various groups of meiofauna (Ankar and Elmgren, 1978). Goodman (1980) calculated a linear relationship between body length and weight, thus enabling the individual dry weight of any harpacticoid of known body length to be estimated (Fig. 9.2).

Table 9.1. Conversion factors used for estimating shell-free dry weight and energy content

Meiofauna	To shell-free dry weight	To ash-free dry weight	To energy
Foraminifera	Ash-free dry wt used	4% of wet wt (Andren et al. 1968)	23.4 $kJ.g^{-1}$ (estimated)
Turbellaria	15% of wet wt (estimated)	90% of dry wt (estimated)	23.4 $kJ.g^{-1}$ (estimated)
Nematoda	23% of wet wt	80% of dry wt	22.2 $kJ.g^{-1}$
Kinorhyncha	20% of wet wt (estimated)	80% of dry wt (estimated)	23.4 $kJ.g^{-1}$ (estimated)
Halicryptus spinulosus	20% of wet wt (wet wt based on dry wt)	80% of dry wt (estimated)	23.4 $kJ.g^{-1}$ (estimated)
Harmothoe sarsi	20% of wet wt (wet wt based on dry wt)	90% of dry wt (estimated)	23.4 $kJ.g^{-1}$ (estimated)
Other polychaetes	20% of wet wt (estimated)		4798 $cal.g^{-1}$ dry wt (Thayer et al., 1968)
Tubifex costatus Other oligochaetes	18.6% of wet wt (Johnson and Brinkhurst, 1971)		5290 $cal.g^{-1}$ dry wt (Johnson and Brinkhurst, 1971)
Clitellio arenarius	18% of wet wt (Ankar, unpublished)		5290 $cal.g^{-1}$ dry wt (Johnson and Brinkhurst, 1971)
Nais elinguis	13% of wet wt (Elmgren, unpublished)		5290 $cal.g^{-1}$ dry wt (Johnson and Brinkhurst, 1971)
Copepoda *Copepod nauplii*	20% of wet wt (wet wt based on dry wt)	80% of dry wt (estimated)	23.4 $kJ.g^{-1}$ (estimated)

Source: Ankar and Elmgren (1978).

UTILIZATION OF MANGROVE FAUNA BY OTHER ANIMALS

Many predators invade the mangrove shore at different periods of the tidal cycle. Man and birds are predators; the former collects *Scylla serrata* and molluscs such as *Cerithidea obtusa*, *Telescopium* spp. and *Geloina* spp. for food.

Fish and crustaceans are known to invade the mangrove shore to consume mangrove invertebrates. These may be sampled quantitatively using gill nets.

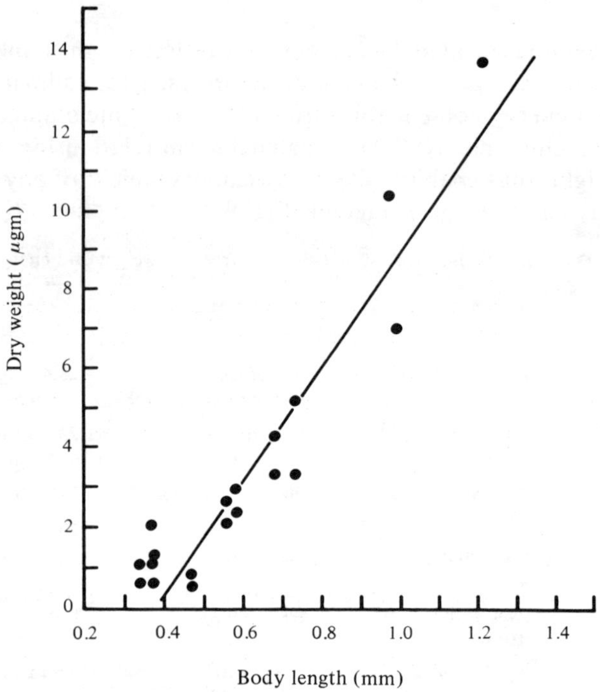

Figure 9.2
The linear relationship between body length and individual dry weight of gravid and non-gravid harpacticoids (Goodman, 1980).

The nets are tied between trees, parallel to the shoreline at low tide. Nets are collected at the next receding tide. Several assistants may be required to remove the great amount of plant debris that typically collects on the net and retrieve the fish. Specimens should be preserved in 4 per cent formalin as soon as possible and formalin is injected into the abdomen to preserve the gut contents of the larger fish specimens.

FOOD WEB

Study of gut contents

Analysis of the stomach contents of the mangrove fauna is a simple way of outlining the food web of the community. Specimens must be preserved in ice or in formalin immediately upon collection. Surface mud samples should be collected for examination under the microscope to identify the organic and inorganic components.

For crabs, the cardiac stomach or proventriculus should be examined. Remove the dorsal part of the carapace by cutting all around the edge with a small pair of scissors. The cardiac stomach is visible just below the cardiac region of the carapace. For details of crab anatomy see Warner (1977) or any book on invertebrate zoology. The cardiac stomach may be removed by cutting at the anterior oesophagus and at the posterior mid-gut. Cut away the dorsal wall of the cardiac stomach and flip over the entire contents on to a slide or Sedgewick rafter cell, adding a small quantity of water to spread the material. The volumetric composition of each item in the stomach is quantified by the method of eye estimation (Pillay, 1953). Random sampling is facilitated by using an eye-piece grid. Twenty to thirty specimens of each species should be examined. The items in the stomach should be identified as fully as possible and placed under various food categories. Empty or fully digested stomach contents should be ignored. The hind gut or rectum may be examined to see the digested state of food items.

In gastropods, the oesophagus and foregut (which in some species is bulged to form a crop), or the stomach, may be examined.

Tropic relationships and energy flow

The ultimate goal of a mangrove ecosystem study should be to find out the trophic relationship of the mangrove benthic macrofauna and meiofauna and the energy circuit of the system.

The living system is divided into various identifiable categories according to their trophic function: primary producers, herbivores, carnivores, detritivores and, finally, decomposers which consume the dead organic matter and liberate nutrients for re-use by plants (Mann, 1972, 1982).

Estuaries, coastal inlets, channels and streams

ZOOPLANKTON

Few studies have been carried out on zooplankton community structure or its biomass in mangrove estuaries or inlets.

Tundisi et al. (1973), in their study of a mangrove estuary in Brazil, found the zooplankton dominated by copepods and their larval stages, with the occasional abundance of other meroplankton taxa. Lee (1977), in a brief study of a mangrove inlet north of Port Klang, Malaysia, found that the zooplankton were also dominated by copepods and their larval stages. They constituted 60 per cent of the zooplankton community. A sergestid, *Acetes* and brachyuran larvae comprised 11 and 10 per cent respectively of the zooplankton.

Sampling techniques for the study of zooplankton are given in Edmondson and Winberg (1971).

NEKTON

Over thirty species of fish are known to live in the mangrove inlets of the Selangor Coast of Malaysia and in the other regions the number is considerably larger. Sampling with modern fishing gear is a problem in this environment. The presence of fallen logs and large branches on the bottom precludes the use of mini-trawlers to collect nekton. Gill nets and cast nets are suitable. Sampling should be carried out during both spring and neap tidal periods.

BENTHOS

Holme and McIntyre (1971) should be consulted in planning a study of the benthos.

The adjacent shallow waters

It is a documented fact that large catches of prawns are landed off mangrove shores. Many species of fin fish are also abundant off mangrove-fringed shores. The aim of the following section is to aid the investigations of the secondary productivity of the adjoining shallow waters. Methods for the study of zooplankton, nekton and benthos are available in IBP Handbooks. Emphasis in this section is devoted to the study of food webs in coastal waters. Such studies could produce evidence on the utilization of mangrove-derived organic matter (both primary and secondary products) by consumers in the sea.

ASSESSMENT OF FOOD OF FISHES

Methods of collection

Gear selection is important in sampling a fish population that is representative of the ecosystem. Bagenal (1978) should be consulted in planning a study.

The study of stomach contents has been the usual way of determining the nature of the diet of the species concerned, and of analysing the various trophic relationships between predators and prey in the food web of a marine biological community. Most workers base their conclusions on the study of stomach contents or, more rarely, of the entire gut of the captured fish. As digestion is less advanced in the stomach, the identification of its contents is therefore more satisfactory. Several methods for identifying and quantifying stomach contents have been employed by different workers. Most of these methods are subjective and vary between investigators. For comparative results of different species or of the same species in different studies, the methods used should be consistent and easy to apply. Because it is not always possible to examine the gut contents in the field, the entire fish is usually preserved in 4 per cent formalin. For the larger specimens, 4 per cent formalin is injected into the abdominal cavities. Windell and Bowen (1978), Berg (1979) and Hyslop (1980) give general reviews of the various methods of investigating the food of fishes. Some of the methods are discussed below.

DETERMINATION OF THE FOOD COMPOSITION

Volumetric methods

The cardiac contents of the stomach should be washed into a petri dish and examined under the microscope, and the volumetric composition of the food items determined in different ways depending on their size. Large food items (e.g. crabs) are sorted into taxonomic categories and the displacement of individual items measured in a partially filled graduated cylinder. A series of different sizes of graduated cylinders may be used to achieve accuracy in estimation of the volume of food items of various sizes. Water surrounding food items must be blotted away before displacement is measured.

Where direct volume estimation is impractical, e.g. where small- or medium-sized particles are prevalent in the stomach, indirect volumetric analysis can be employed. This is done by comparing food items with small blocks or geometric solids of known volume (Hyslop, 1980). Where the food consists of very small particles such as plant detritus or diatoms, the stomach contents should be placed on a slide or a Sedgewick rafter cell and examined under a monocular microscope fitted with a square grid in the eyepiece. Several random fields of view are taken and the number of squares of the grid occupied by each type of food item is counted. In this method, the food items are presumed to be of uniform thickness so that areas can be interpreted as units of volume (Pillay, 1953).

Numerical analysis

This method does not give a good picture of the importance of different food organisms because of large differences in the sizes of individual items. The numerical percentages of food groups is valid if the food is not fragmented during intake and if the particle dimensions do not differ too much. The method is valid when used for zooplankton feeders, which feed on particles of relatively uniform size (Berg, 1979). The main disadvantage of such a method is that amorphous materials, like detritus, cannot be quantified. The method simply involves counting the number of individuals of each food type in each stomach. They are summed to five totals for each kind of food item in the whole sample. A grand total for all the items can then be obtained. Since not all food items appear whole (especially diatoms), only 'significant' wholes are counted. For colonial organisms, counts are made of the number of individuals constituting the colony.

Gravimetric method

The different food items are sorted, identified and the dry weight of each category is determined (Windell and Bowen, 1978). Values for the various kinds of items (each food type) are summed and the results are expressed as a percentage of the weight of the total food in all the samples. The method is tedious as it involves weighing all the food items, including those which occur in very small quantities.

DETERMINATION OF STOMACH FULLNESS

The extent of the feeding intensity is determined by the degree of distension of the stomach and the amount of food it contains. Though the estimate is arbitrary, it is a suitable indication of the feeding intensity and habits of the fish. One method of estimation is Pillay's method (1953). The condition of the stomach is classified as follows: (1) Gorged (swollen with food); (2) Full; (3) Three-quarters full; (4) Half full; (5) A quarter full; (6) A little; (7) Empty.

Tham (1950) used the following to describe the intensity of feeding: E = 'empty' when the stomach is not distended and contained little food; M = 'medium' when the stomach is slightly distended with food; and, H = 'heavy' when the stomach is fully distended with food.

ONTOGENETIC GROWTH PROGRESSION

In order to study the changes in the diet of the fish as it grows (ontogenetic food progression), its length is, of course, required. Usually the standard length taken is from the tip of the snout to the end of the caudal peduncle. It is not advisable to take the total length because some specimens may have damaged caudal fins.

DATA PRESENTATION

For each species, data should be presented as follows:

Date
Number of specimens
Length (mm)
Fullness of stomach
Food items
1.
2.
3.

The data should be averaged to give the per cent composition of each food item of all the specimens examined for the month. From the table, one can compare the variations in the diet with respect to the time or length of fish. If one is interested in the 'main food' or 'dominant' food, it involves the determination of the food type which is both numerically and volumetrically the chief constituent (food item which scored the highest) of all stomachs examined. It is expressed as the percentage of the total number of stomachs examined. The various percentage composition of the diet of a particular species can be presented in the form of pie charts.

The pooled data of all the species examined may be presented as follows:

FOOD ITEMS

Species
1.
2.
3.
Total

The data may represent the average percentage volume/numerical composition of all the specimens of a particular species. From the table, one can determine the relative frequency of occurrence of the various food items.

The relative importance of each food item may also be assessed. This is done by adding up the total for each food item A, then obtaining a grand total B of all the food items. $A/B \times 100\%$ gives an estimate of the relative importance of each food item.

FOOD WEB STUDY THROUGH ANALYSIS OF $\delta\ ^{13}C$ RATIOS

Stable carbon isotope ratios have been recently used in the study of marine food webs (Fry and Parker, 1979; McConnaughey and McRoy, 1979). The $\delta\ ^{13}C$ ratio of a plant is primarily determined by isotope effects in the photosynthesis fixation of carbon dioxide (Park and Epstein, 1961). Plants which fix CO_2 into three-carbon acids via the Calvin cycle are called C_3 plants, and have more negative ratios, −22 to −35 ‰. C_4 plants (e.g. phytoplankton) that fix CO_2 into four-carbon acids have $\delta\ ^{13}C$ values of −8 to −18 ‰.

The above ratios do not change once the plant dies or is eaten. Animals eating plants and detritus reflect the $\delta\ ^{13}C$ ratio of the material they have been assimilating. The $\delta\ ^{13}C$ ratios of mangrove plants (e.g. *Rhizophora*, *Bruguiera*, *Sonneratia* and *Avicennia*) fall in the range −26 to −29 ‰ and that of phytoplankton, −15 to 25 ‰, in Malaysia (Gearing et al., 1980). Therefore, by analysing the $\delta\ ^{13}C$ ratios of consumer organisms in the mangrove and the adjacent shallow-water ecosystems, it would be possible to trace the food web and, further, evaluate the role of mangrove plant detritus in the diet of commercially important species.

IMMUNOLOGICAL METHODS FOR FOOD WEB ANALYSIS

Immunological methods as outlined by Feller et al. (1979) are other means used to trace food webs.

References

ANDRASSY, I. 1956. Die Rauminhalts- und Gewichtsbestimmung der Fadenwurmer (Nematoden). *Acta Zool. Hung.*, Vol. 2, pp. 1–15.

ANDREN, L.; LINDGREN, P. E.; NYHOLM, K. G.; OLSSON, I. 1968. Marinbiologi. *Goteborgs Vattenvardsanl. Unders for havsutslapp, 1966–1967.* Vol. II, No. 7, pp. 127–154.

ANKAR, S.; ELMGREN, R. 1978. *The Benthic Macro- and Meiofauna of the Asko Lansert Area* (*Northern Baltic Proper*). Trosa (Sweden), Asko Laboratory. (Publication No. 11.)

ANSELL, A. D.; MCLUSKY, D. S.; STIRLING, A.; TREVALLION, A. 1978. Production and Energy Flow in the Macrobenthos of Two Sand Beaches in Southwest India. *Proc. Roy. Soc. Edinburgh*, Vol. 76B, pp. 269–96.

BAGENAL, T. (ed.) 1978. *Methods for Assessment of Fish Production in Fresh Waters*, 3rd ed. Oxford, Blackwell Scientific Publications, 365 pp. (IBP Handbook, No. 3.)

BANNER, A. H.; BANNER, D. M. 1966. *The Alpheidae Shrimp of Thailand.* Bangkok, The Siam Society. 167 pp. (Monograph Series, No. 3.)

BERG, J. 1979. Discussion of Methods of Investigating the Food of Fishes, with References to a Preliminary Study of the Prey of *Gobiusculus flavescens* (Gobiidae). *Marine Biol.*, Vol. 50, pp. 263–73.

BERRY, A. J. 1964. Faunal Zonation in Mangrove Swamps. *Bull. Natn. Mus. St. Singapore*, Vol. 32, pp. 90–8.

——. 1972. The Natural History of West Malaysian Mangrove Faunas. *Malay. Nat. J.*, Vol. 25, pp. 135–62.

BRANDT, R. A. M. 1968. Description of New Non-marine Mollusks from Asia. *Arch. Moll.*, Vol. 98, pp. 213–89.

CAMMEN, L. M. 1979. The Macro-infauna of a North Carolina Salt Marsh. *Amer. Midl. Nat.*, Vol. 102, pp. 244–53.

CAMPBELL, B. M. 1967. The Australian Sesarminae (Crustacea: Brachyura): Five Species of *Sesarma* (*Chiromanthes*). *Queensland Museum Memoirs*, Vol. 15, pp. 1–20.

CAMPBELL, B. M.; GRIFFIN, D. J. G. 1966. The Australian Sesarminae (Crustacea: Brachyura): Genera *Helice*, *Helograpsus* nov., *Cyclograpsus* and *Paragrapsus*. *Queensland Museum Memoirs*, Vol. 14, pp. 127–74.

CRANE, J. 1975. *Fiddler Crabs of the World. Ocypodidae: Genus* Uca. Princeton, N.J., Princeton University Press. 736 pp.

CRISP, D. J. 1971. Energy flow Measurements. In: N. A. Holme and A. D. McIntyre (eds.), *Methods for Study of Marine Benthos*, pp. 197–279. (IBP Handbook, No. 16.)

DAY, J. H. 1967. A Monograph of the Polychaeta of Southern Africa. London, British Museum (Natural History). 878 pp.

DE CONINCK, L. 1965. Systématique des nématodes. In: P. O. Grasse (ed.), *Traite de zoologie*, Vol. 4, pp. 586–726.

EDMONDSON, W. T.; WINBERG, G. G. (eds.). 1971. *A Manual on Methods for the Assessment of Secondary Production in Fresh Waters.* (IBP Handbook No. 17.)

ELMGREN, R. 1973. Methods of Sampling Sublittoral Soft-bottom Meiofauna. *Oikos Supplement*, Vol. 15, pp. 112–20.

FAUCHALD, K. 1977. The Polychaete Worms. Definitions and Keys to the Orders, Families and Genera. *Natural History of Los Angeles.* Los Angeles. 188 pp.

FAUVEL, P. 1953. *The Fauna of India Including Pakistan, Ceylon, Burma and Malaya. Annelida. Polychaeta.* Allahabad, The Indian Press. 507 pp.

FELLER, R. J.; TAGHON, G. L.; GALLAGHER, E. D.; KENNY, G. E.; JUMARS, P. A. 1979. Immunological Methods for Food Web Analysis in a Soft-bottom Benthic Community. *Marine Biol.*, Vol. 54, pp. 61–74.

FREY, R. W.; BASAN, P. B.; SCOTT, R. M. 1973. Techniques for Sampling Salt Marsh Benthos and Burrows. *Amer. Midl. Nat.*, Vol. 89, pp. 228–34.

FRITH, D. W. 1977. A Preliminary List of Macrofauna from a Mangrove Forest and Adjacent Biotopes at Surin Island, Western Peninsular Thailand. *Research Bull.* (Phuket Marine Biological Center), Vol. 17, pp. 1–4.

FRITH, D. W.; TANTANASIRIWONG, R.; BHATIA, O. 1976. Zonation of Macrofauna on a Mangrove Shore, Phuket Island. *Research Bull.* (Phuket Marine Biological Center.), Vol. 10, pp. 1–36.

FRY, B.; PARKER, P. L. 1979. Animal Diet in Texas Seagrass Meadows: $\delta\ {}^{13}C$ Evidence for the Importance of Benthic Plants. *Est. Coastal Marine Science*, Vol. 8, pp. 499–509.

GEARING, J.; RODELLI, P. M.; SASEKUMAR, A.; MARSHALL, N. 1980. Initial Findings from Stable Carbon Isotope Ratios in West Coast Mangrove Areas of Peninsular Malaysia. *Asian Symp. Mangrove Environ. and Research. Aug. 1980*. Kuala Lumpur, University of Malaya.

GEARLACH, S. A. 1958. Die mangroveregion tropischer Küsten als Lebensraum. *Z. Morph. u Okol. Tiere*, Vol. 46, pp. 636–730.

GOODMAN, K. S. 1980. The Estimation of Individual Dry Weight and Standing Crop of Harpacticoid Copepods. *Hydrobiologia*, Vol. 72, pp. 253–9.

GRAY, J. S. 1971. Sample Size and Sample Frequency in Relation to the Quantitative Sampling of Sand Meiofauna. In: N. C. Hulings (ed.), *Proceedings of the First International Conference on Meiofauna*, Washington, D.C., Smithsonian Institute, pp. 191–7.

GRODZINSKI, W.; KREKOWSKI, R. Z.; DUNCAN, A. (eds.). 1975. *Methods for Ecological Bioenergetics*. Oxford, Blackwell Scientific Publications. (IBP Handbook, No. 24.)

HALL, D. N. F. 1962. *Observations on the Taxonomy and Biology of Some Indo-West-Pacific Penaeidae* (*Crestacea, Decapoda*). London, Her Majesty's Stationery Office. (Fishery Publication, No. 17.)

HAMOND, R. 1969. Methods of Studying the Copepods. *J. Quekett Microsc. Club*, Vol. 31, pp. 137–49.

HOLME, N. A.; MCINTYRE, A. D. (eds.). 1971. *Methods for the Study of Marine Benthos*. Oxford, Blackwell Scientific Publications. (IBP Handbook No. 16.) 334 pp.

HULINGS, N. C. (ed.) 1971. *Proceedings of the First International Conference on Meiofauna*. Washington, D.C., Smithsonian Institution. (Smithsonian Contribution to Zoology, No. 76.)

HULINGS, N. C.; GRAY, J. S. 1971. *A Manual for the Study of Meiofauna*. Washington, D.C., Smithsonian Institution. (Smithsonian Contribution to Zoology, No. 78.)

HYSLOP, E. J. 1980. Stomach Contents—A Review of Methods and Their Application, *J. Fish. Biol.*, Vol. 17, pp. 411–29.

JANSSON, B. O. 1967. The Significance of Grain Size and Bore Water Content for the Interstitial Fauna of Sandy Beaches. *Oikos*, Vol. 18, pp. 311–22.

JOHNSON, M. G.; BRINKHURST, R. O. 1971. Benthic Community Metabolism in the Bay of Quinte and Lake Ontario. *J. Fish. Res. Bd. Canada*, Vol. 28, pp. 1715–25.

LANG, K. 1948. *Monographie der Harpacticoides*. 2 vols. Lund, Hukan Ohlssons.

——. 1965. Copepoda Harpacticoidea from the California Pacific Coast. *Kungliga Svenska Ventenskapsakademiens Handlinger*, Vol. 10, pp. 1–560.

LEE, Y. S. 1977. Preliminary Investigations in Primary Production and Zooplankton Standing Stock in a Polluted and Unpolluted Mangrove Inlet. Kuala Lumpur, University of Malaya, Department of Zoology. (B.Sc.(Hons.) thesis.)

MCCONNAUGHEY, T. MCROY, R. P. 1979. Food Web Structure and the Fractionation of Carbon Isotopes in the Bering Sea. *Mar. Biol.*, Vol. 53, pp. 257–62.

MACINTOSH, D. J. 1979. Predation of Fiddler Crabs (*Uca* spp.) in Estuarine Mangroves. *Mangrove and Estuarine Vegetation in Southeast Asia.* (Biotropica Special Publication, No. 10.)

MANN, K. H. 1972. The Analysis of Aquatic Ecosystems, pp. 1–14. In: R. H. Clark and R. J. Wooton (eds.), *Essays in Hydrobiology*. University of Exeter.

——. 1982. *Ecology of Coastal Waters. A Systems Approach.* Oxford, Blackwell Scientific Publications. 322 pp.

MUNRO, I. S. R. 1955. *The Marine and Freshwater Fishes of Ceylon.* Canberra, Department of External Affairs.

NICHOLAS, W. L. 1975. *The Biology of Free-living Nematodes.* Oxford, Clarendon Press.

PARK, R.; EPSTEIN, S. 1961. Carbon Isotope Fractionation during Photosynthesis. *Geochim. Cosmochim. Acta*, Vol. 21; p. 110.

PILLAY, T. V. R. 1953. Studies on the Food, Feeding Habits and Alimentary Tract of the Grey Mullet, *Mugil tade* Forskal. *Proc. Natn. Inst. Sci. India*, Vol. 19, pp. 777–823.

RICKER, W. E. (ed.). 1968. *Methods of Assessment of Fish Production in Fresh Waters.* Oxford, Blackwell Scientific Publications. (IBP Handbook, No. 3.)

ROSEWATER, J. 1970. The family Littorinidae in the Indo-Pacific. Part I: The Subfamily Littorininae. *Indo-Pacific Mollusca*, Vol. 2, pp. 417–506.

RUSSEL, H. D. 1963. *Notes on Methods for Narcotization, Killing, Fixation and Preservation of Marine Organisms.* Woods Hole, Mass., Marine Biological Laboratory. 70 pp.

SASEKUMAR, A. 1974. Distribution of Macrofauna on a Malayan Mangrove Shore. *J. Anim. Ecol.* Vol. 43, pp. 51–69.

SERENE, R.; LUNDOER, S. 1974. Observations on the Male Pleoplod of the Species of *Ilyoplax*

Stimpson with a Key to the Identification of the Species. *Research Bull.* (Phuket Marine Biological Center), Vol. 3, pp. 3–10.

SERENE, R.; SOH, C. L. 1970. New Indo-Pacific Genera Allied to *Sesarma* Say 1817 (Brachyura, Decapoda, Crustacea). *Treubia,* Vol. 27, pp. 387–416.

SMITH, J. L. B. 1950. *The Sea Fishes of Southern Africa.* South Africa, Central News Agency Ltd.

SOH, C. L. 1978. On a Collection of Sesariminae Crabs (Decapoda, Brachyura, Grapsidae) from Hong Kong. *Memoirs of the Hong Kong Natural History Society,* Vol. 13, pp. 9–22.

TARJAN, A. C. 1980. *An Illustrated Guide to the Marine Nematodes.* Gainesville, Institute Food and Agricultural Science, University of Florida. 206 pp.

TEAL, J. M. 1962. Energy Flow in the Salt Marsh Ecosystem of Georgia. *Ecology,* Vol 43, pp. 614–24.

THAM, A. K. 1950. *The Food and Feeding Relationships of the Fishes of Singapore Straits.* London, Colonial Office. (Fishery Publication.)

THAYER, G. W.; SCHAAF, W. E.; ANGELOVIC, J. W.; LACROIX, M. W., 1973. Caloric Measurements of Some Estuarine Organisms. *Fish. Bull.,* Vol. 71, pp. 289–96.

TUNDISI, J.; TUNDISI, T. N.; KUTNER, M. B. 1973. Plankton Investigations in a Mangrove Environment. VII: Further Investigations on Primary Production, Standing Stock of Phyto- and Zooplankton and some Environmental Factors. *Int. Revue Ges. Hydrobiol.,* Vol. 58, pp. 925–40.

TWEEDIE, M. W. F. 1935. Notes on the Genus *Ilyoplax Stimpson* (Brachyura: Ocypodidae). *Bull. Raffles Mus.* (Singapore), Vol. 10, pp. 53–61.

——. 1936. On the Crabs of the Family Grapsidae in the Collection of the Raffles Museum. *Bull. Raffles Mus.* (Singapore), Vol. 12, pp. 44–70.

——. 1937. On the Crabs of the Family Ocypodidae in the Collection of the Raffles Museum. *Bull. Raffles Mus.* (Singapore), Vol. 13, pp. 140–70.

——. 1938. A New Scopimerina Crab from the Malay Peninsula. *Bull. Raffles Mus.* (Singapore), Vol. 14, pp. 198–202.

——. 1940. New and Interesting Malaysian Species of *Sesarma* and *Utica* (Crustacea, Brachyura). *Bull. Raffles Mus.* (Singapore), Vol. 16, pp. 88–113.

——. 1950. Grapsoid Crabs from Labuan and Sarawak. *Museum Journal,* Vol. 5, pp. 338–69.

——. 1954. Notes on Grapsoid Crabs from the Raffles Museum, Nos. 3, 4 and 5. *Bull. Raffles Mus.,* Vol. 25, pp. 118–28.

VAN BENTHEM-JUTTING. 1977. *Systematic Studies on the Non-marine Mollusca of the Indo-Australian Archipelago.* Amsterdam, Linaeus Press.

VERWAY, J. 1930. Einiges über Biologie Ost-Indischer Mangrovekrabben. *Treubia,* Vol. 12, pp. 167–261.

WARNER, G. F. 1969. The Occurrence and Distribution of Crabs in a Jamaican Mangrove Swamp. *J. Anim. Ecol.,* Vol. 38, pp. 379–89.

——. 1977. *The Biology of Crabs.* London, Elek Science. 202 pp.

WELLS, J. B. J. 1976. Keys to aid in the Identification of Marine Harpacticoid Copepods. (Available from Professor J. B. J. Wells, Department of Zoology, Victoria University of Wellington, Wellington, New Zealand.)

WIESER, W. 1953. Reports of the Lund University Chile Expedition 1948–1949. 10. Free-living Marine Nematodes. I: Enoploidea. *Lunds Universitets Arsskrift,* Vol. 49, pp. 1–155.

——. 1954. Reports of the Lund University Chile Expedition 1948–1949. 10. Free-living Marine Nematodes. II: Chromadoroidea. *Lunds Universitets Arsskrift,* Vol. 50, pp. 1–148.

——. 1956. Reports of the Lund University Chile Expedition 1948–1948. 10. Free-living Marine Nematodes. III: Axonolaimoidea and Monohysteroidea. *Lunds Universitets Arrskrift,* Vol. 53, pp. 1–115.

WINDELL, J. F.; BOWEN, S. H. 1978. Methods for Study of Fish Diets Based on Analysis of Stomach Contents, pp. 219–26. In: T. Bagenal (ed.), *Methods for Assessment of Fish Production in Fresh Waters.* (IBP Handbook No. 3.) Oxford, Blackwell Scientific Publications.

Bibliography

ABELE, L. G. Taxonomy, Distribution and Ecology of the Genus *Sesarma* (Crustacea, Decapoda, Grapsidae) in Eastern North America with Special Reference to Florida. *The American Midland Naturalist,* Vol. 90, 1973, pp. 375–86.

BERRY, A. J. Molluscs Colonizing Mangrove Trees with Observations on *Enigmonia rosea* (Anomiidae). *Proc. Malac. Soc. Lond.*, Vol. 41, 1975, pp. 589–600.

BROOM, M. J. Size-selection, Consumption Rates and Growth of the Gastropods *Natica maculosa* (L.) and *Thais carinifera* (L.) Preying on the Bivalve, *Annadara granosa* (L.). *J. Exp. Mar. Biol. Ecol.*, Vol. 56, 1982, pp. 213–33.

——. Structure and Seasonality in a Malaysian Mudflat Community. *Estuarine, Coastal and Shelf Science*, Vol. 15, 1982, pp. 135–50.

FRITH, D. W.; CLIFFORD, B. F. Observations on Fiddler Crabs (Ocypodidae: Genus *Uca*) on Surin Island, Western Peninsular Thailand, with Particular Reference to *Uca tetrogonon* (Herbst). *Research Bull.* (Phuket Marine Biological Center), Vol. 18, 1977, pp. 1–14.

HOULIHAN, D. F. Respiration in Air and Water of Three Mangrove Snails. *J. Exp. Mar. Biol. Ecol.*, Vol. 41, 1975, pp. 143–61.

HYNES, H. B. N. The Food of Fresh-water Sticklebacks (*Gasterosteus aculeatus* and *Pygosteus pungitius*), with a Review of Methods used in Studies of the Food of Fishes. *J. Anim. Ecol.*, Vol. 19, 1950, pp. 36–58.

LUNDPER, S. A Checklist of the Brachyura in the References Collection of PMBC, Thailand. *Research Bull.* (Phuket Marine Biological Center), Vol. 4, 1974, pp. 1–11.

MACINTOSH, D. J. Quantitative Sampling and Production Estimates of Fiddler Crabs in a Malaysian Mangrove. *Mar. Res. Indonesia*, Vol. 18, 1977, p. 59.

——. Some Responses of Tropical Mangrove Fiddler Crabs (*Uca* spp.) to High Environmental Temperatures. In: D. S. McLusky and A. J. Berry (eds.) *Proc. 12th European Marine Biol. Symp.* Oxford, Pergamon Press, 1978.

——. The Ecology and Energetics of Mangrove Fiddler Crabs (*Uca* spp.) on the West Coast of the Malay Peninsula. Kuala Lumpur, University of Malaya, Department of Zoology. 332 pp. (Ph.D. thesis.)

MORTON, B. The Diurnal Rhythm and Feeding Responses of the Southeast Asian Mangrove Bivalve, *Geloina proxima* Prime 1864 (Bivalvia: Corbiculacea). *Forma et Functio*, Vol. 8, 1975, pp. 405–18.

——. The Biology and Functional Morphology of the Southeast Asian Mangrove Bivalve, *Polymesoda* (Geloina) *erosa* (Solander, 1786). (Bivalvia: Corbiculidae.) *Canadian J. of Zoology*, Vol. 54, 1976, pp. 482–500.

——. The Biology, Ecology and Functional Aspects of the Organs of Feeding and Digestion of the S.E. Asian Mangrove Bivalve, *Enigmonia aenigmatica* (Mollusca: Anomiacea). *J. Zool. Lond.*, Vol. 179, 1976, pp. 437–66.

NAKASONE, Y. Crab Zonation in the Yuhi River, Okinawa Island. *Jap. J. Ecol.*, Vol. 27, 1977, pp. 61–70.

ONO, Y. On the Ecological Distribution of Ocypodid Crabs in the Estuary. *Memoirs. Series E (Biology)*, Vol. 4, 1965, pp. 1–60. Faculty of Science, Kyushu University.

SERENE, R. Notes on the Brachyura of the Marine Fauna of Thailand. *App. Sci. Res. Corp. of Thailand*, Vol. 18, No. 1, 1966, pp. 1–65.

SERENE, R.; SASEKUMAR, A. Rediscovery of Three Species of Brachyura in Malaysia. *Fed. Mus. Journal*, Vol. 16, 1971, pp. 75–82.

TAN, W. H. Some Singapore Shipworms (Family: Tereinidae). *J. Singapore National Academy of Science*, Vol. 2, 1970, pp. 1–13.

ZAIKA, V. E.; MAKAROVA, N. P. Specific Production of Free-living Marine Nematodes. *Mar. Ecol. Prog. Ser.*, Vol. 1, 1979, pp. 153–8.

10

Determination of estuarine animal biomass

Iver M. Brook
Division of Biology and Living Resources
Rosenstiel School of Marine and Atmospheric Science
University of Miami
4600 Rickenbacker Causeway
Miami, Florida 33149
United States

Introduction

Information on the abundance and distribution of the organisms found in a nearshore estuarine community is the objective of many types of sampling programme. Selection of gear and procedures to accomplish this are a necessary prerequisite for the gathering of these data. Factors which each investigator must consider include: substrate, kinds and sizes of organisms of primary interest, water depth, weight of gear, transportation, sample preservation in the field and sample storage. In many instances, use of traditional gear and techniques will not yield replicate samples nor allow the investigator to express the results on a per-unit-area basis. This chapter provides information on types and uses of benthic and water-column samplers that offer quantification and replication in projects requiring standard information (e.g. species kinds and numbers) and estimates of biomass per unit area.

Quantitative sampling of benthic communities using suction dredges

Traditional methods for quantitatively sampling benthic communities focus on corers, grab samples and simple hand collection using a variety of tools. However, differences in community and substrate type tend to result in data that are difficult to replicate, weak in comparative potential and which frequently cannot be expressed as a reasonable estimate of unit-area biomass. For example, in the nearshore shallow-water environment, the presence of benthic surface debris, roots or hard bottoms limit the use of many types of traditional gear and procedures.

In one of our sampling surveys in a variety of sediments that precluded the use of traditional benthic sampling methods, we developed and used a lightweight portable suction dredge (Brook, 1979). Many other suction devices have been described since 1927 and may be used for quantitative sampling in mangrove estuaries. Several are listed in Holme and McIntyre (1971).

Suction sampling devices can be divided into two classes: airlift and water-powered. Water-powered devices are usually based on the Venturi principle where the low pressure caused by high velocity water flow develops suction at the collecting end of the sampler. Heavy samplers require powerful shipboard winches (Kaplan et al., 1974); others are self-propelled, needing separate boats (Evink, 1973; Allen and Hudson, 1970), while many are lightweight and portable (Brett, 1964; Brook, 1979). Finally, certain dredges may require a diver using scuba gear to be most effective. A reference list is provided in Table 10.1 which classifies various suction samplers, and supplies some other pertinent information. The balance of this section will be confined to a discussion of the uses of suction samplers and dredges that have been used in mangrove estuaries. These range in size from being fully portable to large equipment requiring a small dedicated support boat.

FLUID TRANSFORMER

During 1979 and 1980 a suction dredge, powered by a May's fluid transformer[1] (Figs. 10.1 and 10.2), was used for quantitative sampling in a survey of the benthic communities of Everglades National Park, Florida, the largest mangrove estuary in the continental United States. While the device had been used to sample difficult substrates (Brook, 1979), this was the first extended use in a mangrove dominated community. Sediments where samples were taken included fine particulate peat, peat debris, bedrock with pockets of weathered soil and vegetation, mixed peat and sand over bedrock, and upland sediments with calcareous algae which overlay fossil peat. The mangrove drainage basin empties into the Gulf of Mexico at Ponce de Leon Bay where samples were collected in well-compacted fine siliceous sands, and various peat-and-sand mixtures, some of which contained vegetation. Water depths ranged from 1 m to greater than 3 m.

The usual procedure is to use the dredge operated by a diver. Occasionally, a 3–5 m aluminium pole (made from electrical conduit) is attached by stainless steel clamps to the PVC pipe which serves as the collecting end of the dredge, which permits use from the work boat. In the highly turbid waters of this estuary, it is not usually possible to see the bottom, even in the shallowest waters, from the work boat. Because of this, the earlier samples were taken with the operator in the water, examining the bottom to note the effectiveness of the dredge on the particular substrate. In view of the objectives of the study, and selection of 1-mm mesh to separate the organisms, each replicate consisted of twelve sample probes of the collecting end (7 cm in diameter, 38.2 cm^2) or a total of 458.4 cm^2. Three replicates were taken at each station. However, if polychaetes, amphipods, molluscs or some other group of organ-

1. U.S. Patent 3664768.

TABLE 10.1. Features of various suction samplers

Type	Power source	Number to operate	Driver required	Remarks	References
Water, fluid transformer	3.5 hp, 2-cycle Homelite pump	2	E[1]	Lightweight, 113.6 kg including pump used from 50.2 m to 10 m; large or small replicates, depending on suction end aperture can be used with frame	Brook (1979)
Suction	Manual	2	Y	Lightweight, designed for hard-bottom sampling using scraper and frame	Gulliksen and Deras (1975)
Airlift	Scuba tank	2	Y	Large volume sample; 45 minutes per replicate	Christie and Allen (1972)
Water, venturi	38 hp, 4-cycle gasoline pump	2	N	Separate boat for pump; special separator used for large replicates	Evink (1973)
Water, venturi	16 hp, 4-cycle gasoline engine	2	N	Heavy; 52 kg (empty) 730 cm^2, winch required	Thayer et al. (1975)
Water, fluid transformer	3.5 hp, 2-cycle pump	1	N	Modified for sediment surface sampling (400 cm^2)	Pickral and Odum (1977)
Water, venturi			N	Heavy; 40 or 90 kg, plus core; winch required	Kaplan et al. (1974)
Pump	3.5 hp, 4-cycle		N	Pump and collecting bag on boat; used from 0.2 m to 30 m; weight 37 kg	Proniewski et al. (1978)
Airlift	Scuba tank	1 or 2	Y	Multiple replicates; can replace collecting bag underwater; not to be used less than 2 m depth; weight 25 kg	Hiscock and Hoare (1973)
Water	3 hp gasoline	2	Y	Uses quadrat, 3.2 mm mesh; used to 15 m, more efficient with larger pump	Brett (1964)
Fan	0.5 hp submersible electric	2	Y	Can only take one replicate without being raised and new filter bag inserted; battery on boat	Emig and Lienhart (1971)

1. May be operated either with or without a driver.

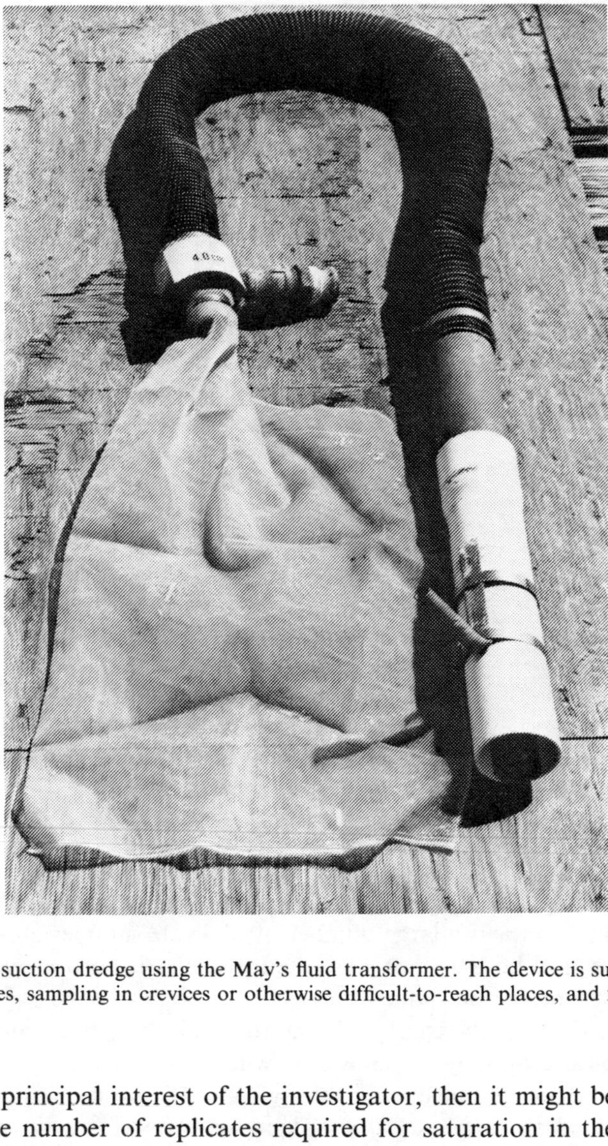

Figure 10.1
Small, portable suction dredge using the May's fluid transformer. The device is suitable for small samples, sampling in crevices or otherwise difficult-to-reach places, and for good replication.

isms are the principal interest of the investigator, then it might be possible to determine the number of replicates required for saturation in the number of species collected by use of a cumulative curve or other appropriate statistical techniques.

This dredge can be modified in many ways in order to suit the needs of the investigator. Mesh size in the collecting bag can be varied depending upon the size of the organism. In the previously described sampling, collecting bags were made from mesh with 600 μm openings and sorted through a 1 mm sieve. Clay, mud-clay mixtures and fine flocculent peat tend to coat the collecting bag mesh, which reduces water flow and suction due to back pressure. Use of a larger bag usually alleviates this problem. As a general rule, bag size should

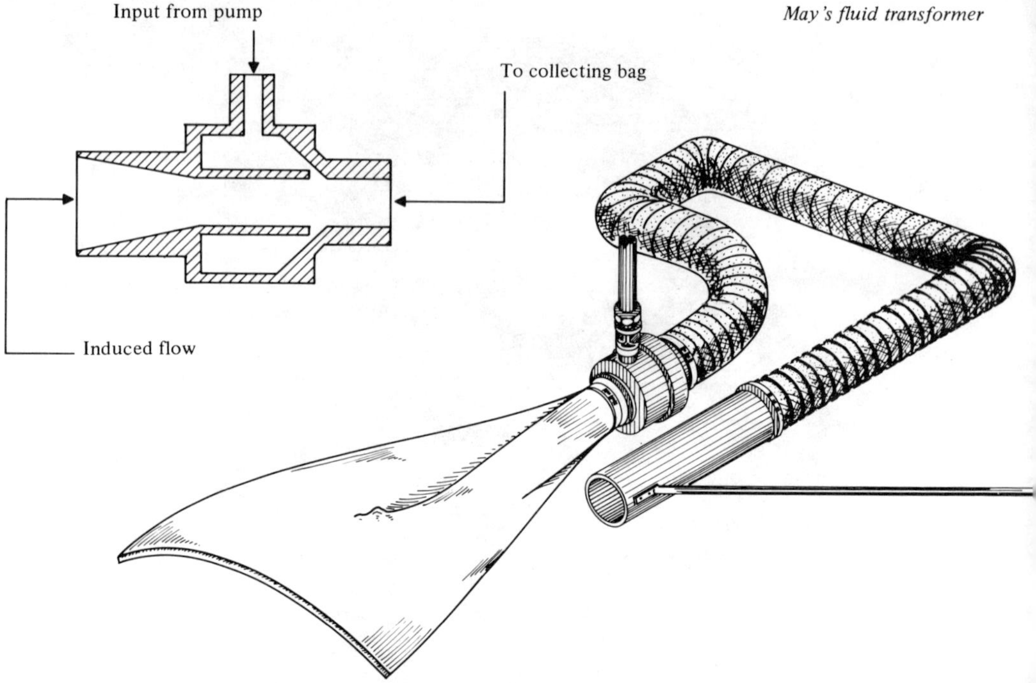

Figure 10.2
Schematic diagram of the portable suction dredge using the May's fluid transformer. The transformer is shown in the cut-away on the left.

increase as mesh size decreases. Since each complete replicate is contained in the collecting bag, the time involved in removing the bag from the dredge could become a factor if a large number of replicate samples are taken. Originally stainless steel hose clamps were used, but elastic cord proved quicker in changing bags and is also quite secure.

Samples are usually emptied from the collecting bags into one-gallon (four-litre) plastic jugs by an assistant, while the operator takes additional replicates. All material not filtered through the 600 μm mesh of the collecting bag is preserved in buffered 10 per cent sea-water formalin solution, and samples from each station should be marked and stored away from the handling area on the work boat. A replicate, consisting of twelve probes taken to a depth of 10 cm in soft mangrove peat, can usually be contained in a single one-gallon jug. Examination of peat cores 25 cm deep showed that most of the macroinvertebrates were found in the top 3 cm of substrate. Two disadvantages in using the suction dredge are: first, the limitation imposed by the aperture of the transformer on the maximum size of the organism collected; and, second, the inability to preserve the integrity of a core in order to determine the vertical distribution of organisms in the sediment.

The fluid transformer has not been produced in quantity, but it can be easily machined from solid-bar PVC in two sections, and held together by stainless-steel bolts. Since there are no moving parts, the only possible malfunction occurs when the pump intake picks up course sediment or debris which plug the circular orifice within the transformer. In this case, the transformer can be disassembled by removing the stainless-steel bolts, and can usually be done in the field using a single hex-head wrench.

Originally a 5-h.p. Homelite XLS-SA pump was used to power the dredge, but later a Homelite XLS-1 1/2 (3.5 hp) was substituted without loss of efficiency, driving a fluid transformer with an aperture of 4.6 cm. The latter pump weighs only 11 kg against 17.2 kg for the 5-h.p. pump. The pump is connected to the transformer by standard 1½-inch (3.8 cm) fire hose. For easy use in the field, all high-pressure connections were aluminium 'Kam-Lock' quick-connect fittings. The intake hose should be protected to keep debris from clogging the transformer and the intake hose must be well above the substrate. While the transformer aperture was 4.6 cm, the collecting end of the sampler had an opening of 7 cm and could have been made larger without noticeable loss of efficiency.

In constructing the May's fluid transformer, it is possible to change the size according to need by proportionately reducing or increasing the dimensions of the apparatus.

DETRITUS SAMPLING

The May's fluid transformer can also be used to produce the suction necessary for the effective sampling of benthic detritus using a sampler box placed over the sediment. Pickral and Odum (1977) describe construction and use of the sampler in great detail. The transformer is mounted in a box (20 × 20 × 40 cm) constructed of aluminium sheet and open at the bottom. To operate the sampler from a boat, a five-metre aluminium pole is mounted on the top of the chamber. The fluid transformer is powered by a small pump (3.5 h.p.) in the boat and the intake hose is covered by mesh equal to the size of the chamber mesh (0.25 mm) to exclude detritus in the water column from mixing with the sample.

In this case, the chamber is pushed into the mud bottom and the pump started, which produces suction within the chamber. All the material is retained by a collecting bag (35 × 35 × 0.25 mm mesh). Material finer than 0.25 mm was allowed to filter through the collecting bag. The authors state that samples may consist of a composite from several locations, retaining all material in a single bag, or that discrete replicates can be taken.

As in the previous case, it is not possible to retain detritus or organisms smaller than the mesh size of the collecting bag. The aperture of the transformer also places a maximum size limit on the detrital materials organisms to be collected.

Slightly modified, the device has been used to remove dead blades of grass and fragments from a sea-grass bed. As described, it is an efficient, quantitative method for sampling organisms and detritus from a mangrove estuary since neither organisms nor detritus are damaged when passing through the transformer.

HEAVY-DUTY, BOAT-MOUNTED DREDGES

Evink (1973) describes a suction dredge system (Fig. 10.3 and 10.4) modified from equipment described by Allen and Hudson (1970). The advantage of this gear, used in conjunction with a sorter/separator, is its ability to obtain large samples from all types of benthic substrates including oyster bars. Whereas the May's apparatus requires a 3.5-h.p. motor-pump, the Evink system requires a larger (38 h.p.) boat-mounted motor-pump to develop the necessary suction. This is a significant disadvantage unless the research project requires the type of data and information obtainable from large samples.

In operation, a one-square-metre quadrat, made from aluminium frame

Figure 10.3
Large, portable boat-mounted suction-dredge system. The photograph shows the motor-pump boat with intake hose over the bow and high-pressure hoses leading to a venturi-operated suction dredge being used by two researchers. This dredge system is suitable for obtaining large samples of benthic materials or samples from relatively deep crevices or holes.

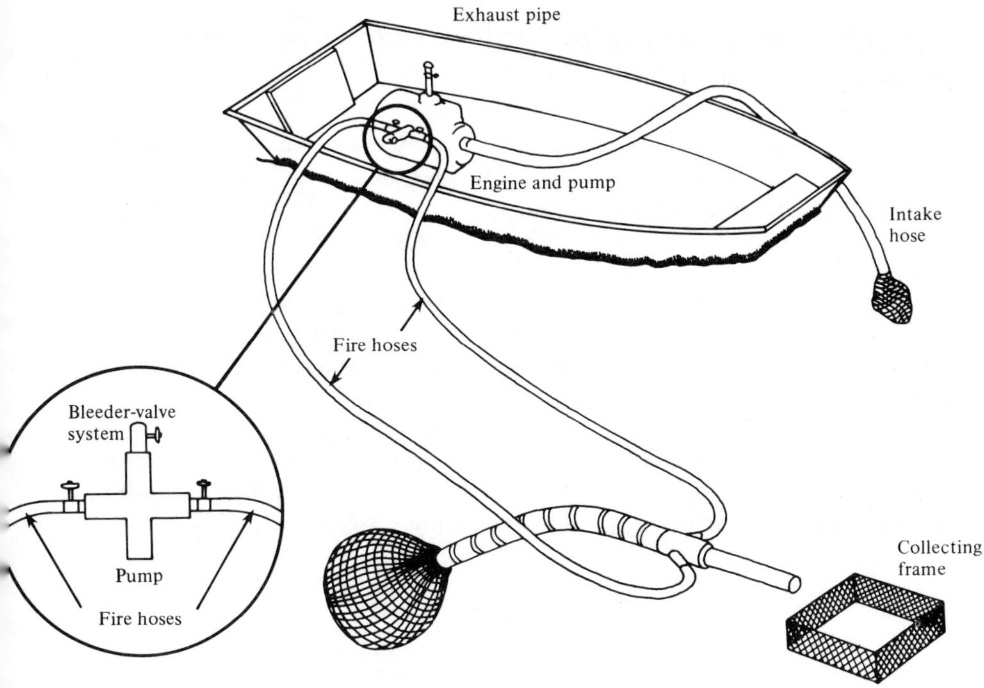

Figure 10.4
Schematic diagram of the large, portable boat-mounted suction dredge showing the basic valve system, the venturi-driven collector and the collecting frame used to demark the sampling area.

0.61 m deep, is pushed into the sediment. All material within the frame quadrat is removed to the desired depth (limited only by the depth of the frame walls) or to bedrock when it is present. Because of the large volume of benthic materials obtained, a special sorting and separating system is required (Fig. 10.5). Called a bubble separator, the device consists of a 30–50-gallon (120–200 litres) can into which the bulk sample is placed. Through a system of perforated tubes and diffusers on the bottom of the can, water and air (bubbles) are introduced in such a manner as to mix and distribute the materials within the water column of the separator. To achieve controlled sorting and separation (based on size and/or density of the materials), the rate of water-flow and air pressure are adjusted; as the water-flow and air pressure are increased, larger and heavier materials are brought to the surface where they overflow across a spill lip. Sieves placed under the spill lip collect the materials for preservation and/or analysis. It should be noted that the sorter/separator has many other related applications in estuarine ecology where mixed, diverse benthic samples require separation and sorting in a relatively short period of time.

Quantitative sampling of shallow-water nekton by drop-net quadrat

Usually, if the aim of a mangrove estuary study includes an examination of its trophic structure, a quantitative estimate of biomass and a monitoring of changing species assemblages of pelagic and demersal organisms, then an intensive and quantitative sampling of the nekton is required. Odum (1970) obtained a qualitative fish and crustacean inventory of the North River of Everglades National Park by using a 'heterogenous assemblage of methods' which included hook-and-line fishing, a 10-foot (3-metre) otter trawl trap, a roller-beam trawl, set lines and a trammel net. Samples were taken in the river, the river mouth and in associated ponds. The topography of the estuary prevented use of a single quantitative method. In selecting a method to sample fishes, an investigator should examine tides, water depth, consistency of the bottom, openness of the water, the presence of snags or other obstructions, such as oyster bars. If the water body is shallow (less than 1.5 m) and is free

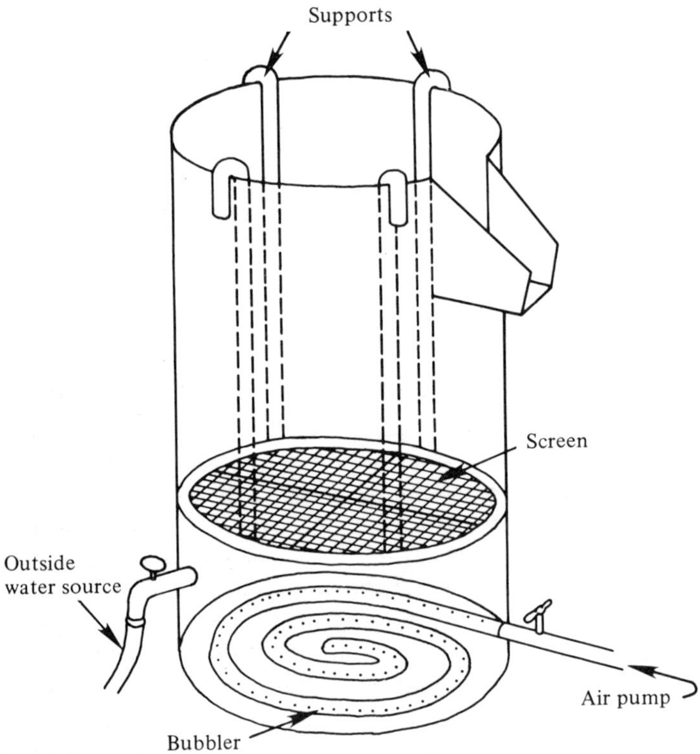

Figure 10.5
Schematic diagram of a bubble sorter/separator used for rapid separating and sorting of bulk benthic samples. Sorting and separation is achieved by controlling the water flow and air pressure lifting materials of different sizes and densities to the overflow lip

from obstructions with a bottom that permits reasonable footing, then the use of a drop-net as a quantitative sampling tool is made possible.

The drop-net, or trap, separates a measured area of a body of water by a wall of mesh. Once the drop is completed, the area can be sieved until all fishes or crustaceans larger than a minimum size, determined by the mesh of the walls and the seine, are removed. Table 10.2 lists several such devices and pertinent information on characteristics. The subsequent section describes portable, easily assembled, relatively inexpensive drop-nets which have been used in south Florida mangrove estuaries with relative success.

Table 10.2. Features of various forms of drop-net sampler

Area	Mounting	Size of mesh	Depth of net	Remarks	References
1 m^2	Aluminium tripod	Sheet metal	1.0 m	Used in Baltic for small fishes; portable.	Aneer and Nellbring (1977)
9 m^2	Aluminium tripod	4.8 mm	1.5 m	Dropped simultaneously with two others for replicate samples of bottom and cryptic seagrass fishes; portable.	Gilmore et al. (1976)
16 m^2	Plastic pipe	1.6 m	1.8 m	Used in mangrove estuary, in conjunction with boat-mounted suction dredge; easily disassembled; net weighted with chain; portable.	Seaman et al. (1973)
1,012 m^2 or 252 m^2	Pilings	19 mm stretch		Heavy framing, not really portable; net heavily weighted with chain.	Hellier (1958)
118 m^2	Pilings	19 mm stretch			Hoese and Jones (1963)
Variable to 360 m^2	Wooden stakes	3.2 mm	1.8 m	Easily disassembled; portable.	Brook (1977)
25 m^2	Floating frame	1 cm		Towed to sampling site; pursed by attending boat; cannot be reset at sampling site.	Moseley and Copeland (1969)

A large semi-permanent drop-net was made of $\frac{1}{8}$-inch (3.2 mm) 'Ace' mesh in three sections: the two side panels were 40 × 1.8 m, and the end panel was 15 × 1.8 m. The net is capable of making an enclosure of 12 × 35 m, but the usual sampling area was 12 × 30 m (360 m^2). The nets are supported by twenty-one 7-foot × $1\frac{1}{4}$-inch (2.1 × 3.2 cm) posts. The nets are placed in the water, and each post is inserted through four metal rings sewed to the vertical seam of the net, and then forced into the sediment. Nets are kept above the water prior to a drop by a pin which supported the rings. Pins on each post are connected by a dacron line which stretches less than other materials. In practice, the lines were arranged so that one investigator was responsible for triggering two sides of the enclosure by pulling the pin line, with the third side handled by one other person. At each post, an additional 0.5 kg weight was added to insure a rapid fall after pin removal. After dropping, the open end of the net was closed by moving two posts to the centre and making sure that the mesh overlapped. The short end-posts also were moved to seal the sides of the quadrat. Finally, the lead line was staked to the bottom to prevent any organisms from leaving the enclosed area by moving under the net. In practice, triggering and sealing the enclosure took no more than five minutes. Because of the side of the enclosure, edge effects are reduced. In some cases, schools of fish (e.g. *Atherinomorous stipes*) were split by a drop.

The net was erected several hours prior to a scheduled set and held clear of the water to permit unimpeded passage of fishes through the quadrat. Fishes and crustaceans were removed from the enclosure by seining. The first pass through the enclosure was made with a 2-inch (5.1-cm) bar mesh seine to remove all of the larger fishes that were trapped when the net was dropped. This first pass usually captured the large crustaceans as well. Six hauls of a 50-foot (12.7-m), $\frac{5}{16}$-inch (7.9 mm) 'Delta' mesh seine weighted by 13 feet (4 m) of chain at the centre, removed all of the larger organisms. A final clean-up of the area consisted of three passes with a 26-foot (8 m) $\frac{1}{8}$-inch (3.2 mm) seine. All organisms were preserved in 10 per cent formalin-sea-water solution. A description of the procedure is provided by Brook (1977).

An advantage of the system described is that the gear can be easily stowed, transported and operated without special equipment and under the conditions

Figure 10.6
Small portable drop-net prior to release of the nets

outlined in the introduction will provide quantitative data on shallow-water nekton. It is also rugged enough to be left in place for sequential sampling to determine changing population distributions, ontological development of resident organisms and time of arrival of migratory species, among other similar objectives.

A smaller (16 m^2), fully portable drop-net (Fig. 10.6) has been used to obtain coverage of large diverse areas and for the purposes of replication (Seaman et al., 1973). Although the sampling area is significantly smaller than the larger semi-permanent drop-net, the portability enables it to be used many times within a relatively short period of time. In practice, two or three drop-nets would be set up and allowed to remain in place for several hours prior to dropping the net. Following enclosure, two persons would enter the net with a fine-mesh beach seine and 'sweep' the enclosure until all animals had been removed. The technique has proved to be useful for estimating biomass of macro-invertebrates and the smaller fishes in an otherwise difficult-to-sample environment.

References

ALLEN, D. M.; HUDSON, J. 1970. *A Sled Mounted Suction Sampler for Benthic Organisms.* 5 pp. (U.S. Fish and Wildlife Service Special Scientific Report—Fish, No. 614.)

ANEER, G.; NELLBRING, S. 1977. A Drop-trap Investigation of the Abundance of Fish in Shallow Water in the Asko Area, Northern Baltic Proper. In: B. Keegan, P. Ceidigh and P. Boaden (eds.) *Biology of Benthic Organisms. Eleventh European Mar. Biol. Symp. 1976*, pp. 21–30. Galway, Ireland.

BRETT, C. E. 1964. A Portable Hydraulic Diver-operated Dredge-sieve for Sampling Subtidal Macrofauna. *J. Mar. Res.*, Vol. 22, pp. 205–9.

BROOK, I. M. 1977. Trophic Relationships in a Seagrass Community (*Thalassia testudinum*) in Card Sound, Florida. Fish Diets in Relation to Macrobenthic and Cryptic Faunal Abundance. *Trans. Am. Fish. Soc.*, Vol. 106, pp. 219–28.

BROOK, I. 1979. A Portable Suction Dredge for Quantitative Sampling on Difficult Substrates. *Estuaries*, Vol. 2, pp. 54–8.

CHRISTIE, N.; ALLEN, J. 1972. A Self-contained Diver-operated Quantitative Sampler for Investigating the Macrofauna of Soft Substrates. *Transactions of the Royal Society of South Africa*, Vol. 4, pp. 299–307.

EMIG, C.; LIENHART, R. 1971. Principe de l'aspirateur sous-marin automatique pour sediments meubles. *Vie et Milieu* (supplement), Vol. 22, pp. 573–8.

EVINK, G. L. 1973. *The Role of Mangrove Ecosystems: Biomass and Diversity of Benthic Macroinvertebrates of Fahka Union and Fahkahatchee Bays, Florida.* Bureau of Sport Fisheries and Wildlife, NTIS, Department of Commerce. Springfield, Va., 73 pp. (No. PB-231 749.)

GILMORE, R. G.; KULCZYCKI, G. R.; HASTINGS, P. A.; MAGLEY, W. C. 1976. Studies of the Indian River Lagoon and Vicinity. In: D. K. Young (ed.), Indian River Coastal Zone Study. Third Annual Report to Harbor Branch Consortium. 187 pp. (Unpublished MS.)

GULLIKSEN, B.; DERAS, K. M. 1975. A Diver-operated Suction Sampler for Fauna on Rocky Bottoms. *Oikos*, Vol. 26, pp. 246–9.

HELLIER, T. 1958. The Drop-net Quadrat, A New Population Sampling Device. *Publ. Inst. Mar. Sci. Univ. Texas*, Vol. 5, pp. 165–8.

HISCOCK, K.; HOARE, R. 1973. A Portable Suction Sampler for Rock Epibiota. Helgolander wiss. Meeresunters, Vol. 25, pp. 35–8.

HOESE, H. D.; JONES, R. S. 1963. Seasonality of Large Animals in a Turtle Grass Community. *Publ. Inst. Mar. Sci. Univ. Texas*, Vol. 9, pp. 37–47.

HOLME, N. A.; MCINTYRE, A. D. 1971. *Methods for the Study of Marine Benthos*. Oxford, Blackwell Scientific Publications, 334 pp. (IBP Handbook, No. 16.)
KAPLAN, E. H.; WELKER, J. R.; KRAUSE, M. G. 1974. A Shallow Water System for Sampling Macrobenthic Infauna. *Limnol. Oceanogr.*, Vol. 19, pp. 346–9.
MOSELEY, F. N.; COPELAND, B. J. 1969. A Portable Drop-net for Representative Sampling of Nekton. *Contr. Mar. Sci. Univ. Texas*, Vol. 14, pp. 37–45.
ODUM, W. E. 1970. Pathways of Energy Flow in a South Florida Estuary. Coral Gables, Fla. School of Marine and Atmospheric Science, University of Miami. 162 pp. (Ph.D. thesis.)
PICKRAL, J. C.; ODUM, W. E. 1977. Benthic Detritus in a Salt Marsh Tidal Creek. In: M. Wiley (ed.), *Estuarine Processes*, Vol. II, pp. 280–92. New York, Academic Press.
PRONIEWSKI, F.; DELALANDE, P.; ELKAIM, B. 1978. Utilisation d'un nouveau modèle de suceuse à macrobenthos sur les fonds estuariens. *J. Rech. Oceanogr.*, Vol. 3, pp. 3–7.
SEAMAN, W.; ADAMS, C. A.; SNEDAKER, S. C. 1973. Biomass Determination in Shallow Estuaries; Technique Evaluation and Preliminary Data. In: S. Snedaker and A. Lugo (eds.), The Role of Mangrove Ecosystems in the Maintenance of Environmental Quality and a High Productivity of Desirable Fishes. Report to Bureau of Sport Fisheries and Wildlife. (Unpublished MS.)
THAYER, G. W.; WILLIAMS, R. B.; PRICE, T. J.; COLBY, D. R. 1975. A Large Corer for Quantitatively Sampling Benthos in Shallow Water. *Limnol. Oceanogr.*, Vol. 20, pp. 474–80.

11

Role of diatoms in the mangrove habitat

Keith E. Cooksey
Division of Biology and Living Resources
Rosenstiel School of Marine and Atmospheric Science
University of Miami
4600 Rickenbacker Causeway
Miami, Florida 33149, United States

Introduction

Considering the major role that microalgae are known to play in aquatic ecosystems, it is surprising that so little work has been published on either the planktonic or benthic algal components of the mangrove environment. A possible reason for this may lie in the fact that most of the major oceanographic institutions are situated in temperate waters, beyond the limits of mangrove growth. In recent years, however, workers in Brazil and India have begun to redress the balance. Tundisi and his collaborators (Tundisi and Texeira, 1968; Tundisi et al., 1973) have described the seasonal variations in the planktonic communities and the primary production of the phytoplankton component. They found that diatoms formed the major fraction of the phytoplankton, especially in the summer months when *Skeletonema costatum* was particularly abundant. Dinoflagellates and other flagellates were less obvious. Krishnamurthy (1971) studied phytoplankton dynamics by means of their photosynthetic pigment content in an area adjacent to the Bay of Bengal, where he too found that diatoms formed the major part of the phytoplankton, although he mentioned several organisms besides *Skeletonema*, and also some dinoflagellates.

There have been no reports, so far as I am aware, which specifically concern the epibenthic diatoms of the mangrove environment. The work by Sprogis (1975) concerned the diatom populations of Card Sound, Biscayne Bay, Florida, United States, an area surrounded by mangroves, but her studies did not include the swamps themselves. Similarly, Wood (1963) studied bays which received mangrove drainage waters, but he did not venture into the mangrove areas. A review of the benthic diatoms by Round (1971) does not mention the mangrove ecosystem, although the periphyton referred to as a major component of the fringing mangrove system by Lugo and Snedaker (1975) certainly contain epiphytic diatoms (Cooksey, personal observation). Nelson (1947) and Moul and Mason (1957) suggest the possible importance of the contribution of benthic microflora in shallow water. Pomeroy (1959), working in a salt marsh, provides evidence to support this suggestion. In recent years there have been several papers comparing the contributions of the epibenthic and planktonic microalgae to the productivity of various water masses (e.g. Sheldon and Boylen, 1975; Kairesalo, 1980), but none have concerned a mangrove environment. It is very likely that both components of the microalgal populations are important in the overall productivity of the swamp.

One feature of the mangrove habitat that contributes considerably to its complexity is the large quantity of leaf and other vascular plant detritus that accumulates, and is subsequently exported while in varying degrees of microbial degradation. In undergoing microbial decay, this material generates a pool of soluble organic materials (Fell et al., see Chapter 15). Much of the work on this aspect of mangrove microbiology has been concerned with the soluble humic fraction of the plant litter. Smayda (1970) found that the coloured mangrove drainage waters from Phosphorescent Bay, Puerto Rico, affected the growth of planktonic diatoms in a variable fashion. In some instances, water from the bay was toxic to the test algae, whereas other times growth was enhanced.

Smayda ascribed the variable phenomenon to 'patchiness'. Prakash and Rashid (1969) and Prakash et al. (1973), on the other hand, showed that low molecular-weight fractions derived from mangrove leachates consistently stimulated the growth rate of several diatoms. The precise mechanism for this stimulation is not known, but it could concern the chelation of trace metals, although Prakash and Rashid believe that the humic compounds may act directly on microalgal physiological or biochemical processes. Humic compounds are not the only ones found in the mangrove waters. We described the promotion of heterotrophic diatom growth by extracts of local mangrove sediments (Cooksey and Cooksey, 1978). Humic compounds are not capable of supporting the growth of algae; thus we were dealing with a different class of materials.

The organisms concerned in these experiments were various species of *Amphora coffeaeformis*, diatoms known to be capable of heterotrophic growth on some sugars and glutamate (Cooksey and Chansang, 1976). Although no analyses of the mangrove leachates were reported, there were qualitative differences in the leachates as seen by diatom growth bioassay. Early in the year, just after the beginning of the spring rains, material in the leachates was growth-inhibitory, whereas the growth-promoting materials were not seen until later in the year when fungal decay of recently fallen leaves had begun. Recent work (Miller and Cooksey, 1980) has shown that in a laboratory model system where mangrove leaves were undergoing degradation by the fungus *Phytophtora vesicula*, compounds affecting the growth of *A. coffeaeformis* were also produced. Early in the process, a simple sugar (probably glucose) and glutamate were produced in sufficient quantities to support the heterotrophic growth of the diatom, whereas, later in the experiments, leaf-leachates inhibited growth. Again we have no explanation concerning the inhibitory mechanism.

The production in the mangrove swamp of compounds that affect microalgal growth, and the subsequent export of these compounds to adjacent bays and estuaries, deserves further study. It is by no means certain that the compounds to which I have alluded influence only the growth and metabolism of microalgae. Interactions, especially chemosensory ones, with more highly developed organisms are entirely possible. I believe that diatoms, because of their ease of manipulation in the laboratory, will prove to be convenient assay organisms for such substances.

Methods for the study of diatoms in the laboratory and field

MICROSCOPY

Diatoms range in size from a few to several hundred micrometres, thus a microscope capable of a magnification of at least $\times 400$ is needed for their study. Most classical studies were done using bright field illumination but I prefer to use phase-contrast optics. Nomarski interference contrast illumination provides exceptional internal detail of living organisms.

Sample preparation

Living cells from cultures or plankton tows can be examined in wet mounts. Motile organisms can be arrested with a few microlitres of 1 per cent formaldehyde. Cell counts can be made in standard haemocytometers or with electronic particle counters such as those manufactured by the Coulter Company, Hialeah, Florida. Cells in sediments are much harder to examine; however, a convenient procedure to examine motile benthic organisms is as follows. Sediment from the samplings is spread evenly in a petri dish. The surface of the sediment is covered with microscope slide cover-glasses and the dish covered. After several days the motile organisms will have migrated to the surface of the sediment and often will have attached themselves to the surface of the cover glass, which can then be examined microscopically. This procedure also works well as an enrichment step in the initial cultivation of sediment diatoms. Settling vessels and the use of the inverted microscope are covered in standard oceanographic texts (Lund et al., 1958).

Cells in sediments can be examined also, after removal of the sediment by a combination of oxidant and acid treatment (Zoto et al., 1973). Hydrogen peroxide (28 per cent) followed by concentrated nitric acid has been used (Van der Werff, 1955). The cleaned diatom frustules can then be washed with water and dried down on a microscope slide for permanent mount in hyrax.

CULTURE

Before diatom culture is attempted, it is recommended that Volume 1 of *Handbook of Phycological Methods* (Stein, 1973) be consulted. This is a highly detailed and practical account of the state of the art in algal culture.

Medium

I recommend the use of the artificial sea-water medium of Provasoli et al. (1957) and the natural sea-water medium F/2 of Guillard and Ryther (1962). There are many other media (McLachlan, 1973), but these two allow the culture of most brackish and marine diatoms. Many of the problems associated with the adhesion of diatoms to culture vessels during growth studies can be solved by lowering the calcium content of the medium (Cooksey and Chansang, 1976; Cooksey and Cooksey, 1980).

Light and temperature

Light requirements for diatoms vary considerably, but usually organisms can be grown at between 3,000–5,000 lux from cool-white fluorescent tubes (approximately 25–50 cm from a 40-watt tube). On occasion, organisms will grow with considerably less illumination, especially in the presence of utilizable organic substrate (Cooksey and Chansang, 1976). Many tropical organisms grow in the range of 25–30 °C. Few grow at higher temperatures unless isolated from near geothermal vents.

Axenic cultures

Although many ecological experiments with diatoms can be performed with cultures containing bacteria, critical physiological experiments can be performed only with axenic cultures. Methods for the production of such cultures have been reviewed (Hoshaw and Rosowski, 1973). I have had particular success with the antibiotic chloramphenicol in liquid culture and, with a solution of penicillin and streptomycin placed in a trench cut in an agar–sea-water medium, contained in a petri dish. Diatom cultures containing bacteria are then streaked at right angles to the trench and are thus subjected to a gradient in antibiotic concentration as it diffuses into the agar.

Physiological properties of cultured organisms

Axenic cultures can be tested for their ability to metabolize and grow on simple organic molecules such as sugars, organic acids or amino acids. A requirement for vitamins can also be investigated, as well as the need for certain forms of combined nitrogen (NH_4^{2-}, NO_3^-, amino acids). It is informative to perform these studies with light-(mixotrophy) and dark-incubated cultures (heterotrophy).

Growth rates are most conveniently measured by counting cells with time using a haemocytometer. Division rates can be followed also by measuring increases in chlorophyll pigments (Eppley and Sloan, 1966; Parsons and Strickland, 1963; Hansmann, 1973), cellular protein (Lowry et al., 1951) and packed cell volume (Sorokin, 1973).

Diatoms as assay organisms

As mentioned in the introduction, diatoms can be used as assay organisms to detect and measure growth-modifying compounds, and to investigate the nutrient status of samples of natural water. A technique for the investigation of mangrove-associated compounds has been described in detail (Cooksey et al., 1976; Cooksey and Cooksey, 1978; Cooksey and Chansang, 1976). I have also used the assay to estimate the copper-ion binding capacity of mangrove leachates by measuring the relief of growth inhibition they provided when combined in the medium that contained levels of copper toxic to the diatom *Coscinodiscus*.

References

COOKSEY, K. E.; CHANSANG, H. 1976. Isolation and Physiological Studies on Three Isolates of *Amphora* (Bacillariophyceae). *J. Phycol.*, Vol. 12, pp. 455–60.

COOKSEY, B.; COOKSEY, K. E. 1980. Calcium is Required for Motility in the Diatom *Amphora coffeaeformis. Plant Physiol.*, Vol 65, pp. 129–31.

COOKSEY, K. E.; COOKSEY, B. 1978. Growth-influencing Substances in Sediment Extracts from a Subtropical Wetland: Investigation using a Diatom Bioassay. *J. Phycol.*, Vol. 14, pp. 347–52.

COOKSEY, K. E.; COOKSEY, B.; EVANS, P. M.; HILDEBRAND, E. L. 1976. Benthic Diatoms as Contributors in the Carbon Cycle in a Mangrove Community. In: G. Persoone and E. Jaspers (eds.) *Proc. 10th European Symp. Marine Biol.*, Vol. 2. pp. 165–78. Wetteren (Belgium), Universa Press.

EPPLEY, R. W.; SLOAN, P. R. 1966. Growth Rates of Marine Phytoplankton: Correlation with Light Absorption by Cell Chlorphyll *a*. *Physiol. Plant.*, Vol. 19, pp. 47–59.

GUILLARD, R. R. L.; RYTHER, J. H. 1962. Studies of Marine Planktonic Diatoms. 1. *Cyclotella nana*. Hustedt and Detonula *confervacea* (Clere) Gran. *Can. J. Microbiol.*, Vol. 8, pp. 229–39.

HANSMANN, E. 1973. Pigment Analysis. In: J. R. Stein (ed.), *Handbook of Phycological Methods*, Vol. 1, pp. 359–68. London, Cambridge University Press.

HOSHAW, R. W.; ROSOWSKI, J. R. 1973. Methods for Microscopic Algae. In: J. R. Stein (ed.), *Handbook of Phycological Methods*, Vol. 1, pp. 53–67. London, Cambridge University Press.

KAIRESALO, T. 1980. Comparison of *In Situ* Photosynthetic Activity of Opiphytic, Pipielic and Planktonic Algal Communities in an Oligotroptic Lake, Southern Finland. *J. Phycol.*, Vol. 16, pp. 57–62.

KRISHNAMURTHY, K. 1971. Phytoplankton Pigments in Porto Novo Waters (India). *Int. Revue. Ges. Hydrobiol.*, Vol. 56, pp. 273–82.

LOWRY, O. H.; ROSEBROUGH, N. J.; FARR, A. L.; RANDALL, R. J. 1951. Protein Measurement with the Folin-Phenol Reagent. *J. Biol. Chem.*, Vol. 193, pp. 265–75.

LUGO, A. E.; SNEDAKER, S. C. 1975. Properties of a Mangrove Forest in Southern Florida. In: G. Walsh, S. Snedaker and H. Teas (eds.) International Symp. *Biology and Management of Mangroves*, Vol. I, pp. 170–212. Gainesville, University of Florida.

LUND, J. W. G.; KIPLING, C.; LECREN, E. D. 1958. The Inverted Microscope Method of Estimating Algal Numbers and the Statistical Analysis of Estimations by Counting. *Hydrobiologia*, Vol. 2, pp. 143–70.

MCLACHLAN, J. 1973. Growth Media-Marine. In: J. R. Stein (ed.) *Handbook of Phycological Methods*, Vol. 1, pp. 25–51. London, Cambridge University Press.

MILLER, C. A.; COOKSEY, K. E. 1980. Glucose Metabolism by the Diatom *Amphora coffeaeformis* and its Ability to Grow Heterotrophically in Nature. Abstract of paper presented at the Annual Meeting of the American Society of Microbiology, held at Miami Beach, Florida.

MOUL, E. T.; MASON, D. 1957. Study of Diatom Populations on Sand and Mud Flats in the Woods Hole Area. *Biol. Bull.*, Vol. 113, p. 351.

NELSON, T. C. 1947. Some Contributions from the Land in Determining Conditions of Life in the Sea. *Ecol. Monogr.*, Vol. 17, pp. 337–46.

PARSONS, T. R.; STRICKLAND, J. D. H. 1963. Discussion of Spectrophotometric Determination of Marine Plant Pigments, with Revised Equation for Ascertaining Chlorophylls and Carotenoids. *J. Mar. Res.*, Vol. 21, pp. 155–63.

POMEROY, L. R. 1959. Algal Productivity in Salt Marshes of Georgia. *Limnol. Oceanogr.*, Vol. 4, pp. 386–98.

PRAKASH, A.; RASHID., M. A. 1969. The Influence of Humic Substances on Coastal Phytoplankton Productivity. In: A. Ayala-Castañares and F. B. Phleger (eds), *Coastal Lagoons, A Symposium*, Paris, pp. 431–8. UNAM-Unesco. (Mem. Symp. Int. Lagunas Costeras.)

PRAKASH, A.; RASHID, M. A.; JEMSEN, A.; SUBBA-RAO, D. V. 1973. Influence of Humic Substances on the Growth of Marine Phytoplanton: Diatoms. *Limnol. Oceanogr.*, Vol. 18, pp. 576–624.

PROVASOLI, L., MCLAUGHLIN, J. J. A., DROOP, M. 1957. The Development of Artificial Media for Marine Algae. *Arch. Mikrobiol.*, Vol. 25, pp. 392–428.

ROUND, F. E. 1971. Benthic Marine Diatoms. *Oceanogr. Mar. Biol. Ann. Rev.*, Vol. 9, pp. 83–139.

SHELDON, R. B.; BOYLEN, C. W. 1975. Factors Affecting the Contribution by Epiphytic Algae to the Primary Productivity of an Oligotrophic Freshwater Lake. *Appl. Microbiol.*, Vol. 30, pp. 657–67.

SMAYDA, T. J. 1970. Growth Potential Bioassay of Water Masses using Diatom Cultures. Phosphorescent Bay (Puerto Rico) and Caribbean Waters. *Helgol. Wiss. Meeresunters*, Vol. 20, pp. 172–94.

SOROKIN, C. 1973. Dry Weight, Packed Cell Volume and Optical Density. In: J. R. Stein (ed.), *Handbook of Phycological Methods*, Vol. 1, pp. 321–43. London, Cambridge University Press.

SPROGIS, J. M. 1975. Changes in Benthic Diatom Assemblages within the Thermal Effluent at Turkey Point, South Biscayne Bay, Florida. (Master's thesis.) Coral Gables, Fla. University of Miami. 85 pp.

STEIN, J. R. 1973. *Handbook of Phycological Methods*, Vol. 1. London, Cambridge University Press, 448 pp.

TUNDISI, J.; TEXEIRA, C. 1968. Plankton Studies in a Mangrove Environment. VII: Size Fractionation of the Phytoplankton: Some Studies on Methods. *Bol. Inst. Oceanogr.* (São Paulo), Vol. 17, pp. 89–94.

TUNDISI, J.; TUNDISI, T. M.; KUTNER, M. B. 1973. Plankton Studies in a Mangrove Environment. VII: Further Investigations of Primary Production, Standing-stock of Phyto- and Zooplankton and Some Environmental Factors. *Int. Revue. Ges. Hydrobiol.*, Vol. 58 pp. 925–40.

VAN DER WERFF, A. 1955. A New Method of Concentrating and Cleaning Diatoms and Other Organisms. *Int. Ver. Theor. Angew. Limnol. Verh.*, Vol. 12, pp. 276–7.

WOOD, E. J. F. 1963. A Study of the Diatom Flora of Fresh Sediments of the South Texas Bays and Adjacent Waters. *Publ. Inst. Mar. Sci. Univ. Texas*, Vol. 9, pp. 237–310.

ZOTO, G.; DILLON, D.; SCHLICHTING, H., Jr. 1973. A Rapid Method for Clearing Diatoms for Taxonomic and Ecological Studies. *Phycologia*, Vol. 12, pp. 69–70.

Part III

Mangrove functions and community processes

12

Mangrove physiology: photosynthesis

Federico Pannier
Universidad Central de Venezuela
Facultad de Ciencias, Escuela de Biología
Departamento de Botánica, Apartado 80.390
Caracas, Venezuela

Photosynthesis

Research on photosynthetic metabolism in mangroves not only allows the interpretation of their growth responses to diverse environmental factors, but also enables an estimation to be made of the primary productivity in the natural community.

Most research conducted on mangrove photosynthesis has been carried out with regard to the last aspect (Golley et al., 1962; Moore et al., 1972, 1973; Carter et al., 1973; Lugo and Snedaker, 1974; Lugo et al., 1974; Hicks and Burns, 1975; Attiwill and Clough, 1980). Nevertheless, there is a level of physio-ecological research on photosynthesis which is of fundamental importance to the understanding of the adaptative mechanisms which allow the individual species to cope with the particular environmental conditions of their habitat.

This research depends on the existence of adequate conditions for experimental work, growth chambers, special instrumentation, facilities for chemical analysis, etc., and has not been backed up until the present, there being a need for more comparative data on the different mangrove species and the main characteristics of the photosynthetic process such as photosynthetic activity, CO_2 transfer resistances, photorespiration, light compensation and biochemical CO_2 fixation pattern. These data must be related to the conditions prevailing in the natural habitat (microclimate, salinity, soil nutrients, etc.).

In order to obtain comparable data on the photosynthetic response of the different mangrove species, it is of fundamental importance to use plants in physiologically optimal conditions. In the case of seedlings, they have to be collected from the same parent plant and grown under controlled conditions and selected for the experiment according to physiological age (for example, plants of ten months counted from the unfolding of their first leaf pair), their morphological status (for instance, having *x* pairs of leaves, being *x* cm in height, with *x* number of shoot or root ramifications, etc.), and healthy appearance (absence of infections, turgid leaves without curved borders, normal leaf pigmentation, etc.).

There must be a sufficiently high number of plants available to allow, in some cases, experiments involving a great number of leaves, and to secure sufficient material for the analytical procedures which require tissue destruction. There should always be available a minimum number of 100–200 plants grown under generally even conditions in greenhouses or, even better, grown under the finely controlled conditions of a growth chamber.

Before starting any type of photosynthetic measurement the physiological adaptation of the plants should be achieved. This pre-adaptation may be reached by maintaining the plants during a period of one to four weeks under uniform conditions, preferably in growth chambers, by simulating the environmental parameters existing in the habitat and taking special care of thermo- and photoperiodical fluctuations.

The measurement itself, if not continuous, has to be done with a constant frequency and a periodicity throughout the experiment, special precaution being taken to quantifying the photosynthetically active radiation, as well as controlling the other parameters of the photosynthetic process.

A synoptical table of the main limiting factors of the photosynthetic process that should be taken in account during the planning of the experiment (Sestak et al., 1971) is given in Table 12.1.[1]

Once the characteristic of the photosynthetic process is established, using plants grown under controlled conditions and with the precautions mentioned, comparative measurements under field conditions should be undertaken in order to obtain an appreciation of the range of variation of the data obtained between the physiological optimum and the ecological optimum of photosynthesis, as illustrated by Figure 12.1 (after Bannister, 1976).

METHODS TO MEASURE PHOTOSYNTHETIC ACTIVITY

These methods can be applied to intact plants grown in the field or under laboratory conditions, permitting an estimation of the photosynthetic activity of the whole plant or of parts of it, by measuring: (a) changes of CO_2 concentration produced in the environment immediate to the plant, recorded continuously (gas analytical methods); (b) changes of the CO_2-partial pressure (physico-chemical pCO_2 methods); and (c) changes in the ^{14}C-fixation rate from the plant enclosure ($^{14}CO_2$-assimilation methods).

Gas analytical method

This method uses a measuring system consisting of an assimilation chamber containing the plant or a plant part, an infra-red gas analyser (IRGA), a CO_2 circulation pump, and a system for conditioning the air before it comes into

1. Table 12.1 summarizes the main factors that affect the rate of photosynthesis. The importance of these factors, their mode of action (during the whole previous ontogeny of the plant, during one or several of its phases or during the measurement only), and the interrelationships amongst them cannot be defined precisely. Nevertheless, the table illustrates that measured rates of photosynthesis must be regarded as the result of numerous effects.

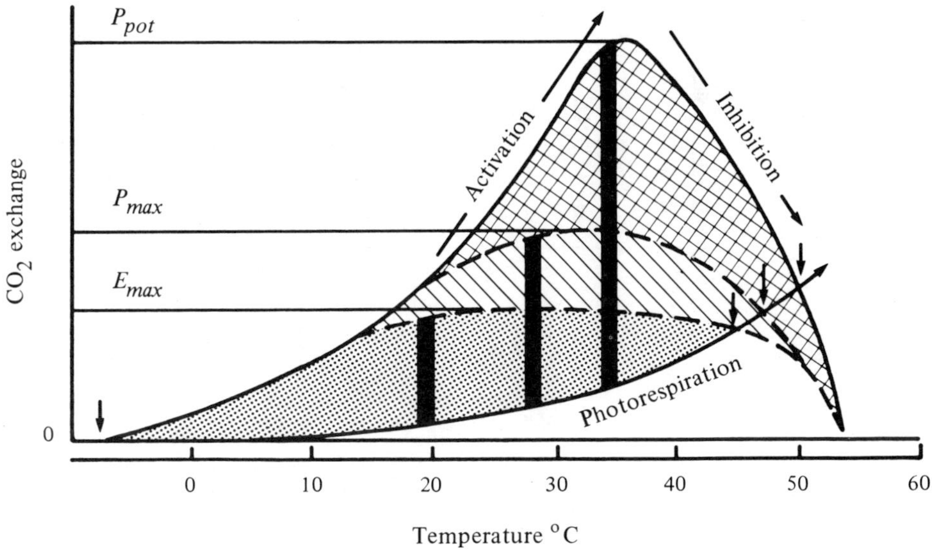

Figure 12.1
Relationships between photosynthesis, respiration, temperature and the physiological and ecological responses of a plant.
P_{pot} = potential physiological maximum of gross photosynthesis (all requirements, including CO_2, at optimum levels).
P_{max} = physiological maximum (CO_2-limiting). E_{max}, ecological maximum (most factors-limiting).
Vertical bars represent optima for net photosynthesis.
Vertical arrows indicate temperature compensation points.

contact with the plant. This method is one most frequently used and the details for mounting and handling the system can be found in the manual of methods of Sestek et al. (1971), as well as in the literature concerning measurements in mangroves (Moore et al., 1972, 1973; Attiwill and Clough, 1980).

The method, despite the complexity of its operation (it needs precise and simultaneous control of a great diversity of environmental parameters during the experiment), represents today the most adequate for obtaining precise data suitable for evaluation by computer. It is therefore recommended that those laboratories engaging in long-term research of mangrove photosynthesis should direct their efforts towards the acquisition of such a system. Once calibrated, it can be employed to study the effect of hydrological (tidal) changes, as well as the effects produced by the chemistry of the substrate on the photosynthetic capacity of the diverse mangrove species, being also able to measure the main components of photosynthesis, such as photorespiration, CO_2 transfer resistances and light- and CO_2-compensation points.

Physico-chemical methods to measure CO_2-partial pressure (pCO_2)

These methods are based on the determination of the CO_2-partial pressure (pCO_2) by measuring changes produced in CO_2-absorbing solutions.

TABLE 12.1 Review of rate-limiting factors of photosynthesis

Factor	Reason	Remarks	References
Genetic factors			
A. Existence of stomatal resistance			
1. In both leaf surfaces	CO_2 diffusion resistance	Error in partitioning diffusion resistances. Gas exchange rates through upper and lower leaf surfaces should be determined separately	Raschke (1958, Chaps. 2, 15, 16); El-Sharkawy and Hesketh (1964); Holmgren (1968); Moreshet, Koller and Stanhill (1968)
2. High resistance in upper surface of the leaf		Pronounced stomatal limitation of gas exchange rate at saturating irradiance. Assimilation chamber not separating the leaf surfaces utilizable. Porometry not reliable	Holmgren, Jarvis and Jarvis (1965); Bertsch and Domes (1969); Domes and Bertsch (1969); Zelitch (1969*a*)
B. Differences in pathways of anabolism and catabolism of carbon substances			
1. Pathways of carbon fixation	C_4-dicarboxylic acid pathway of CO_2 fixation (typical for tropical grasses and related plant species) or the more common Calvin cycle	Upper limits of CO_2 influx as well as the level of saturating irradiance much lower in Calvin cycle plants	Kortschak et al. (1965); Hatch and Slack (1966); Bassham and Jensen (1967); Ludlow and Wilson (1968); Bull (1969); Bjorkman (1970); Hatch (1970); Hatch and Slack (1970)
2. Activity of the carboxylating enzymes			Huffaker et al. (1966); Slack and Hatch (1967); Bjorkman (1968*a*); Wareing et al. (1968); Bjorkman and Gauhl (1969); Eagles and Treharne (1969); Treharne and Cooper (1969); Preiss and Kosuge (1970); Treharne and Eagles (1970); Walker and Crofts (1970)

3. Presence of photorespiration	Photorespiration present in plants with Calvin cycle destroys part of the newly formed photosynthates	High CO_2 compensation. P_G cannot be determined as the sum of $P_N + R_D$. Comparison of P_N with the activity of dark enzymatic reactions of photosynthesis not reliable	Forrester, et al. (1966*a,b*); Moss (1966, 1968); Hesketh (1967); Lake (1967*a, b*); Downton and Tregunna (1968); Johnson and Hatch (1968); Jolliffe and Tregunna (1968); Zelitch (1969*b*); Ludlow (1970); Walker and Crofts (1970)
C. Composition of photosystems and activity of photoreactions			
1. Amount of chlorophyll-a and its its distribution between the in vivo forms, amounts of accessory pigments	Rate of radiant energy fixation, formation of $NADPH_2$ and ATP	P_N usually better correlated with the amount of chlorophyll-a than total chlorophylls	Butler (1965); Deroche and Costes (1966); Schmid and Gaffron (1967); Briantais (1968); French et al. (1968); Hager (1969)
2. Amount of limiting components of electron transfer chain (cytochrome f, plastocyanin, P 700, ferrodoxin, compounds Z, X, etc.)		Potential indicators of photosynthetic activity	Hill (1965); Boardman (1967, 1968, 1969, 1970); Levine (1968, 1969); Fork and Amesz (1969); Cheniae (1970)
3. Arrangement of pigments and electron carriers, ultra-structure of thylakoid membrane			Park (1966); Bogorad (1967); von Wettstein (1967); Frey-Wyssling (1968)

TABLE 12.1—*continued*

Factor	Reason	Remarks	References
4. Variation in activity of individual reactions of the electron transfer chain			Hoffman and Miller (1966); McNaughton (1967); Olson (1970)
D. Shape, structure and number of chloroplasts and their position in the cells	CO_2 diffusion resistance in mesophyll	CO_2 diffusion in liquid phase may be changed with light-dependent chloroplast movements in cells. Light distribution within the leaf	Laetsch and Stetler (1965); Haupt (1966, 1969); Homann and Schmid (1967); Mousseau (1967); Thomas et al. (1967); Laetsch (1968, 1969); Laisk (1968, 1970); Leech (1968); Mousseau and Bourdu (1968)
E. Leaf structure			
1. Leaf anatomy	CO_2 diffusion resistance	Affected by growing conditions and pretreatment	McClendon (1962); Bjorkman and Holmgren (1963); El-Sharkawy and Hesketh (1965); Wilson and Cooper (1967, 1969*a*, *b*, *c*, *d*); Pearce et al. (1969); Hesketh and Baker (1970)
2. Optical properties	Efficiency of energy absorption: factor C_1	May change even during slight wilting; affected deposits on the leaf surface	Gates et al. (1965); Loomis (1965); Bjorkman (1968*b*)
Environmental and internal factors			
F. Energy supply			
1. Radiation 400–700 nm, its quality and duration	Energy supply to photochemical reactions: factor N	Effect of E of different sources of radiation and of their filtration (e.g. by water). Effect of irradiance during growing period	Talling (1961, Chap 19); Kriedman et al. (1964); Hiesey and Milner (1965); Mousseau (1966, 1967, 1968); Loach (1967); Mousseau et al. (1967); Pearce et al. (1967); Eagles and Treharne (1969); Hatch, et al. (1969); Larcher (1969*c*); Wilson and Cooper (1969*a*, *d*)

2. Radiation 400–700 nm, flux density and duration	Photoinhibition—precise causes uncertain		Klein and Bogorad (1964); Kok et al. (1965)
3. Radiation 400 nm, and 700 nm, its quality and duration	Changes in ultrastructure and thus changes in other factors; factors E and N	Change in leaf energy balance when using different light sources	
4. Photoperiod during development and immediate photoperiod history	Formative effects on plant ontogenesis as well as on their photosynthetic apparatus; factor P		Cockshull (1966); Sestak (1966); Park and Drury (1967); Sironval and Englert (1967); Thomas et al. (1967)
G. CO_2 supply			
1. CO_2 concentration in the ambient air	CO_2 gradient; CO_2 concentration near sites of carboxylation; CO_2 dependent control of stomatal opening	Control of CO_2 concentration in the assimilation chamber according to outlet concentration preferable	Hesketh (1963); Holmgren and Jarvis (1967); Bishop and Whittingham (1968); Heath and Orchard (1968); Holmgren (1968); Fock et al. (1969); Wilson and Cooper (1969*a*)
2. Ventilation	Effect on diffusion resistance in the boundary layer, thus on transfer of CO_2, O_2, water vapour and heat	Decrease in photosynthetic rate and large CO_2 gradients in the assimilation chamber at low ventilation rates	Avery (1966, Chap 2); Takakura (1966); Nevins and Loomis (1970*a*)
H. Leaf and air temperatures	Rate of chemical reactions; effect on transpiration rate and thus interrelationships with factor I; also influence on other factors	Difficulties in measuring the actual leaf temperature	Kozlowski and Keller (1966, Chap. 16, 17); Kriedemann (1968*a*, *b*); Hesketh and Baker (1969); Hew et al. (1969); Hofstra and Hesketh (1969*a*, *b*); Larcher (1969*c*); Pisek et al. (1969); Wilson and Cooper (1969*c*); Woledge and Jewiss (1969); Mooney and Harrison (1970); Treharne

TABLE 12.1—*continued*

Factor	Reason	Remarks	References
I. Plant water relations			
1. Water content	Rate of translocations of photosynthates; effect on factor J	Sufficient water supply to the object difficult at high irradiances. Leaf energy balance important	Clements (1964); Roberts (1964); Slavik (1966); Crafts (1968); Wardlaw (1967, 1968); Larcher (1969*c*); Packer (1966)
2. Water potential in chloroplasts	Enzymatic activity in gels; supply of water to photo-system II and enzymatic reactions of CO_2 fixation		Santarius and Ernst (1967); Santarius and Heber (1967); Crafts (1968)
3. Water potential along the diffusion paths (also presence of water in liquid phase in intercellular spaces)	CO_2 diffusion resistances; H_2O dependent control of stomata, and thus factor G		Brix (1962); El-Sharkawy and Hesketh (1964); Crafts (1968); Kriedmann (1968*a*); Statyer (1969); Troughton and Slatyer (1969); Troughton (1969)
J. Concentration of photosynthates and translocation rate	Equilibrium of chemical reactions; osmotic potential	Effect of photosynthate accumulation in detached objects (leaves, discs, segments)	King et al. (1967, Chap. 9); Neale and Incoll (1968); Haapala (1969); Hofstra and Nelson (1969*a*, *b*)
K. Soil moisture	Water potential in the plant (factors I, J)	Lack of oxygen supply to roots in fully watered potted plants	Sestak and Vaclavik (1965); Kozlowski and Keller (1966); Negisi (1966)
L. Mineral nutrition and contents of mineral elements in the tissue	Effects of factors E (e.g. nitrogen, phosphorus), I (including permeability of membranes, e.g. potassium), C, J, etc.	Equilibrated and steady mineral nutrition during cultivation important	Clements (1964); Keller and Koch (1964); Murata (1965); Ozbun et al. (1965); Kozlowski and Keller (1966); Bourdu et al. (1967); Hartt and Burr (1967); Keller (1968); Meinl (1969); Ryle and Hesketh (1969); Natr (1970); Nevins and Loomis (1970*b*); Walker and Crofts (1970)
M. Pathological state	Effect of other factors	Effect of latent disease (e.g. viruses)	Livne (1964); Scott and Smillie (1966); Harding et al. (1968)

N. Endogenous diurnal cycle	Stomatal CO_2 diffusion resistance; rate of translocation of assimilate	Comparative measurements should be done at the same time of day. Influence on prolonged measurements	Gates (1965); Kluge (1968); Hoffman and Miller (1966); Kozlowski and Keller (1966); Kortschak and Forbes (1969); Jones and Mansfield (1970)
O. Seasonal cycle	Effect on other factors	All comparative measurements on plants of the same age	Kozlowski and Keller (1966); Shiroya et al. (1966); Bamberg et al (1967); Gordon and Larson (1968); Larcher (1969*a*)
P. Development stage of the plant	Effect on other factors	Photoperiodic influences. Existence of vegetative cycle with two peaks of photosynthetic activity, the second often at the time of flowering	Thorne (1965); Sweet and Wareing (1966); Kortschak and Forbes (1969)
R. Age of leaf	Effect on other factors	Complex factor. Also differences in area of one leaf blade	Sestak and Catsky (1967); Hardwick et al. (1968); Kriedemann (1968*b*); Pearce et al. (1968); Wada (1968); Walker and Waygood (1968); Wilson and Cooper (1969*b*)
S. Age of meristem when leaf primordia were produced	Effect on other factors	Effect of initial growth rate of seedlings	Bormann (1965)
T. Effect of growing conditions, adaptation and pretreatment	Effect on other factors	Also adaptation to the quality of light source	Tranquillini (1964); Loach (1967); Mousseau et al. (1967); Bjorkman (1968*a*, *b*); Duncan and Hesketh (1968); Hesketh (1968); Holmgren (1968); Eagles and Treharne (1969); Wilson and Cooper (1969*c*, *d*) and other papers cited in the previous sections

Source: Sestak et al. (1971).

The estimation of the CO_2-partial pressure produced can be made: (a) through measurement of the pH of the solution by colorimetry (Catsky and Slavik, 1958); (b) by means of the measurement of the changes of conductivity produced in a sodium hydroxide solution absorbing CO_2 (Thomas, 1933; Stocker and Vieweg, 1960); and (c) by means of continuous measurement of pH of diluted solutions of bicarbonate (Zunker and Kreeb, 1970).

The above-mentioned methods do not reach the extreme sensitivity of the gas analytical method, and require precise temperature control. Nevertheless, these inconveniences are compensated for by the ease of setting it up, at reasonable cost, in every experimental laboratory.

Because of its simplicity of construction and the possibility of measuring photosynthetic rates in the field, the potentiometric method developed by Zunker and Kreeb (1970) is described below.

The components of the system, as shown in Figure 12.2, can be easily constructed. The procedure for measurement is as follows:

1. Aspirate an air-control sample in open circuit until stabilization of pH of the reaction solution is attained (five to ten minutes).
2. Introduce the plant organ inside the photosynthetic chamber.
3. Measure the actual environment parameters: air temperature, radiation and relative humidity, inside and outside the photosynthetic chamber, by means of thermocouples, radiometer and psychrometer.
4. Five minutes after introducing the plant part, circulate air for two minutes and read off pH and temperature.
5. Take the plant part out of the chamber, and recirculate air in open circuit—prior to the next measurement—until pH stabilization of the absorbing solution is reached.

The pair of values obtained, that is the pH value reached by the reaction solution, at the given temperature, after the control measurement, and the recorded value with the plant inside the chamber, are used for calculating the photosynthetic rate, taking into consideration: (a) the total air volume enclosed in the system; (b) the leaf area, or the leaf weight of the part which had been exposed; and (c) the exposure time. For calculating the CO_2 concentrations in mg/1 of air, based on the pH differences observed, the equation of Catsky–Sestak is employed, where $0.934 \lg P = a\text{-pH}$, a being the temperature. In order to calculate the CO_2 concentration of the air from the pH data measured in the absorbing bicarbonate solution at different temperatures, the tables contained in Sestak et al. (1971) may be used.

The experimenter must seek ways of improving the microclimatic conditions inside the photosynthetic chamber and of developing criteria to define the minimum exposure time as well as the number of measuring points required to obtain acceptable mean values. In each case he must take into account the particularities of the plant material employed.

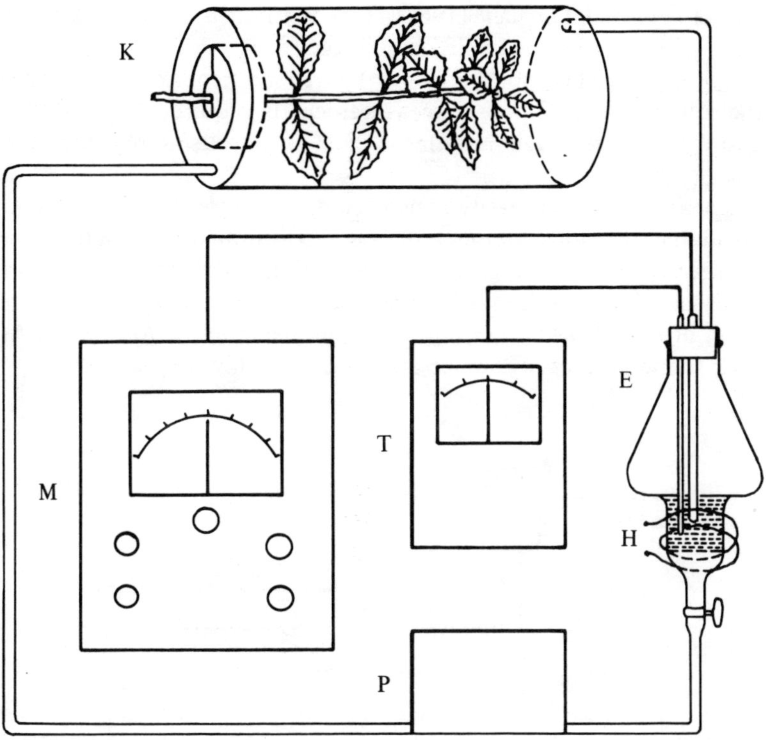

Figure 12.2
Potentiometric device for field measurement of photosynthetic rates (after Zunker and Kreeb, 1970).
K = photosynthetic chamber made of lucite no greater than 400 to 500 cc capacity.
E–H = reaction chamber made from an Erlenmeyer flask of 50 cc to which a filter funnel with a porous glass bottom and a wrench is soldered. This chamber contains all CO_2-absorbing solution which is composed of: 84.0 mg $NaHCO_3$ (= sol. 0.001N) + 7.460 KCl (= sol. 0.099N) brought to a litre with distilled water.
P = membrane pump actioned by 6-V batteries.
T = temperature sensor. M = portable pH-meter with glass electrode.

Methods of measuring the rate of CO_2 fixation

The principle on which the methods are based consists of exposing the plant tissues inside an enclosure to a stream of air containing a known amount of $^{14}CO_2$, followed by the measurement of the radioactivity of the $^{14}CO_2$ incorporated into the exposed material.

The methods have the advantage of being simple, speedy in their operation and allowing one to collect a great number of data on the rate of photosynthetic fixation of CO_2 in individuals of the same species, corresponding to their age and to the environmental conditions.

Method of Austin and Longden (1967). This method is most suitable for use with plants grown under controlled conditions.

The $^{14}CO_2$ supplied to the tissue has to be generated and circulated inside the apparatus each time a measurement is required. This allows only one measurement every ten minutes, which is unsatisfactory for large experiments.

The apparatus itself is easily constructed, and depending on the counting device available, the measurement of the incorporated radioactivity can be done by direct plate counting or by scintillation counting of the previously digested tissue. This method has been successfully employed for the determination of the CO_2 fixation rate of young seedlings of *Rhizophora mangle* grown in a saline gradient (Pannier and Kochanovsky, unpublished).

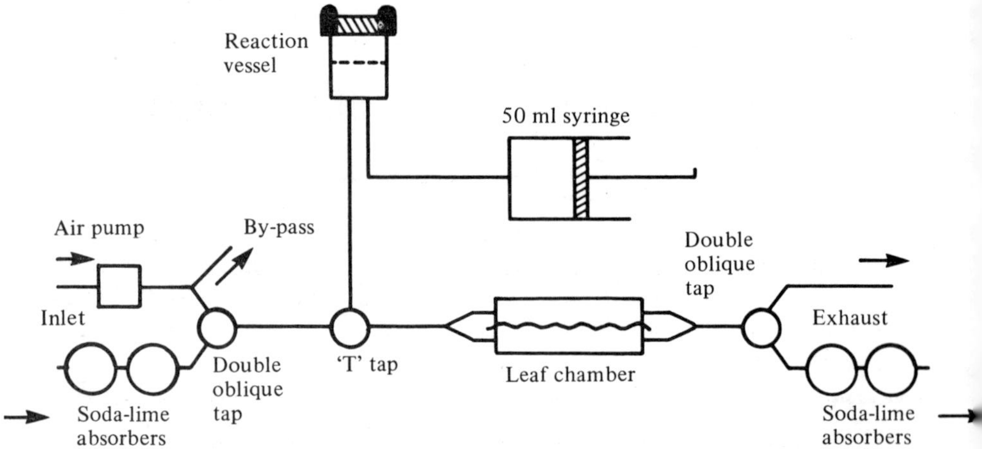

Figure 12.3
Apparatus used for measuring photosynthesis: Austin and Longden's method.

The operation of the apparatus is as follows (see Figure 12.3):

1. A disc of filter paper, having a known amount of ^{14}C labelled K_2CO_3 of known activity, is placed in the reaction vessel which is then closed with a suba-seal-serum cap.
2. Air is drawn over the soda-lime absorber, through the reaction vessel and into the 50-ml syringe, and expelled to waste via the leaf chamber, which is opened at this stage.
3. With a hypodermic syringe, sufficient normal perchloric acid is applied to the filter paper in the reaction vessel to liberate the $^{14}CO_2$. Sixty seconds is allowed for this reaction to proceed to completion.
4. Fifty ml of air is drawn across the soda-lime absorbers, through the reaction vessel and into the syringe.
5. The taps are turned so that the air containing the $^{14}CO_2$ is isolated in the

reaction vessel and syringe while ordinary air is pumped through the apparatus to exhaust at the rate of 200 ml per minute.

6. A leaf is placed in the leaf chamber and allowed to equilibrate for two minutes while air is passing through the leaf chamber as in 5, above.
7. The flow of air from the pump is interrupted for fifteen seconds while 50 ml of air containing $^{14}CO_2$ is passed through the leaf chamber. For these fifteen seconds, and for the following thirty seconds, the air from the cell is directed through the soda-lime absorbers on the exhaust side of the apparatus. Thus, the $^{14}CO_2$ not used in photosynthesis is absorbed in soda-lime and does not contaminate the atmosphere.
8. After the thirty second period, the leaf is taken out of the chamber and discs cut from it with a cork borer.

   The discs can be directly counted, after having been fixed to aluminum planchets and dried in the oven at 100 °C for ten minutes, with an end-window counter, or preferably, digested for counting with a scintillation detector. Since the latter procedure has a higher counting efficiency, it is recommended.

The following procedure for scintillation counting was used by Shimishi (1969).

Fill a series of counting vials with 2 ml of the CO_2 absorber, NaOH 0.25N, and close immediately.

The cut leaf disc is placed in a microtube filled with 1 ml of chromic acid, and the tube itself is then placed in the counting vial. The closed counting vials are then exposed for thirty minutes to an autoclave at 120 °C. During this time the $^{14}CO_2$ previously contained in the leaf tissue should be absorbed by the NaOH solution.

After cooling the vials in ice water, 8 ml of the standard scintillation fluid adequate for counting aqueous solutions ('Aquasol', or PPO + POPOP according to the indications of Patterson and Greene, 1965) are added. A few drops of Triton X–100 facilitate mixing.

A counting efficiency of 65–70 per cent can be attained.

The following calculation was made:

$$\mu\text{g assimilated } CO_2 = \frac{\text{cpm leaf disc}}{\text{cpm } CO_2\text{-generating paper disc}} \times \frac{\mu M CO_2 \text{ applied to}}{\text{the generating disc}} \times 44$$

This value has to be referred to leaf area utilized, time of exposure of the leaf disc to the $^{14}CO_2$-pulse in order to obtain the final expression: mg $CO_2 cm^{-2} min^{-1}$.

Method of Shimshi (*1969*). This method represents a substantial improvement of the method of Austin and Longden (1967) because it allows a considerable reduction in the time of the $^{14}CO_2$-(pulse and consequently increases the stability of the environmental conditions inside the leaf-chamber. $^{14}CO_2$ is supplied from cylinders to the leaf surface inside a microchamber mounted on steel crucible tongs, and the required amount is injected automatically.

Details of the procedure of preparing and filling the air cylinder with $^{14}CO_2$ as well as the construction details of the micro-chamber may be seen in the original paper. The method was developed for field work, and 30–40 measurements per hour can be realized.

Method of Johnson et al. (1979). This method has the advantage of allowing simultaneous measurement of photosynthesis and transpiration under laboratory, as well as field, conditions using an airstream containing tritium THO and $^{14}CO_2$ which diffuses into the leaf each according to their own concentration gradients. Thus, photosynthetic rates in terms of CO_2 incorporation can be correlated with the gas-exchange efficiency of the leaf as a function of stomatal conductance. The description of the apparatus, called a double diffusion or isotope porometer, is given in Figure 12.4. More details about the theories and data evaluation can be found in the original paper.

This method was successfully applied to seedlings of *Rhizophora mangle* in the controlled chambers of the Plant Science Laboratories of the University of California, Riverside (results to be published).

METHODS OF MEASURING RESISTANCE TO CO_2 TRANSFER

Due to the great structural diversity of the leaves of the different mangrove species which is manifested in the foliar symmetry, structure and distribution of stomata, position of the photosynthetic tissue and the internal accumulation of water, the measurement of the resistances opposing the diffusion of CO_2 molecules from the outer leaf environment to the sites of reaction in the chloroplasts is needed to interpret the photosynthetic response of a particular species.

Theoretical consideration on the transfer processes of CO_2 molecules during their diffusive pathway can be found in Gaastra (1959), Jarvis (1971), Monteith (1963), Moreshet et al. (1968), Lake (1967*a*, *b*), Lake and Slatyer (1970), Cowan and Farquhar (1977).

Of the components of the diffusion pathway, the most adequate resistances to be directly measured are the stomatal resistance (r_s) and the mesophyll resistance (r_m), which also includes the carboxylation resistance (t_x), that is, the limitations due to concentrations and activities of carboxylating enzymes to the CO_2 incorporation into the chloroplast.

Methods of measuring stomatal resistance to CO_2

Methods using inert gases These methods are based on the principle of gas diffusion porometry, consisting of exposing one face of a leaf to a physiologically inert gas, such as helium (Gale et al., 1967) or radioactive argon, ^{41}A, (Moreshet et al., 1968) in order to establish a concentration gradient through the leaf, and measuring the corresponding gas efflux which leaves other leaf face.

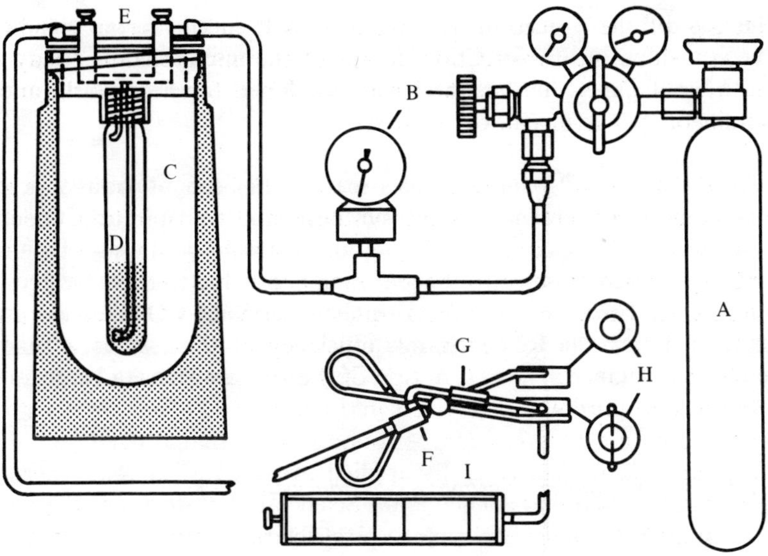

Figure 12.4
Double isotope porometer. Compresses air cylinder (lecture bottle size). A contains the $^{14}CO_2$ supply in air. Gas regulator (B) with a sensitive pressure gauge ensures precise pressure control and gas-flow regulation. Insulated flask temperature bath (C), approximately 1,000 cm^3 in volume, maintains constant temperature (most simply that of ice) during operation for the tritiated water supply (D). The insulated flask (C) has a wide mouth fitted with a plexiglass cover (E) which has been threaded to hold a 30 cm^3 culture tube (D) bearing the liquid THO supply in a gas-tight seal. The top of the plexiglass cover is also fitted with two needle valves at the exit and entrance ports in the gas-flow line of the culture tube to provide additional safety controls. Labelled air is bubbled through the THO supply to achieve THO vapour saturation with a known specific activity. The flow line leads from the exit port of the culture tube through small diameter teflon tubing to the leaf-chamber clamp. At the leaf-chamber clamp a finger-release spring valve (G) is used to start and stop the flow of isotopes through the leaf chamber (H). A 2-cm section of 0.1 mm capillary tubing (G) between the finger-release valve and the chamber jaws controls the flow rate of the isotope gas mixture at approximately 2 cm^3s^{-1} with the system pressurized to 26.7 kPa. Further adjustments in pressure result in linearly proportional changes in the flow rate. The chamber jaws H of the leaf clamp are made of plexiglass. The upper jaw has an open top that allows maximum illumination. The lower jaw is fitted with an 'O' ring 1.26 cm in diameter which maintains a space of approximately 0.15 cm between the plant tissue and the entrance and exhaust isotope gas ports when the chamber is clamped on a leaf. The chamber volume is thus 0.180 cm^3. Plant tissue inside the 'O' ring is exposed to $^{14}CO_2$ and THO vapour when the gas flow is initiated with the finger-release spring valve. A line leads from the lower chamber jaw, which during operation is sealed against the plant leaf, to the isotope trap (I). The trap has two compartments, the first contains a desiccant to remove excess THO vapour and the second a strong base to remove excess $^{14}CO_2$ from the airstream passing out the exhaust. Not shown are two small thermistors, one mounted in the THO supply culture tube and the other in the leaf chamber, for monitoring temperatures critical in data analysis.

The gas-diffusion resistances measured with specific gases can be converted to equivalent values for CO_2. Nevertheless, these methods, besides being difficult to handle, have limitations, such that CO_2 isotope methods are always to be preferred.

Method using the $^{14}CO_2$-isotope This method has been previously described (Johnson et al., 1979) for its use in measuring photosynthetic CO_2-fixation rates.

It allows a quantitative expression of CO_2-conductance, or its reciprocal resistance, to be obtained by redefining the traditional equation for gaseous exchange, including the following relationship: $C = F/\Delta$, where F is $^{14}CO_2$ uptake (units: $s^{-1}cm^{-1}s^{-1}$) and Δ is $^{14}CO_2$ airstream density (units: $s^{-1}cm$) (see Johnson et al., 1979).

Method for measuring mesophyll resistance to CO_2

In the same manner as for the quantitative determination of r_s, the measurement of the rate of $^{14}CO_2$-fixation to the leaf tissue by means of the double diffusion porometer (described by Johnson et al., 1979) permits an estimation of the value corresponding to the mesophyll resistance, through the formula: $r_m = \Sigma r_{CO_2} - r_s - r_c$, but only if light saturation for photosynthesis is attained.

A useful alternative for the researcher having a gas analytical system for continuous recording of photosynthetic rates is to calculate the mesophyll resistance (r_m) from the data of CO_2 exchange and the total resistance for water-vapour diffusion measured in the plant enclosure.

For more details the reader is recommended to consult references indicated in Sestak et al. (1971), and for the specific case of mangroves the calculation procedure stated in the papers of Moore et al. (1972, 1973).

METHOD FOR THE MEASUREMENT OF PHOTORESPIRATION BY MEANS OF THE DETERMINATION OF THE COMPENSATION CONCENTRATION OF CO_2 [1]

Although there is no convincing evidence for the existence of a photorespiratory mechanism in mangroves, which would indicate the presence of a biochemical carbon pathway different from the C_3, it is convenient to carry out a screening inside the high species diversity of mangroves and their accompanying species, taking special account of the salinity of the substrate and of the physiological age of the individuals.

The most appropriate method for doing this is to measure the capacity of illuminated leaves, exposed in a closed system with air deprived of CO_2, to enhance the concentration of CO_2 to levels of 50 ppm.

1. Treguna et al. (1961), Meidner (1967), Hesketh (1967).

Assuming that the relation between photosynthetic rate (P_N) and the concentration of CO_2 of the air is linear, $T = CO_2$ efflux/Δ P_N/CO_2, the origin of the produced CO_2 could be attributed to the respiration process which occurs simultaneously with photosynthesis in the light.

An essential requirement for employing the method is to have a sufficiently highly sensitive instrument to measure CO_2, such as the IRGA. The measurement must be performed under conditions of light saturation, having taken the precaution of checking for possible leaks in the system.

A very convenient improvement of the method which allows the handling of a great number of samples in the shortest time has been described by Goldsworthy and Day (1970). It consists of exposing the leaves by floating them on water in petri dishes which are enclosed in large transparent, CO_2-impermeable Mylar-bags, which have been previously flushed with nitrogen and exposed to saturating light for one hour. After this, the CO_2 accumulated inside the bag is expelled manually, by pressing the air through the measuring cuvette of the IRGA, the maximum deflection obtained being recorded.

Measurement performed in this way with seedlings of *Rhizophora mangle* showed rapid increase of CO_2 levels exceeding 200 ppm per hour (Pannier and Pannier, unpublished).

METHOD FOR MEASURING THE LIGHT COMPENSATION POINT[1]

Determination of the light compensation point, i.e., the light intensity in which photosynthetic CO_2-incorporation ceases, is very important for estimating the adaptive capacity of the physiological developmental stages of the plants, or of leaves belonging to the canopy of the same individual, to the light conditions prevailing in the habitat. In addition, it is important for defining the minimum light needed to cultivate seedlings under laboratory conditions, as well as gaining some idea of the amount of light needed for development in the natural forest shade.

Working with *Avicennia*, Attiwill and Clough (1980) found significative adaptive differences between leaves belonging to the same canopy. Shadow leaves showed always a higher light-compensation point than sun leaves.

A recommended low-cost and simple field method is the colorimetric method of Alvik (1939), further improved by Pavletvic and Lieth (1958). Leaves enclosed over a solution of bicarbonate containing an appropriate indicator produce, according to the light intensity, a change of colour in the solution, which can be visually or colorimetrically estimated and compared against a control.

These colour changes can also be calibrated in absolute pCO_2 values when an IRGA is available (Lieth, 1960).

1. Alvik (1939), Lange (1956), Pavletvic and Lieth (1958).

DETERMINATION OF BIOCHEMICAL CO_2-FIXATION PATHWAYS

Research to establish a CO_2-fixation pathway of the C_4 type for mangrove species (specifically for *Rhizophora mucronata, Sonneratia acida, Aegiceras majus* and *Excoecaria agallocha*) has been reported by Joshi et al. (1979, 1976).

Exposing leaves to $^{14}CO_2$ for times varying between a few minutes and several hours, the above-mentioned authors were able to find aspartate in the corresponding chromatograms, as the initial product of photosynthetic carbon assimilation. The appearance of this compound, precursor for the synthesis of alanine and other amino acids, was only observed in plants grown under saline conditions and not in fresh water. This fact, as well as the discovery of an increase in quantity and activity of PEP carboxylase, and in the case of plants of *Aegiceras majus* of an inhibition of the malic dehydrogenase activity, has suggested that the mangroves studied are C_4 plants, even though they do not have the corresponding Kranz anatomy of this plant type. The ecological implications that such an adaptative photosynthetical pathway to saline environmental stress could have, make it worth-while to extend the research to other mangrove species, so as to dissipate contradictory results, such as those obtained with seedlings of *Rhizophora mangle* cultured in saline substrate (Pannier and Kochanovsky, unpublished) which indicate no enhancement of PEP carboxylase activity, or the fact that the carbon isotope ratio $^{13}C/^{12}C$, reported for sixteen species belonging to fourteen genera of Australian mangroves, are better fitted to the typical values of C_3 plants (Andrews, personal communication). In view of the fact that the biochemical methodologies employed to trace the specific CO_2-fixation pathway are not in the scope of this book, the reader is referred to the following important papers: Bassham and Calvin (1957), Kortschak et al. (1965), Hatch and Slack (1966, 1970), Black et al. (1973); and to the following reviews: Coombs (1971), Hatch et al. (1971), Bjorkman (1973), Black (1973).

Respiration

The determination of the respiration rate in mangroves is important because of its value as an indicator of the energy requirement for the maintenance of the different physiological processes of the plant (growth, translocation, glandular excretion, etc.), as well as for quantifying adaptative responses to the environment.

Comparative measurement performed on different mangrove species (Brown et al., 1969; Chapman, 1962*a,b*) have shown that the transpiration rates of young seedlings transferred from sea-water to distilled water undergo a remarkable increase. This fact, as well as the observed capacity of seedlings of *Rhizophora* to respire under anaerobical conditions, has been interpreted as physiological adaptations in stages previous to, and during their implantation in, the natural habitat.

Pannier (1962), studying on a comparative basis the oxygen consumption of tissues constituting the viviparous seedlings of *Rhizophora mangle*, found a relationship between the differences in respiratory activity of the different tissues and the existence of an osmotic threshold in the translocation pathway connecting the viviparous seedling with its parent. Chapman (1944) and Troll and Dragendorff (1931) presented experimental evidence that the contribution made by the pneumatophores of *Avicennia* and *Sonneratia* to the CO_2 pool located in the intercellular species of the aerenchyma is partially due to respiratory activity of the cells of their own tissues. This finding has importance in relation to the problem of internal diffusion of gases and this should be investigated.

Measurement of the respiration rate of the absorbing nutritional rootlets of *Rhizophora racemosa* cultivated in a saline gradient showed their dependence on the previous conditions in which the plants were maintained. Thus, evidence for a compensating effect of respiration at specific salinities by duration of the light period and dependence on temperature has been found (Pannier and Pannier, 1971).

Values on leaf dark respiration for several mangroves are provided by Moore et al. (1972, 1973) indicating higher and more fluctuating values for *Avicennia germinans* than for *Rhizophora* and *Languncularia*.

It is interesting to mention in this connection that the productivity measurements of mangrove communities as performed by Hicks and Burns (1975) showed an increase of respiratory activity of the species with increasing salinity of the substrate, according to the following species' specific response:

Laguncularia < *Avicennia* < *Rhizophora*

Future research on the respiration process in mangroves should be oriented towards those energy-requiring processes linked with root growth, ion uptake and ion translocation. It is necessary to use techniques for continuous measurement of oxygen consumption by intact organs avoiding the use of detached plant parts.

The changes occurring during the development of the embryos of the different mangrove species, the mobilization of assimilation products in the viviparous seedlings as well as in the adult plant, and the processes of leaf and hypocotyl abcission should be also correlated with the respiratory activity of the corresponding tissues.

Special importance should be attached to the study of the metabolic pathways in tissues having active respiration (glandular tissues, absorbing roots, developing buds, etc.) in order to determine the response to changes of endogeneous and exogenous salt levels.

METHODS

In general, the methods to measure respiratory activity require instruments in common use in laboratories and which can be easily adapted.

Direct analysis method

This method can be employed for the determination of CO_2 evolution by plant seedlings in closed systems, and consists in taking air samples which are then analysed in a Haldane apparatus (Haldane and Graham, 1933). This method was employed by Brown et al. (1969) in their studies on the respiration metabolism of Australian mangroves.

Gas analytical method

This method permits continuous measurement of CO_2 evolved, but its applicability is restricted to the availability of the gas analyser.

The use of the IRGA can be simplified, and its efficiency in measuring small quantities of respiratory CO_2 can be enhanced, by applying the method described by Clegg et al. (1978), which makes use of the IRGA released from the normally employed system for measuring gas samples injected into a flowing carrier gas stream.

Carbon dioxide absorption methods

Methods based on this principle differ only in the means of measuring the CO_2 concentration in the absorbing solution, after contact with the gas sample to be measured.

In its simplest form, the absorbed CO_2 is calculated from the amount of acid required to neutralize the absorbing solution of NaOH or KOH, in the presence of an indicator.

A more elaborate way is measuring the CO_2 through the electric conductivity changes of the absorbing solution. The changes of specific conductance can be measured with any conductimeter, after the establishment of a calibration curve, as indicated by Wollum and Gomez (1970). A device for the continuous conductometric measurement of absorbed CO_2 which is very appropriate for root respiration studies has been described by Nieman (1950).

Warburg manometric method

This method was employed by Chapman (1962*a, b*) and Pannier and Pannier (1971) for respiration measurements in *Rhizophora* seedlings and roots, respectively.

It was described in detail by Umbreit et al. (1957) and Sestak et al. (1971), but had the disadvantage of measuring only tissue slices, root tips, etc., which did not necessarily correspond to the physiological conditions existing in the intact plant. Nevertheless, the method is very convenient for studying the metabolic pathways of tissues in relation to the problem of salt tolerance (Porath and Poljakoff-Mayber, 1964).

Polarographic method

An improved polarographic method well suited for root-respiration measurements has been developed by Armstrong (1967).

The employment of the Clark oxygen sensor for measuring oxygen consumption of intact root systems has been described by Steiner (1968). The

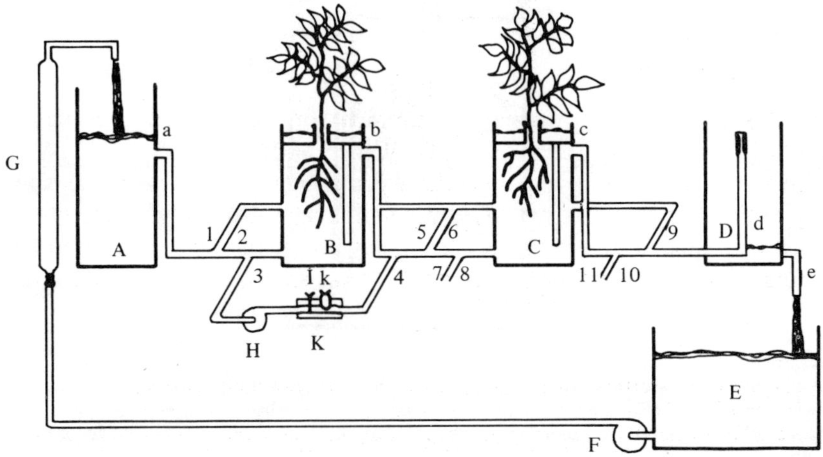

Figure 12.5
A scheme of the apparatus (longitudinal section) for growth and measurement of oxygen uptake by root systems. It consists of four plastic vessels (A), (B), (C), and (D) and a tank underneath (E). The tank is large to guarantee a constant composition of the nutrient solution. The solution is pumped out of the tank by a small pump (F) to a flowmeter (G). This flowmeter is used to control the volume of solution circulating. The solution falls free into vessel (A), filling it up to the outlet pipe (a). It flows through valve 2 to vessel (B), which is filled up to outlet pipe (b). By valves 5 and 7 it goes to vessel (C), also filling this up to outlet pipe (c). After flowing through valve 10, the solution enters vessel (D) by stand pipe (d). It overflows around the stand pipe and then it drains back into the tank by outlet pipe (e).

In this circuit there are three places of intense aeration of the nutrient solution: the free fall at the inlet in vessel (A), the overflow down in a thin layer around stand pipe (d) and where the solution drains back into the tank (E).

Vessels (B) and (C) have a covering disc of plastic inside (as shown in Fig. 12.6), placed just above the outlet. This disc is loose and rests on a plastic ring attached to the wall of the vessel. Between the disc and the ring there is a rubber ring.

Stand pipe (d) has a variable height and acts as an overflow, hence, the level of the solution in the vessel (A), (B) and (C) can be adjusted a few cm above the disc by the stand pipe in vessel (D). The solution above the disc is a water seal to prevent undesired aeration from the air into the nutrient solution beneath the disc. This is only important during the measurement (see Fig. 12.6).

In the middle of the disc there is a hole on which a tube is placed (f). The stem of a plant can be placed in this tube, preferably free, hence, without any supporting material such as foam plastic. The plant can be kept in position by hanging with string. Contact of fresh air with the nutrient solution in this tube must be prevented by a small rubber or weak plastic cap around the stem and around the top of the tube (g).

Near this wide tube there is a small hole with a small pipe underneath the disc (h). This gives an open connection between the nutrient solution underneath and above the disc. This pipe is a buffer to prevent aeration of the solution in the vessel during readings. It has to be long enough to prevent solution above the disc mixing with solution below, if there are small fluctuations in water volume through fluctuations in pump capacity. This is also only important during the measurement (Fig. 12.6).

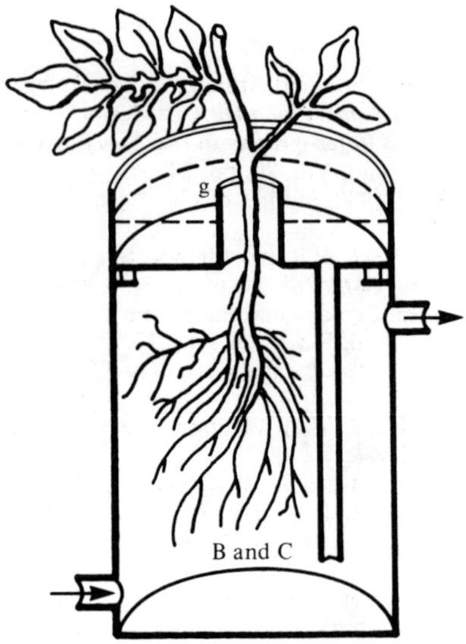

Figure 12.6
Detail of a growing vessel (longitudinal section). The oxygen consumption of a plant (e.g. in vessel (B)) can be measured by isolating it in a known volume of nutrient solution, whose oxygen depletion can be measured over a certain period. During this time aeration from outside must be excluded.

Valves 1 and 6 have to be opened and valves 2 and 5 to be closed. The circulation of aerated nutrient solution through vessel (C) will be continued, but there is a bypass for vessel (B).

A pipe connects valve 3 and 4 (Fig. 12.5). A small pump (H) is placed in this pipe and also a hollow block of polished methyl acrylate (K). In this block a hole with a screw thread is made for attaching the oxygen sensor (k) and a smaller one to hold a thermistor (1) for temperature estimation.

A polished methyl acrylate block is used because it is essential to examine the teflon membrane of the oxygen sensor. If there is an air-bubble underneath, new KCl and new membrane have to be placed on it.

By opening valves 3 and 4 and switching on the motor of pump (H), the nutrient solution in vessel (B) will be mixed and circulated past the oxygen sensor.

apparatus (Figs. 12.5 and 12.6) seems well suited to performing measurements on mangrove seedlings maintained under cultivation, and for this the entire original description is set out in the figure caption.

The oxygen meter gives the oxygen content directly in ppm O_2 so that the consumption by the whole root system through time can be read. During this period there is no aeration in vessel B. Oxygen deficiency prevents readings being taken over a period longer than, say, ten minutes. This, of course, depends on the size and kind of plant and on the volume of the vessel. Usually the system—vessel B together with the connecting pipes—contains 3–4 l of nutrient solution. Knowing the exact volume of solution in vessel B, the oxygen consumption of the whole root system can thus be calculated.

It is possible to have replicate vessels like B and C in a single apparatus, but care must be taken that the pipes are wide enough to avoid unequal levels of solution in the vessels.

References

ALVIK, G. 1939. Ueber die Assimilation und Atmung einiger Holzgewaechse im westnorwegischen Winter. *Meddel*, No. 22, *Vestlandets forstlige Forsoksstation Bergen*, Vol. 6, pp. 1–266.

ARMSTRONG, W. 1967. The Use of Polarography in the Assay of Oxygen Diffusion from Roots in Anaerobic Media. *Physiol. Plant*, Vol. 20, pp. 240–53.

ATTIWILL, P. W.; CLOUGH, B. F. 1980. Studies of Gas Exchange in the White Mangrove. *Photosynthetica*, Vol. 14, pp. 40–7.

AUSTIN, R. B.; LONGDEN, P. C. 1967. A Rapid Method for the Measurement of Rates of Photosynthesis Using $^{14}CO_2$. *Ann. Bot.*, Vol. 31, pp. 245–53.

BANNISTER, P. 1976. Physiological Ecology and Plant Nutrition. In: S. B. Chapman (ed.), *Methods in Plant Ecology*, p. 536. Oxford, Blackwell Scientific Publications.

BASSHAM, J. A.; CALVIN, M. 1957. *The Path of Carbon in Photosynthesis*. Englewood Cliffs, N.J., Prentice Hall.

BLACK, C. C. 1973. Photosynthetic Carbon Fixation in Relation to Net CO_2 Uptake. *Ann. Rev. Pl. Physiol.*, Vol. 24, pp. 253–86.

BLACK, C. C.; CAMPBELL, W. H.; CHEN, T. M.; DITTRICH, P. 1973. The Monocotyledons, Their Evolution and Comparative Biology. III: Pathways of Carbon Metabolism Related to Net Carbon Dioxide Assimilation by Monocotyledons. *Quart. Rev. Biol.*, Vol. 48, pp. 299–313.

BJORKMAN, O. 1973. Comparative Studies on Photosynthesis in Higher plants. In: A. C. Giese (ed.) *Photophysiology*, Vol. 8, p. 63. Academic Press, New York.

BROWN, J. M. A.; OUTRED, H. A.; HILL, C. F. 1969. Respiratory Metabolism in Mangrove Seedlings. *Plant Physiol.*, Vol. 44, pp. 287–94.

CARTER, M. R. et al. 1973. *Ecosystems analysis of the Big Cypress Swamp and Estuaries*. Athens, Georgia, U.S. Environmental Protection Agency.

CATSKY, J.; SLAVIK, B. 1958. Eine neue Anwendung der CO_2-bestimmung nach Kauko zu Assimilationsmessungen. *Planta*, Vol. 51, pp. 63–9.

CHAPMAN, V. J. 1944. Cambridge University Expedition to Jamaica. *J. Linn. Soc. Bot.* (London) Vol. 52, pp. 407–533.

——. 1962*a*. Respiration Studies of Mangrove Seedlings, Part I. *Bull. Mar. Sci. Gulf. Caribb.*, Vol. 12, No. pp. 137–67.

——. 1962*b*. Respiration Studies of Mangrove Seedlings, Part. II. *Bull. Mar. Sci. Gulf Caribb.*, Vol. 12, No. 2 pp. 245–63.

CLEGG, M. D.; SULLIVAN, C. Y.; EASTIN, J. D. 1978. A Sensitive Technique for the Rapid Measurement of Carbon Dioxide Concentrations. *Plant Physiol.*, Vol. 62, pp. 924–6.

COOMBS, J. 1971. The Potential of Higher Plants with the Phosphopyruvic Acid Cycle. *Proc. Roy. Soc. Ser. B.*, Vol. 179, pp. 221–35.

COWAN, I. R.; FARQUHAR, G. C. 1977. Stomatal Function in Relation to Leaf Metabolism and Environment. *Symp. Soc. Expl. Biol.*, Vol. 31, pp. 471–505. Cambridge, Cambridge University Press.

GAASTRA, P. 1959. Photosynthesis of Crop Plants as Influenced by Light, Carbon Dioxide, Temperature and Stomatal Diffusion Resistance. *Meded. Lanbouwhogesch. Wageningen*, Vol. 59, pp. 1–68.

GALE, J.; POLJAKOFF-MAYBER, A.; KAHANE, I. 1967. The Gas Diffusion Porometer Technique and its Application to the Measurement of Leaf Mesophyll Resistance. *Israel J. Bot.*, Vol. 16, pp. 187–204.

GOLDSWORTHY, A.; DAY, P. R. 1970. A Simple Technique for the Rapid Determination of Plant CO_2 Compensation Points. *Plant Physiol.*, Vol. 46, pp. 850–1.

GOLLEY, F. B.; ODUM, H. T.; WILSON, R. F. 1962. The Structure of Metabolism of a Puerto Rican Red Mangrove Forest in May. *Ecology*, Vol. 43, pp. 9–19.

HALDANE, J. S.; GRAHAM, J. T. 1933. *Methods of Air Analysis*. 4th ed. London, Charles Griffin.

HATCH, M. D.; SLACK, C. R. 1966. Photosynthesis by Sugar Cane Leaves. A New Carboxylation Reaction and the Pathway of Sugar Formation. *Biochem. J.*, Vol. 101, pp. 103–11.

——. 1970. Photosynthetic CO_2-fixation Pathways. *Ann. Rev. Pl. Physiol.*, Vol. 21, pp. 144–62.

HATCH, M. D.; OSMOND, C. B.; SLATYER, R. O. (eds.). 1971. *Photosynthesis and Photorespiration*. New York, Wiley Interscience.

HESKETH, J. D. 1967. Enhancement of Photosynthetic CO_2 Assimilation in the Absence of Oxygen, as Dependent of Species and Temperature. *Planta*, Vol. 76, pp. 371–74.

HICKS, D. B.; BURNS, L. A. 1975. Mangrove Metabolic Response to Alteration of Natural Freshwater Drainage in South-western Florida Estuaries. In: G. Walsh, S. Snedaker and H. Teas (eds.) *Proc. Int. Symp. Biol. Management of Mangroves*, pp. 238–55. Gainesville, University of Florida.

JARVIS, P. G. 1971. The Estimation of Resistances to Carbon Dioxide Transfer. In: Z. Sestak, J. Catsky and P. G. Jarvis (eds.), *Plant Photosynthetic Production. Manual of Methods.* pp. 556–631. The Hague, Dr W. Junk, Publisher.

JOHNSON, H. B.; ROWLANDS, P. G.; TING, I. P. 1979. Tritium and Carbon-14 Double Isotope Borometer for Simultaneous Measurement of Transpiration and Photosynthesis. *Photosynthetica*, Vol. 13, pp. 409–18.

JOSHI, G. V. 1976. *Photosynthesis under Saline Conditions. Final Report on the P.L. 480 Project.* Kolhapur (India), Shivaji University. 195 pp.

JOSHI, G. V., KARKAR, M. D.; GOWDN, C. A.; BHOSALE, L. 1974. Photosynthetic Carbon Metabolism and Carboxylating Enzymes in Algae and Mangrove under Saline Conditions. *Photosynthetica*, Vol. 8, pp. 51–2.

KORTSCHAK, H. P.; HARTT, C. E.; BURR, G. 1965. Carbon Dioxide Fixation in Sugarcane Leaves. *Plant Physiol.*, Vol. 40, pp. 209–14.

LAKE, J. V. 1967*a*. Respiration of Leaves during Photosynthesis. I: Estimates from an Electrical Analogue. *Austral. J. Biol. Sci.*, Vol. 20, pp. 487–93.

——. 1967*b*. Respiration of Leaves during Photosynthesis. II: Effects on the Estimation of Mesophyll Resistance. *Austral. J. Biol. Sci.*, Vol. 20, pp. 495–9.

LAKE, J. V.; SLATYER, R. O. 1970. Respiration of Leaves during Photosynthesis. III: Respiration Rate and Mesophyll Resistance in Turgid Cotton Leaves with Stomatal Control Diminished. *Austral. J. Biol. Sci.*, Vol. 23, pp. 529–35.

LANGE, O. L. 1956. Zur Methodik der kolorimetrischen CO_2- bestimmung nach Alvik. *Ber. Deutsch. Bot. Ges.*, Vol. 69, pp. 49–60.

LIETH, H. 1960. Über den Lichtkompensationspunkt der Landpflanzen. I. *Planta*, Vol. 54, pp. 530–54

——. 1960. Über den Lichtkompensationspunkt der Landpflanzen, II. *Planta*, Vol. 54, pp. 555–76.

LUGO, A. E.; SNEDAKER, S. C. 1974. The Ecology of Mangroves. *Ann. Rev. Ecology and Systematics*, Vol. 5, pp. 40–64.

LUGO, A. E.; EVINK, G.; BRINSON, M. M.; BROCE, A.; SNEDAKER, S. C., 1974. Diurnal Rates of Photosynthesis, Respiration and Transpiration in Mangrove Forests of South Florida. In: F. B. Golley and E. Medina (eds.), *Tropical Ecology*. New York, Springer Verlag.

MEIDNER, H. 1967. Further Observations on the Minimum Intracellular Space Carbon Dioxide Concentration (I) of Maize Leaves, and the Postulated Roles of 'Photo-respiration' and Gycollate Metabolism. *J. Expl. Bot.*, Vol. 18, pp. 177–85.

MONTEITH, J. L. 1963. Gas Exchange in Plant Communities. In: L. T. Evans (ed.), *Environmental Control of Plant Growth*, pp. 95–112. New York, Academic Press.

MOORE, R. T.; MILLER, P. C.; ALBRIGHT, D.; TIEZEN, L. L. 1972. Comparative Gas Exchange Characteristics of Three Mangrove Species during the Winter. *Photosynthetica*, Vol. 6 No., 4, pp. 387–93.

MOORE, R. T.; MILLER, P. C.; EHLERINGER, J.; LAWRENCE, W. 1973. Seasonal Trends in Gas Exchange Characteristics of Three Mangrove Species. *Photosynthetica*, Vol. 7, No. 4, pp. 387–94.

MORESHET, S.; KOLLER, D.; STANHILL, G. 1968. The Partitioning of Resistances to Gaseous Diffusion in the Leaf Epidermis and the Boundary Layer. *Ann. Bot. N. S.*, Vol. 32, pp. 695–701.

NIEMAN, R. H. 1950. Salt Activated Respiration in Excised Roots of the Garden Pea. *Bot. Gaz.*, Vol. 118, pp. 1–13.

PANNIER, F. 1962. Estudios fisiologicos sobre la viviparia de *Rhizophora mangle* L. *Acta Cientif. Venezolana*, Vol. 13, pp. 184–97.

PANNIER, F.; PANNIER, R. F. 1971. Compensacion del incremento respiratorio de raices de *Rhizophora mangle* inducido por elevadas concentraciones salinas, por los parametros luz y temperatura. *Resum. IV. Symp. Latinoamer. Fisiol. Veg. Lima, Peru.*

PATTERSON, M. S.; GREENE, R. C. 1965. Measurement of Low Beta Emitters in Aqueous Solution by Liquid Scintillation Counting of Emulsions. *Analyt. Chem.*, Vol. 37, pp. 854–7.

PAVLETVIC, Z.; LIETH, H. 1958. Der Lichtkompensationspunkt einiger immergruener Pflanzen im winter und im Fruehjarh. *Ber. Deutsch Bot. Ges.*, Vol. 71, pp. 309–14.

PORATH, E.; POLJAKOFF-MAYBER, A. 1964. Effect of Salinity on Metabolic Pathways in Pea Root Tip. *Israel J. Bot.*, Vol. 13, pp. 115–21.

SESTAK, Z.; CATSKY, J.; JARVIS, P. G. 1971. *Plant Photosynthetic Production. Manual of Methods.* The Hague, Dr W. Junk, Publishers. 800 pp.

SHIMSHI, D. 1969. A Rapid Field Method for Measuring Photosynthesis with Labelled Carbon Dioxide. *J. Epl. Bot.*, Vol. 20, pp. 381–401.

STEINER, A. A. 1968. Apparatus for Measurement of Oxygen Consumption by the Root System of a Plant. *Plant and Soil*, Vol. 28, pp. 173–6.

STOCKER, O.; VIEWEG, G. H. 1960. Die Darmstaedter Apparatur zur Mommentanmessung der Photosynthese unter oekologischen Bedingungen. *Ber. Deutsch. Bot. Ges.*, Vol. 73, pp. 198–208.

THOMAS, M. D. 1933. Precise Automatic Apparatus for Continuous Determination of Carbon Dioxide in Air. *Ind. Eng. Chem., Anal. ed.*, Vol. 5, pp. 193–8.

TREGUNA, E. B.; KROTKOV, G.; NELSON, C. D. 1961. Evolution of Carbon Dioxide by Tobacco Leaves During the Dark Period Following Illumination with Light of Different Intensities. *Can. J. Bot.*, Vol. 39, pp. 1045–56.

TROLL, W.; DRAGENDORFF, O. 1931. Über die Luftwurzeln von *Sonneratia* L. und ihre biologische Bedeutung. *Planta*, Vol. 13, pp. 311–473.

UMBREIT, W. W.; BURRIS, R. H.; STAUFFER, J. F. 1964. *Manometric Techniques.* 4th ed. Minneapolis, Burgess.

WOLLUM, A. G.; GOMEZ, J. E. 1970. A Conductivity Method for Measuring Microbially Evolved Carbon Dioxide. *Ecology*, Vol. 51, pp. 155–6.

ZUNKER, E.; KREEB, K. 1970. Untersuchungen zur potentiometrischen Messung der Photosyntheserate. *Ber. Deutsch. Bot. Ges.*, Vol. 83, pp. 245–57.

13

Mangrove physiology: water relations

Federico Pannier
Universidad Central de Venezuela
Facultad de Ciencias, Escuela de Biología
Departamento de Botánica, Apartado 80.390
Caracas, Venezuela

Research on the water relations of mangroves is of key importance in understanding the role of water in the physiology of growth and nutrition of individual species, as well as in the analysis of the water utilization efficiency of the community as a whole.

A quantitative approach to the water balance, taking into consideration the measurement of its main components in the context of the substrate–plant–air continuum, has to be undertaken. This means, as was duly pointed out by Miller et al. (1975), that to obtain comparable data of the water relations of mangroves at the individual level, one must first establish a model of water flow from the substrate to the leaf, and to the atmosphere, and of its regulation by leaf water potential changes. Such models of water flux, like that developed by Waring and Running (1976) for conifers, should be of great value in analysing water relationships in mangrove trees.

Nevertheless, it must be emphasized that mangroves have to be considered as a special group of halophytes, which, despite the high water-retaining capacity of the saline substrate on which they live, can maintain a good water balance, due to diverse regulating mechanisms such as stomatal behaviour, osmotic adjustment, succulence and salt excretion.

Since these regulating mechanisms have not been sufficiently studied as to their significance for the global water balance, and considering that even data on the different components of water flux are still very scarce, much research remains to be done before a comparative analysis of the water use of the individuals belonging to different species can be undertaken. However, an important methodological approach has been achieved by Miller et al. (1975).

The methods for studying water relations in mangroves are: (a) measurement of the components of water status; (b) measurement of transpiration; (c) stomatal measurements; (d) measurement of water uptake by the roots; and (e) measurement of water flow in the xylem.

Since methods have been described in detail in several well-known methodology manuals on plant water relationships (see Slavik, 1974; Kozlowsky, 1968–76; Eckardt, 1965), references to such publications, as well as comments on their application to specific mangrove research topics, will be given.

Measurement of the components of water status

Knowledge of the water status of plant tissues represents a valuable means of understanding the physiological responses of mangroves at the species level, in relation to the changing climate and substrate conditions of their habitat.

Its description should be approached by the measurement of the widest range of its components, that is: (a) water content, and water saturation deficit; (b) total water potential; (c) osmotic potential; and (d) turgor pressure. Nevertheless, despite few exceptions of global measurements of water status in mangroves (Miller et al., 1975) most research has been oriented towards measurement of an isolated component, such as osmotic potential, for which a general response pattern for the limited number of examined species has been well established (Harris and Lawrence, 1917; Walter and Steiner, 1936; Navalkar, 1940; Bharucha and Navalkar, 1942; Biebl and Kinzel, 1965; Scholander, 1968).

The significance of water status, extremely well discussed by Hsiao (1973) and Richter (1976), goes far beyond its use as an indicator of internal physiological water stress, and should constitute a tool for the establishment of evolutionary and adaptative differences in plants of extreme habitats, to which group mangroves belong.

It seems, therefore, advisable to direct more attention in the future towards obtaining precise information on the patterns of diurnal, seasonal and developmental changes of leaf water status for a broad range of mangrove species. These data should be employed first for absolute comparisonal purposes of the behaviour between the individual species, and gained under strictly controlled environmental conditions (growth chamber experiments) before going into the complex analysis of the individual response to the environmental interactions existing in the natural habitat.

Concomitantly, the link between water status and metabolism, through the measurement of changes in the rates of synthesis of carbohydrates and nitrogen compounds, and the link with hormone and enzyme activities, have clearly to be established, as well as the effect on the regulation of physiological control mechanisms, such as stomatal control, salt uptake and excretion (see also corresponding sections of this chapter). Measurement of differences of water status between tissues involved in internal water movement to sinks—such as represented by growing viviparous embryos (Pannier, 1962) or to root primordia in developing hypocotyles—should be also of outstanding physiological interest.

MEASUREMENT OF WATER SATURATION DEFICIT AND RELATIVE WATER CONTENT

Water saturation deficit (WSD) is expressed by the formula of Stocker (1928, 1929) as:

$$\text{WSD} = \frac{(\text{fully turgid weight} - \text{fresh weight})}{(\text{fully turgid weight} - \text{dry weight})} \times 100$$

and has been extensively measured as an index of water deficit in plants, despite the claims of Slatyer (1960) and Slatyer and MacIllroy (1961) as to its insensitivity during measurements performed at high values of water potential (γ).

However, the possibility of measuring the water potential (see method below) at the same time as determining the WSD, overcomes this incertainty since the values of γ can be plotted against those of WSD (examples, in Miller et al., 1965; Jones et al., 1980) and the resulting type of curve evaluated.

Knowing the actual weight, and using the data for full turgor weight obtained after exposure of the plant or plant part to optimal water saturation for 24–48 hours, and for dry weight, after drying the plant material to constant weight in an oven at 85 °C, the terms can be substituted in the equation, being referred to dry weight of the sample, and preferably, when working with leaves, to the leaf area.

According to the plant parts used, the following general considerations may be made: (a) mangrove branches, even when collected by cutting underwater, and subsequently exposed to water vapour saturation exhibit visible wilting symptoms very rapidly, so that exposure times should not exceed a few hours; and (b) isolated mangrove leaves with their petioles in water inside saturation chambers under diffuse light should not be exposed longer than twenty-four hours to air saturation, because of visible damage occurring (black spots in *Rhizophora*, apical injury in *Avicennia*).

Nevertheless, the adequate time exposure conditions for (a) and (b), according to the behaviour of the species and of its corresponding light compensation intensity, has still to be worked out.

Although leaf discs of mangroves have apparently not been used for measurement of WSD, an attempt to calibrate the 'floating leaf-disc method' developed by Weatherly (1950), taking into account the improvements and the sources of error described by Barrs and Weatherly (1962) and Catsky (1965), should be undertaken.

Using this floating leaf-disc method, the term suggested by Barrs (1968) to express the water deficit should be correctly stated as relative water content (RWC) and the following equation employed:

$$\text{RWC} = \frac{(\text{fresh weight} - \text{dry weight}) \times 100}{(\text{fully turgid weight} - \text{dry weight})}$$

Besides the direct measurement of water content by the determinations of WSD and RWC, a number of indirect methods measuring changes in leaf thickness (Gardner and Ehlig, 1965) circumferences of fruits (Bartholomew, 1927), and trunk diameter (McDougal, 1921; Alvin, 1975) have been generally applied in plants (see reviews by Barrs, 1968; Slavik, 1974; Kramer, 1964).

In mangroves, measurement of leaf thickness has been used to calculate the relative amount of water-storage tissue of several species (Mullan, 1931; Biebl and Kinzel, 1965), and Pannier and Pannier (1975) have been able to correlate growth measurements of *Rhizophora* seedlings with their water content, in order to establish precise reference points for the description of development stages.

MEASUREMENT OF TOTAL WATER POTENTIAL

Two basic types of methodologies to measure total water potential, γ, can be distinguished: the liquid-phase and vapour-phase methods.

Liquid-phase methods

The principle of the liquid-phase method consists in bringing tissue samples into direct contact with a gradient of solutions of known concentration, and finding the concentration where no exchange of water between sample and solution occurs. The equalizing osmotic concentration of this solution is assumed to be equal to the γ of the tissue.

A great diversity of liquid-phase methods, using changes of the properties of the plant samples, as well as of the test solutions employed, have been developed, but because of their laborious and time-consuming exigencies, generally low accuracy and a great number of sources of error (see Barrs, 1968), they have not been applied in mangrove research. Only the following method which, having been applied to mangroves, gives an approximate estimation of the γ of stems will be described.

The 'closed burette technique' (Scholander, 1962). The method measures the water absorption rate from a closed burette applied externally to the woody tissue of a stem. Water absorption against the existing vacuum inside the burette is considered to be driven by the existence of a negative hydrostatic pressure in the xylem, and assuming equivalence between hydrostatic pressure and osmotic potential, and knowing that xylem water of mangroves is practically deprived of solutes, an estimation of γ can be obtained.

The procedure consists of removing, by means of a cork borer, a disc of bark, and placing on the exposed surface a metal or plastic button with a 5 mm bore and an 'O' ring tightly fixed to the xylem with a hose clamp. After filling the bore with water, the xylem is scooped out with a sharp drill which passes through the water, the shavings are flushed out and a burette connected. Air-bubbles inside the burette are evacuated by forcing them out with a large syringe applied to the open end of the burette. Rate of water absorption is then determined and the burette closed, enclosing a known volume of air. The rate of water absorption is again observed. After absorption stops, the gas volume is read, water is admitted to the burette and gas volume again read. From the reduction in the gas volume the pressure can be calculated.

Values measured by this method in *Rhizophora mangle*, *Avicennia nitida*, and *Languncularia racemosa* range between 0.4 and 0.7 bar.

An improved methodology based on it and called by Scholander the 'delta pressure technique' (described in the same paper) is more sensitive due to the utilization of a microburette and to smaller cuts to the xylem tissue.

Plotting water-absorption rates (vol/min) against the calculated pressures, and extrapolating to a control value close to ambient pressure, an estimation of actual stem pressure = stem γ is obtained.

An extention of the use of the method to measure γ of stem in other mangrove species, as well as its possible calibration against the thermocouple psychrometric method for measurement γ *in situ* with sensing elements implanted in the stems (Michel, 1977, 1979), is suggested.

Vapour-phase methods

Vapour-phase methods make use of the existing quantitative relationship between water potential of a solution γ, and the vapour pressure above it: $\gamma = (\mathrm{RT/M})\,(\ln p/p_0)$.

By exposing the tissue in a closed space near (i.e. not in contact with) the solution of known γ, or preferably having the possibility of using electric transducers in sealed chambers without any salt solution but only the tissue sample, the point of relative vapour-pressure equilibration, and therefore the equivalent γ, can be measured.

As in the case of liquid-phase methods, a great number of techniques are available, whose suitability for use in the field and laboratory, accuracy and reality is discussed by Barrs (1968) and Slavik (1974). From these methods, the most suitable for mangrove research are undoubtedly: (a) the direct pressure-bomb method; (b) the psychometric method; and (c) the gasometric 'null-point compensation' method.

Pressure-bomb method. This method (Scholander et al., 1964, 1965, 1966) was specifically developed for the determination of xylem γ in mangroves, but has found general acceptance as a useful field method for estimating water potential in woody plants.

The procedure consists in fixing leafy twigs inside a closed steel chamber, with the cuts outside, and continuously increasing the gas (nitrogen) pressure, until sap begins to exude from the cut surface. The amount of pressure needed to yield the liquid is considered equivalent to the γ of the xylem before the twig was cut off the tree.

The errors involved in the use of this method have been well discussed by numerous workers (Tyree et al., 1978) and it is generally recommended (Kaufman, 1968; Frank and Harris, 1973) that calibration curves be established for each species and age, and that the values obtained be compared with known leaf water potentials obtained through psychrometric methods.

Psychrometric method. The principle underlying the psychrometric method is to measure, by means of a thermocouple junction or a thermistor enclosed in a gaseous system maintained at constant temperature, the wet bulb depression which reaches equilibration with the plant sample.

From the diverse types of measuring instruments available, the most appropriate for accurate leaf γ determination in the field is the thermocouple psychrometer of Spanner (1951) which has been commercially developed by Wescor Inc., Logan, Utah. This instrument has been very useful for measuring leaf-water potential in mangroves (Miller et al., 1975; Pannier and Pannier, 1973) and is recommended, despite certain disadvantages such as sometimes very extended equilibration times, and surface-salt contamination effects when using salt-exuding leaves which could result in a steep decline of the response of the instrument. A very valuable critical discussion concerning the interpretation of foliar γ values obtained by the Spanner-psychrometer has been provided by Oertil et al. (1975).

For measurement of γ in roots a special root psychrometer chamber was developed by Fiscus (1972), and for stem γ a microsensing element has been described by Michel (1977, 1979); but the adequacy of both devices for measurement in mangroves has not yet been proved.

Gasometric 'null-point compensation' method. For those laboratories having the facility to use expensive gas-analytical equipment for continuous measurement of photosynthesis and transpiration rates, this method, by the inclusion of an ingenious compensating valve system, is able to measure the equilibrating vapour pressure of the external atmosphere at which net flux of water vapour between the plant and the surrounding air attains zero. Details of the measuring system can be found in the original paper of Koller and Samish (1964).

MEASUREMENT OF OSMOTIC POTENTIAL

The osmotic potential component π of total water potential has been regarded, long before the general use of total water potential γ, as an indicator of internal tissue water deficit, as well as for establishing gradients from substrate to plant and between different plant sites, knowing the nature of the water-flow.

As with the preceding description of methods for γ determination, only a few techniques can be successfully applied for the measurement of π in cell sap obtained from mangrove tissues; the cryscopic method, the psychrometric method and the improved refractometric method are the most relevant here, since the limiting plasmolysis method by direct microscopical observation, as employed initially by Von Faber (1913) in his studies on mangroves, has been proved to give inconsistent values.

Cryoscopic method

This method uses the determination of the freezing-point depression ΔT, as a measure of the concentration of dissolved solutes in the cell sap, and for the procedure several cryoscopic devices described by Walter (1931), Slavik (1974) and Kreeb (1965) can be used. It is also possible to perform these determinations on any of the clinical cryoscopes ('osmolimeters') commercially available, so long as care is taken to dilute the sap samples down to the rather narrow measuring range for which these instruments are usually constructed.

In mangroves, the extraction of the minimum amount of cell sap required (about 2 ml) for measurement is generally very tedious, especially with *Rhizophora* leaves because of its mucilagenous consistency, and the use of a pressure bomb or hydraulic oil press is strongly recommended. Sometimes, when the cell sap has a high carbohydrate content and a very viscous consistency, such as in the case of the cotyledon body of *Rhizophora*, freezing point determination of undiluted cell sap is simply not possible. In this case, an alternative method, the tensiometric thermoelectric method described by Van Andel (1952), has been successfully applied (Pannier, 1962).

Psychrometric method

This method, in the form of the Wescor Thermocouple Psychrometer, is very adequate for easy direct determination of π, since the cell sap of previously frozen or boiled leaf tissue can be squeezed directly on to small filter-paper discs, which are handled in the same way as leaf discs used for the determination of water potential. A great number of samples (20–30 per hour) can be handled with a single sample chamber.

Freezing of the plant tissue, prior to squeezing, can be performed in sealed test-tubes which, after exposure to liquid nitrogen, may be stored in a freezer at -20 °C until required for determination.

Dilution errors in the osmotic potential data obtained from sap of thawed tissue have to be considered for very precise measurements (see Boyer and Potter, 1973; Tyree, 1976), but they can be neglected for coarse field studies.

Improved refractometric method

This method (Shimshi and Livne, 1967) is very well suited for field measurements, and combines the advantage of the use of very small amounts of cell sap obtained by squeezing the tissue with an adequate hand hydraulic press (described in the original paper) with the direct measurement of its electrical conductivity.

In assuming that the osmotic potential is the sum of the concentrations of sugars and the electrolytes present in the cell sap, the former is detected by measuring the refractive index n_D and comparing the data with tables expressing the n_D values for known sucrose solutions; the latter, by measurement of the electrical conductivity of the same sample, using a specially designed platinum microsensor

MEASURE OF TURGOR PRESSURE

Turgor pressure, P, is a property sustained by the rigid cell wall of plant cells, which is directly involved in the mechanical support of plant tissues, besides being an important parameter in the regulation of water and ion transport, and of growth.

Relatively well studied at the cellular level (see Zimmerman, 1977) because of the possibility of its direct measurement in adequate biological material, such as giant algal cells (*Nitella*, *Chara*, *Valonia*), and more recently in bladder cells of the crassulacean *Mesembryanthemum crystallinum* (Steudle et al., 1977), turgor pressure is extremely difficult to measure in higher plant tissues, despite some rare cases of techniques employed to obtain values from the phloem of trees (Buttery and Boatman, 1966; Hammel, 1968). Measurements performed with the pressure-bomb technique (Scholander et al., 1964) in mangroves have shown that the xylem sap has to be considered in these plants as being in a state of tension (i.e. negative hydrostatic pressure) instead of compression as in most intensely transpiring plants, and it has been assumed that for this reason the tissues develop negative turgor, a characteristic generally shown by plants with very low water potential γ, a fact which has been rejected by Tyree (1976) as a fallacy.

Data on turgor potential of leaves of Florida mangroves was obtained by Miller et al. (1975) indirectly by calculation, i.e. by subtracting the measured values of osmotic potential from the also-measured data of total water potential. These data showed in all the three species studied a decreasing pattern of turgor pressure potential with increasing water deficit, decaying from the initial positive value to zero, with *Laguncularia* being the only species to reach negative turgor values.

Since turgor potential analysis has been improved meanwhile (Tyree et al., 1978) the following method should prove useful for further mangrove research.

Calculation of turgor potential from leaf γ and the pressure-volume curve

The method consists basically in calculating the turgor potential P, by substituting in the osmotic equation: $-\gamma = -\pi(\pm)P$ (γ_m not being considered), the values for leaf-water potential being measured by one of the methods described above and the values of the osmotic potential by the following procedure described by Jones and Turner (1980).

Sap from a leaf enclosed in a pressure bomb is squeezed out by increasing gradually the applied pressure in increments of 0.2 MPa.

The squeezed sap volume is collected for 15–25 minutes at each applied pressure and the balance pressure recorded. The squeezed sap is collected on filter-paper discs with vials and the weight increase of the vial (*g*) is taken as a measure of the volume of sap squeezed (ml). The inverse of the balance pressure (y^{-1}) (*y*-axis) is plotted against the cumulative volume (*v*) (*x*-axis), and a least-squares linear regression is fitted to the linear part of the curve. This line represents the relationship between the inverse of the leaf osmotic potential (π^{-1}) and the volume expressed, *V*. The *y*-intercept of this line ($\pi_0{}^{-1}$) gives the inverse of the leaf osmotic potential at full turgor and the *x*-intercept gives the total volume of osmotic water in the leaf (V_{ref}).

The relative osmotic water content (W_{ro}) is calculated accordingly: $W_{ro} = 100\,[100(V_{ref} - V)/(V_{ref}]$.

In addition, from the obtained values the bulk volumetric elastic modulus ε (Philip, 1958), an expression used to define turgor changes responses of the tissue to osmotic stress induced in the environment, can be calculated as the slope of the relationship between leaf-turgor potential and relative osmotic water content:

$$\varepsilon = \frac{\mathrm{d}p}{\mathrm{d}V/V}\ \frac{\Delta P}{\Delta W_{ro}} \times 100$$

where *P* is the turgor potential.

Measurement of transpiration

From the pioneering ecological research on mangrove transpiration of Walter and Steiner (1936) followed by that of Lamberti (1969), Moore et al. (1972, 1973) and Miller et al. (1975), we know that these plants belong, with few exceptions (see also comparative tables of transpiration rates in Adriani, 1965, and Chapman, 1976), to the group of low transpiring halophytes.

This behaviour has been associated mainly with the saline conditions of their substrate, since transpiration is proportionally reduced as salinity increases (Bowman, 1917; Scholander et al., 1962, 1965); but the regulation of the water loss from leaves to the atmosphere by environmental factors prevailing in the habitat has been increasingly recognized (Miller et al., 1975; Pannier and Pannier, 1967). In this context, it is necessary to underline that for comparison of mean transpiration of different mangrove species, the values of the physical evaporation force of the air surrounding the leaf at the precise moment of measurement of the water loss, obtained by a comparable method such as the Piche-Evaporimeter (see Roth, 1961) must be known, as well as the specific character of structures of the leaves to be compared (distribution and size of water-holding tissues, type of distribution of stomata, presence of glands, hairs, etc.). Such analysis will undoubtedly show significant species differences of transpirational behaviour of mangroves, requiring a better knowledge of the mechanism of water loss of the individual, as well as of the community.

Besides the need for a better understanding of the interrelationship between absolute water loss and the mechanism of cuticular transpiration, stomatal regulation and salt regulation, further specific research on mangrove transpiration should consider the following:

1. Comparative measurement of water loss in different developmental stages of the diverse mangrove species.
2. Comparative measurement of water loss by different organs of the same individual (such as pneumatophores, buds, flowers and hypocotyls) related to the total amount of water transpired by the plant.
3. Measurement of transpiration changes (stomatal and cuticular) according to salinity fluctuations of the substrate and the local climatic conditions.
4. Measurement of the time course of transpirational deposition of salts in leaf tissues, in terms of absolute values of transpired water.
5. Measurement of the vertical distribution pattern of leaf transpiration in a canopy, and in accordance to leaf angle position.

METHODS SUITABLE FOR MEASURING WATER LOSS OF ISOLATED LEAVES AND OTHER PLANT PARTS

Methods to measure instantaneous transpiration in situ

Quick-weighing method (Huber, 1927; Stocker, 1929). This method consists in measuring, by means of a torsion balance, the weight loss by a leaf or another plant organ, over very short time intervals. Although extensively used in the past, the method has been found to be very prone to errors introduced by the hydroactive closing of the stomata and by the influence of microclimatical changes in the leaf's environment. Nevertheless, it can be used for leaf-transpiration measurements under certain circumstances, adopting the criteria discussed by Slavik (1974).

For the determination of water loss by plant organs having no stomatal regulation problems, like pneumatophores, hypocotyls, buds and flowers, the method can be still of value.

Method to measure water-retention capacity of detached leaves (Fukuda, 1935; Hygen, 1951, 1953; Bannister, 1964). This method is based on the analysis of the water-loss curves of single detached leaves, obtained by exposing the leaves for several hours to the conditions of their natural environment, and measuring continuously the weight differences by means of a sensitive torsion balance.

Due to its simplicity, this method can serve as a rough estimation of the cuticular and stomatal transpiration resistances of the leaf, and is useful for screening structural differences produced by growing plants on different substrates under controlled conditions, as well as for obtaining evidence of the variation in the water-holding capacities of individuals of the same species under different micro-environmental conditions in their natural habitat.

Gas-analytical method (Decker and Wien, 1960). Employing an infra-red gas analyser adapted to water-vapour measurement, either in a closed or open-air circulation system, instantaneous changes of the water loss can be detected. The method is rather expensive, difficult to handle because of the frequent calibration required, and has its limitations for field-work. If possible, it should be employed simultaneously with the gas analytical method for measuring CO_2 exchange, as described by Parkinson (1968).

Differential psychrometric method (Slayter and Bierhuizen, 1964). This method has been found to be very sensitive and responds well to the continuous measurement of transpiration. It detects differences of absolute humidities between two air-streams: one coming from the plant enclosure, and the other represented by air chosen for control. Because of the need for accurate thermal control of the airstreams the method has limitations for use in the field.

Lithium chloride sensors (Lieneweg, 1955; Gaastra, 1959). Methods employing commercially available humidity sensors for measuring changes of water-vapour pressure in flowing airstreams through changes of electrical resistance of LiCl crystals should be viewed with caution for transpiration measurement, since errors due to variations of the accuracy of the sensor depending on the range of relative humidity can be introduced, once adapted to a well-defined measuring range. This method should be useful for recording the time course of cuticular water loss.

METHODS SUITABLE FOR MEASURING WATER LOSS OF WHOLE PLANTS

Method employing micro-lysimeters (Hesse, 1954*a*, *b*; Hanks and Shawcroft, 1965; Courtin and Bliss, 1971). This method should be very suited to the measurement of water loss from developing mangrove seedlings in their natural habitat. It consists of the continuous and accurate determination, over long periods of time, of weight differences of small plant containers in which seedlings are planted and periodically watered.

Detailed handling of this method for transpiration measurement of mangroves in the field has still to be worked out, taking into account the tidal effects on capillary capacity of the substrate, avoidance of contamination of the vessels by organic material, and correction for the increase of dry matter of the plants due to growth.

Methods for measuring transpiration resistances

Determination of steady-state transpiration rates by gas analysis (Raschke, 1958;

Jarvis, 1971; Jarvis et al., 1967). This method to calculate water-diffusion resistances (r_{H_2O}), mesophyll resistance (r_m) and intracellular resistances (r_{int}) has been successfully employed by Moore et al. (1972, 1973) in their measurement of the gas exchange in Florida mangrove species. The authors used a Siemens 'Sirigor' environmental controlled cuvette and a dewpoint hygrometer in an open system. (For details see their papers.)

Determination of transpiration by non-steady-state methods (Jarvis 1971). The fundamental principle on which the great diversity of methods are based, consists in measuring inside a plant enclosure the increase, in a certain range and during a short time, of the air humidity due to water loss from the plant surface, with the air and leaf temperatures strictly controlled. According to the facilities and the character of the specific research problem, an adequate device has to be chosen.

The following relatively simple method has been developed by Miller et al. (1975) to measure the transpirational behaviour of Florida mangroves. It is based on the calculation of a combined leaf and air boundary conductance value (h_w) obtained during the continuous measurement of water loss under exactly controlled environmental conditions.

Considering the conductance as the reciprocal of the sum of the leaf and air boundary-layer resistances, the following equation for calculating h_w is utilized:

$$h_w = \Delta w/[\Delta t(x_l - x_a)A]$$

where w = weight loss in a time period;
t = length of time period in seconds;
x_l = saturation vapour density at leaf temperature;
x_a = vapour density of the air measured with a hygrometer; and
A = leaf area.

The procedure consists in removing three to five leaves from a previous illuminated branch and weighing them inside a chamber with known air movement every two minutes on a precision balance until reaching constant weight.

During the weighing procedure, leaf temperature, as well as air temperature and vapour density inside the chamber has to be measured by thermocouples and a hygrometer. The rate of air movement through the chamber for calculating air resistances has to be measured, and further, knowing the total exposed leaf area, the conductance of the air boundary layer (h_a) can be calculated.

After reaching constant weight, the leaves are oven-dried to obtain dry weight. The leaf water content is calculated assuming the initial weight as being the turgid weight.

With the data obtained, h_w (see equation above) can be calculated. In a further step, h_w is substituted with the previously calculated value for air boundary layer conductance (h_a), in the following corrected equation for leaf conductance $h_l = h_a \times h_w/(h_a - h_w)$.

The obtained values should be corroborated with measurements employing a diffusion porometer under field conditions. A very suitable apparatus for this is the previously described double diffusion porometer (see Chapter 12) which has been utilized to measure transpirational behaviour of *Rhizophora* seedlings subjected to salinity treatments (Pannier, Mizrachi and Ting, unpublished).

Calculation of transpiration from changes of the leaf energy balance (Miller and Gates, 1967). This method, which needs an integrated knowledge of the different processes involved in heat exchange of the individual leaf, and a very accurate method to measure leaf temperature changes, requires special radiation measuring equipment which is described in detail in the paper of Miller (1972) concerning water balance calculation of a mangrove community. For an excellent review on the relationships between transpiration and energy exchange, the reader is referred to the article by Gates (1976).

Stomatal measurement

Measurement of the differences in stomatal behaviour between mangrove species is important, not only for the evaluation of the transpirational and photosynthetical performances of these plants, but also to gain insight into their adaptation to their natural habitat.

The analysis of stomatal functioning in a continuously changing environment poses, nevertheless, extremely difficult problems because of the complex interactions between the physiological status of the leaf and the whole range of environmental factors (for discussion see Meidner and Mansfield, 1968; Raschke, 1975; Hall et al., 1976).

According to the few data available on stomatal regulation in mangroves, there seems to be a tendancy for the stomata of several Indian mangrove species (Joshi, 1976) to remain open only during the early hours of the day.

This behaviour is apparently not shared by *Avicennia*, which is known as the most intense transpiring genus of mangroves, and has a steadily declining closure movement during the day after a first maximum peak value in the mid morning hours, only to achieve a second peak in the late afternoon (Lewis and Naidoo, 1970).

Interpretation of the second peak by these authors was related to an opening of the stomata due to a gaining of turgor of the guard cells as a result of the relief of soil water tension by the incoming tide. However, this 'tidal effect' on stomatal movement could not be confirmed by Steinke (1979), even employing the same method.

From the values given by Miller et al. (1975) for leaf conductance of three Florida mangrove species, a greater sensitivity of stomatal opening of *Avicennia*, in comparison to *Laguncularia* and *Rhizophora*, can be inferred. The authors interpreted this difference of leaf conductance on the basis of species-specific mechanisms to maintain leaf turgor potential under the prevailing salinity conditions of the substrate.

Nevertheless, due to the scarcity of available data on stomatal movement obtained under diverse experimental conditions by different, not directly comparable methods, there is an urgent need for more research on this aspect, and this should take account of the following.

1. An exact description of the physical environment in which the measurement is performed (laboratory/natural habitat), and the recording of any changes of the individual environmental parameters involved (light intensity and quality, air temperature and humidity, wind velocity, nubosity, as well as soil moisture and salinity).
2. The selection of an adequate measuring method, which will vary according to the kind of research intended (for transpirational, as well as photosynthetical related stomatal measurements, diffusive conductance methods should be always preferred).
3. The use of the leaf material under strict criteria of leaf morphology (succulence, leaf symmetry) leaf physiology (age, water status) and stomatal structure and distribution.

In the context of a more complete description of stomatal behaviour of the different mangrove species growing in their natural habitat, the following specific problems concerning stomatal response need investigation.

1. The effect of differences of relative humidity and temperature on members of the same species in wind-protected and exposed habitats.
2. Light and CO_2-concentration effects in vertical gradients of the canopy.
3. Effects due to primary osmotic potential changes of the substrate.
4. Effect of photo- and thermoperiodic regional differences.
5. Regulation by endogenous hormone levels caused by environmental and phenological changes.

METHODS TO MEASURE STOMATAL APERTURE

Replica methods

Replica methods are based on microscopic examination of impressions of stomata obtained after applying solidifying substances to the leaf surface.

Method of negative impression (Buscalioni and Pollacci, 1901; Wenzel, 1939). The solid replica of the leaf surface can be obtained after applying to the leaf any of the following substances: (a) commercial solution of collodion in chloroform (4 per cent); (b) 5 per cent celloidin solution in ethanol-ether mixture (1 : 2); (c) concentrate solution of cellulose acetate in ethylacetate; (d) colourless nail varnish. A convenient feature of the method is that the replica obtained in the field, adequately identified, can be stored and examined in the laboratory.

However, leaves with deeply immersed stomata, or covered with hairs or scales, are not suitable for the method.

Positive impression method (Sampson, 1961). After having obtained a negative replica of the leaf surface, with fluid silicone rubber, a positive replica is made by painting the negative replica with one of the above mentioned solutions. The stomata impressed in the peeling can be seen exactly as they would be seen directly on the leaf.

Infiltration method

This method (Molisch, 1912; Alvim and Havis, 1954) is based on the measurement of the time required for liquids of low surface tension to infiltrate into the leaf. Widely used in field work, when other more precise methods are not available, this method is very subjective and cannot be applied for comparative purposes between species.

Porometric methods

A great variety of methods to measure the resistance to diffusive flux of water vapour offered by stomata of transpiring leaves is available. A few of them have been described earlier in relation to CO_2-exchange measurements (see Chapter 12) and to transpiration.

A useful overview of the existing methods, allowing one to choose the most adequate according to the circumstances, is given in Table 13.1 (after Slavik, 1974).

Details of the methods may be found in Slavik (1974), Jarvis (1971), as well as in the original papers referring to the design of low-cost mass-flow porometers (Alvim, 1965; Meidner, 1970; Gregory and Pearse, 1934; Heath and Russel, 1951; Raschke, 1965).

Excellent theoretical accounts on stomatal response to environmental and internal factors are given by Burrows and Milthorpe (1976), Hall et al. (1976) and Cowan and Farquhar (1977).

Measurement of water uptake by the roots

Mangroves live in habitats where poor soil aeration and accumulation of toxic substances, as well as high and fluctuating salinity, raises interesting questions regarding the mechanism of root water absorption.

Although a great variety of morphological and anatomical structures, which presumably allow the plants to cope with these adverse environmental conditions, have been described (see Chapman, 1976) very little research in mangroves has been done to explain at all satisfactorily how water enters the plant (Walter and Steiner, 1936; Scholander, 1968).

Experimental evidence gained by the combined measurement of negative hydrostatic pressures in mangrove twigs, with chloride determinations in the expressed sap and the water flow response of root-poisoned plants with carbon monoxide and dinitrophenol (Scholander, 1968) point to a mechanism whereby a separation of freshwater from sea-water is achieved in the root membranes, in the sense of an ultrafiltration process independent of energy requirement and driven by the hydrostatic pressure difference provided by the transpirational stream.

Since new concepts have been developed for analysing driving forces of water movement into, and inside, roots (see reviews by Caldwell, 1976; Greacen et al., 1976; Weatherly, 1970, 1974; Newman, 1974), as well as new approaches to salinity effects on root water absorption (O'Leary, 1974, 1969), an urgent revision of this problem is advised.

Besides physiological aspects, such as the need for an exact localization of the water-flow resistance in the very diversified root system of mangroves, the response of the flow resistance to specific environmental factors, the measurement of diurnal periodicity of root water flow, and of possible links with root metabolism, the ecological significance of root density for the growing shoot and its global water requirement, as well as the search of possible changes of the water-absorption pattern according to the development from the seedling stage to the adult tree, have to be investigated.

Since no specific recommendations concerning adequate methods for water-absorption measurement in mangrove roots can be given, the reader is referred to the rich diversity of porometric devices described in Slavik (1974).

Measurement of water flow in the xylem

Measurement of water flow in the xylem can be a useful means of investigating the diurnal pattern of the linear velocity of the transpiration stream, that is, the translocation velocity of the water contained in the xylematic pathway of the plant.

Application of existing methods should permit the following:

1. Correlation of the changes in internal water velocity with the stomatal opening behaviour of the species.
2. Measurement of internal water flow, independent of stomatal movement, in organs of mangroves bearing mainly lenticels, such as hypocotyls and pneumatophores.
3. Quantitative assessment of the internal water-flow interrelationship between the semiparasitic Loranthaceae commonly growing on mangroves, and their hosts.

So far, only few measurements of water flow in mangroves have been carried out. Pannier (1962), using dye indicators (K-fluoroscein) in an attempt to trace the pathway of water flow in the viviparous seedlings of *Rhizophora*,

TABLE 13.1 Methods for the investigation of stomata

Method	What is measured	Equipment	Advantages and usefulness	Disadvantages and limitations
Pressure infiltration	As in 3	Special, inexpensive equipment with small pressure bomb	Values representing large number of stomata. Useful for field work, also for linear (needle) leaves and both hypostomatous (or epistomatous) and amphistomatous leaves	As in 3(a)
Porometry				
1. Water vapour diffusion porometry	A sum of boundary-layer resistance plus transpiration resistance, i.e. diffusive resistance to water vapour of stomata, intercellular spaces and cuticle	Equipment for measuring rate of change of humidity and air and leaf temperature in leaf chamber	Yields diffusive resistance values for water vapour	Includes the flux of water through the cuticle
2. Diffusion porometry	Diffusive resistance of the leaf to model gases which may be recalculated for CO_2 ($s\ cm^{-1}$)	Usually leaf chambers on both sides of a leaf are used; the concentration of the gas is measured by elaborate and expensive equipment	A useful laboratory method yielding directly values of diffusive resistance	Expensive laboratory equipment needed, only for amphistomatous leaves. Includes the intercellular space resistance through the leaf
3. Viscous flow porometry (mass flow)	Viscous flow conductivity (or resistance) of the leaf usually to the air		Useful for field measurements or sophisticated laboratory devices, which can be adapted for recording laboratory methods	Pressure difference (if high) may cause artifact and change stomatal aperture

(a) Viscous flow measurements	Viscous flow-rate measured, viscous resistance calculated	Leaf chambers, flow-meter			Calculation of diffusive resistances is not simple: the relationship of viscous flow conductivity (resistance) to diffusive conductivity (resistance) also depends on the geometry of the stomatal aperture. Includes the intercellular space resistance through the leaf
(b) Resistance porometers	Viscous flow resistance directly measured	Laboratory equipment with leaf chambers			
(c) Pressure-drop measurements	Viscous conductance (resistance) may be calculated from pressure drop, time and volume of the porometer chamber	Mostly simple devices using leaf chamber(s)	Useful for field measurements		High pressure difference may cause artefacts
Calculation of transpiration resistances and hence stomatal resistance from determination of transpiration rate	Transpiration rate from a leaf surface may be recalculated to the transpiration resistance values	Equipment of gasometric measurements of transpiration rate using leaf chambers. Leaf and air temperatures must be measured simultaneously	Useful with simultaneous gasometric measurements of CO_2 exchange in controlled conditions		

Source: Slavik (1974)

observed an extremely low displacement of the fluorochrome in the xylem vessels, so this method seems not to be appropriate.

An apparatus based on the steady-state heat-flow method originally employed by Vieweg and Ziegler (1960) was developed by Saddler and Pitman (1970) to measure rates of water flow in shoots of woody plants.

Despite its restricted use in laboratory conditions and the difficulty in its calibration in absolute values, the authors were able to follow over a period of more than twenty-four hours diurnal changes of sap flow (expressed as relative values) in five Australian mangrove species. These cyclic changes were very similar to the pattern of balancing pressure values of the corresponding species measured by Scholander et al. (1966). Details of the construction of the apparatus can be found in the original paper.

The feasibility of using radioactive tracer methods in mangroves, which have been employed to measure linear velocity of water movement (Klemm and Klemm, 1964) as well as rates of volume flux of water through vascular bundles of trees (Kline et al., 1970), should also be seriously considered.

References

ADRIANI, M. J. 1956. Der Wasserhaushalt der Halophyten. *Encyclop. Plant Physiol.*, Vol. 3, pp. 902–14.

ALVIM, P. de T. 1965. A New Type of Porometer for Measuring Stomatal Opening and its Use in Irrigation Studies. In: F. E. Eckardt (ed.), *Methodology of Plant Ecophysiology*, pp. 325–9. Paris, Unesco. (Arid Zone Research, 25.)

——. 1975. A new Dendrometer for Monitoring Cambium Activity and Changes in the Internal Water Status of Trees. *Turialba*, Vol. 25 pp. 445–7.

ALVIM, P. de T.; HAVIS, J. R. 1954. An Improved Infiltration Series for Studying Stomatal Opening as Illustrated with Coffee. *Plant Physiol.*, Vol. 29 pp. 97–8.

ATTIWILL, P. M.; CLOUGH, B. F. 1980. Studies of Gas Exchange in the White Mangrove. *Photosynthetica*, Vol. 14, No., pp. 40–7.

BANNISTER, P. 1964. Stomatal Responses of Healthy Plants to Water Deficits. *J. Ecol.*, Vol. 52, pp. 151–8.

BARRS, H. D. 1968. Determination of Water Deficits in Plant Tissues. In: T. T. Kozlowski (ed.), *Water Deficits and Plant Growth*, Vol. I, pp. 235–368. New York, Academic Press.

BARRS, H. D.; WEATHERLY, P. E. 1962. A Re-examination of the Relative Turgidity Technique for Estimating Water Deficit in Leaves. *Austral. J. Biol. Sci.*, Vol. 15, p. 413.

BARTHOLOMEW, E. T. 1926. Internal Decline of Lemons. III: Water Deficit in Lemon Fruit Caused by Excessive Leaf Evaporation. *Amer. J. Bot.*, Vol. 8, p. 102.

BHARUCHA, F. R.; NAVALKAR, B. S. 1942. Studies in the Ecology of Mangroves, III. *J. Univ. Bombay*, Vol. 10, No. 5.

BIEBL, R.; KINZEL, H. 1965. Blattbau und Salzhaushalt von *Laguncularia racemosa* (L.) Gaertn. und anderer Mangrovebaume auf Puerto Rico. *Österr. Bot. Zeitsch.*, Vol. 112, p. 56.

BOYER, J. S.; POTTER, J. R. 1973. Chloroplast Response to Low Leaf Water Potentials. I: Role of Turgor. *Plant Physiol.*, Vol. 51, pp. 989–92.

BOWMAN, H. H. M. 1917. Ecology and Physiology of the Red Mangrove. *Proc. Am. Phil. Soc.*, Vol. 56, pp. 589–672.

BUSCALIONI, L.; POLLACCI, G. 1901. L'applicazione delle pellicole di colodio allo studio di alcuni processi fisiologici della piante et in particolar modo alla traspirazione. *Atti. Inst. Bot. Univ. Pavia*, Vol. 7, pp. 83–95.

BURROWS, F. J.; MILTHROPE, F. L. 1976. Stomatal Conductance in Gas Exchange Control. In: T. T. Kozlowski (ed.), *Water Deficits and Plant Growth*, Vol. 4, pp. 103–52. New York, Academic Press.

BUTTERY, B. R.; BOATMAN, S. G. 1966. Manometric Measurement of Turgor Pressures in Lactiferous Phloem Tissues. *J. Expl. Bot.*, Vol. 17, pp. 283–96.

CALDWELL, M. M. 1976. Root Extension and Water Absorption, In: O. L. Lange et al. (eds.), *Water and Plant Life. Ecological Studies*, Vol. 19, pp. 63–85. New York, Springer Verlag.

CATSKY, J. 1965. Leaf-disc Method for Determining Water Saturation Deficit. *Arid Zone Res.* (Unesco), Vol. 25, p. 353.

CHAPMAN, V. J. 1976. *Mangrove Vegetation.* Vaduz, J. Cramer, 447 pp.

COURTIN, G. M.; BLISS, L. C. 1971. A Hydrostatic Lysimeter to Measure Evapotranspiration under Remote Field Conditions. *Arctic Alpine Res.*, Vol. 3, pp. 81–9.

COWAN, I. R.; FARQUHAR, G. D. 1977. Stomatal Function in Relation to Leaf Metabolism and Environment. *Symp. Soc. Expl. Biol.*, Vol. 31, pp. 471–505.

DECKER, J. P.; WIEN, J. D., 1960. Transpiration Surges in *Tamarix* and *Eucaliptus* as Measured with an Infrared Gas Analyzer. *Plant Physiol.*, Vol. 35 pp. 340–3.

ECKARDT, F. E. (ed.). 1965. *Methodology of Plant Ecophysiology*. Paris, Unesco. (Arid Zone Research, 25.)

FISCUS, E. F. 1972. In Situ Measurement of Root-water Potential. *Plant Physiol.*, Vol. 50, pp. 191–3.

FRANK, A. B.; HARRIS, D. G. 1973. Measurement of Leaf Water Potential with a Pressure Chamber. *Agron. J.*, Vol. 65, pp. 334–5.

FUKUDA, Y. 1935. Über die Hydratur der Pflanzen und eine empirische Formel der Verdunstung und Transpiration. *Pflanzenforschung*, Vol. 19. Jena, G. Fischer Verlag.

GAASTRA, P. 1959. Photosynthesis of Crop Plants as Influenced by Light, Carbon Dioxide, Temperature and Stomatal Diffusion Resistance. *Med. Landbouwhogeschool. Wageningen*, Vol. 59, pp. 1–68.

GARDNER, W. R.; EHLIG, C. F. 1965. Physical Aspects of the Internal Water Relations of Plant Leaves. *Plant Physiol.*, Vol. 40, p. 705.

GATES, D. M. 1976. Energy Exchange and Transpiration. In: L. O. Lange et al. (eds.), *Water and Plant Life. Ecological Studies*, Vol. 19, pp. 137–47. New York, Springer Verlag.

GESSNER, F. 1967. Untersuchungen an der Mangrove on Ost-Venezuela. *Int. Rev. Ges. Hydrobiol.*, Vol. 52, No. 5, p. 769.

GREACEN, E. L.; PONSANA, P.; BARLEY, K. P. 1976. Resistance to Water Flow in the Roots of Cereals. In: O. L. Lange et al. (eds.), *Water and Plant Life. Ecological Studies*, Vol. 19, pp. 86–100. New York, Springer Verlag.

GREGORY, F. G.; PEARSE, H. L. 1934. The Resistance Porometer and its Application to the Study of Stomatal Movement. *Proc. Roy. Soc. London, Ser. B.*, Vol. 114, pp. 477–93.

HALL, A. E.; SCHULZE, E. D.; LANGE, O. L. 1976. Current Perspectives of Steady-state Stomatal Responses to Environment. In: O. L. Lange et al. (eds.), *Water and Plant Life. Ecological Studies*, Vol. 19, pp. 169–88. New York, Springer Verlag.

HAMMEL, H. T. 1968. Measurement of Turgor Pressure and its Pradient in the Phloem of Oak. *Plant Physiol.*, Vol. 43, pp. 1042–8.

HANKS, R. J.; SHAWCROFT, R. W. 1965. An Economical Lysimeter for Evaporation Studies. *Proc. Soil Sci. Soc. Amer.*, Vol. 57, pp. 634–5.

HARRIS, J. A.; LAWRENCE, J. V. 1917. The Osmotic Concentration of the Sap of the Leaves of Mangroves Trees. *Biol. Bull.*, Vol. 32, pp. 202–11.

HEATH, O. V. S.; RUSSEL, J. 1951. The Wheatstone Bridge Porometer. *J. Exptl. Bot.*, Vol. 2, pp. 111–16.

HESSE, W. 1954*a*. Messung der Pflanzentranspiration mit Kleinlysimetern. *Zeitsch. Acker. Pflanzenbau*, Vol. 98, pp. 107–18.

——. 1954*b*. Ergebnisse von Pflanzentranspirationsmessungen mit Kleinlysimetern in Zusammenhang mit metereologischen Einfluessen. *Angew. Metereol.*, Vol. 2, pp. 65–82.

HSIAO, T. C. 1973. Plant Responses to Water Stress. *Ann. Rev. Plant Physiol.*, Vol. 24, pp. 519–70.

HUBER, B. 1927. Zur Methodik der Transpirationsbestimmung am Standort. *Ber. Deutsch Bot. Ges.*, Vol. 45, pp. 611–18.

HYGEN, G. 1951. Studies in Plant Transpiration, I. *Physiol. Plant*, Vol. 6.

——. 1953. Studies in Plant Transpiration, II. *Physiol. Plant*, Vol. 8.

JARVIS, P. G. 1971. The Estimation of Resistances to Carbon Dioxide Transfer. In: Sestak et al. (eds.), *Plant Photosynthetic Production. Manual of Methods*, pp. 556–631. The Hague, W. Junk, Publisher.

JARVIS, P. G.; ROSE, C. W.; BEGG, J. E. 1967. An Experimental and Theoretical Comparison of

Viscous and Diffusive Resistances to Gas Flow through Amphistomous Leaves. *Agr. Meterel.*, Vol. 4, pp. 103–17.

JONES, M. M.; TURNER, N. C. 1980. Osmotic Adjustment in Expanding and Fully Expanded Leaves of Sunflower in Response to Water Deficits. *Austral. J. Plant Physiol.*, Vol. 7, pp. 181–92.

JOSHI, G. V. 1976. *Photosynthesis under Saline Conditions. Final Report on the PL. 480 Project.* Kolhapur (India), Shivaji University. 195 pp.

KAUFMAN, M. R. 1968. Evaluation of the Pressure Chamber Technique for Estimating Plant Water Potential of Forest Tree Species. *Forest Sci.*, Vol. 14, pp. 369–74.

KLEMM, M.; KLEMM, W. 1964. Die Verwendung von Radiosotopen zur kontinuirlichen Bestimmung des Tagesverlaufes der Transpirationstromsgeschwindigkeit bei Baumen. *Flora*, Vol. 154, pp. 89–93.

KLINE, J. R.; MARTIN, J. R.; JORDAN, C. F.; KORANDA, J. J. 1970. Measurement of Transpiration in Tropical Trees with Tritiated Water. *Ecology*, Vol. 51, pp. 1068–79.

KOLLER, D.; SAMISH, Y. 1964. A Null-point Compensating System for Simultaneous and Continuous Measurement of Net Photosynthesis and Transpiration by Controlled Gas-stream Analysis. *Bot. Gaz.*, Vol. 25, pp. 81–8.

KOZLOWSKI. T. T. (ed.). 1968–76. *Water Deficit and Plant Growth*, Vol. 4. New York, Academic Press.

KRAMER, P. J. 1964. *Plant and Soil Water Relationships. A Modern Synthesis.* New York, McGraw-Hill.

KREEB, K. 1965. Determination of the Internal Water Balance (Hydrature) in the Field by Measuring Suction Force and Refractive Index. In: F. E. Eckardt (ed.), *Methodology of Plant Ecophysiology*, pp. 385–91. Paris, Unesco. (Arid Zone Research, 25.)

——. 1966. Die Registrierung des Wasserzustandes ueber die elektrische Leitfahigkeit der Blatter. *Ber. Deutsch. Bot. Ges.*, Vol. 79, pp. 150–62.

LAMBERTI, A. 1969. Contribuicão a conhecimiento da ecologia das plantas de Manguezal de Itanhaeim. *Univ. São Paulo. Fac. Filos. Cienc. e Letras. Bol. 317, Bot.*, Vol. 23, pp. 17–218.

LEWIS, O. A. M.; NAIDOO, G. 1970. Tidal Influence on the Apparent Transpirational Rhythm of the White Mangrove. *S. Afr. J. Sci.*, Vol. 66, pp. 268–70.

LIENEWEG, F. 1955. Absolute und relative Feuchtebestimmung mit dem Lithiumchlorid-feuchtemesser. *Siemens-Zeitschrift*, Vol. 29, pp. 212–82.

MCDOUGAL, D. T. 1921. *Growth in Trees.* (Carnegie Institute, Washington, Publication 307.)

MEIDNER, H. 1970. A Critical Study of Sensor Element Diffusion Porometers. *J. Expl. Bot.*, Vol. 21, pp. 1060–6.

MEIDNER, H.; MANSFIELD, T. A. 1968. *Physiology of Stomata.* London, McGraw-Hill.

MICHEL, B. E. 1977. A Miniature Stem Thermocouple Hygrometer. *Plant Physiol.*, Vol. 60, pp. 645–7.

——. 1979. Correction of Thermal Gradient in Stem Thermocouple Hygrometers. *Plant Physiol.*, Vol. 63, pp. 221–4.

MILLER, P. C. 1972. Bioclimate, Leaf Temperature and Primary Production in Red Mangrove Canopies in South Florida. *Ecology*, Vol. 53, pp. 22–45.

——. 1975. Simulation of Water Relations and Net Photosynthesis in Mangroves in South Florida. In: G. Walsh, S. Snedaker and H. Teas (eds.), Proceedings of *International Symposium on the Biology and Management of Mangroves.* Gainesville, University of Florida.

MILLER, P. C.; HOM, J.; POOL, D. K. 1975. Water Relations of Three Mangrove Species in South Florida. *Oecol. Plantarum*, Vol. 10, No. 4, pp. 351–67.

MOLISCH, H. 1912. Das offen- und geschlossensein der Spaltoffnungen veranschaulicht durch eine neue Methode (Infiltrationsmethode). *Z.* Bot., Vol. 4, pp. 106–12.

MOORE, R. T.; MILLER, P. C.; ALBRIGHT, D.; TIESZEN, L. L.; 1972. Comparative Gas Exchange Characteristics of Three Mangrove Species During the Winter. *Photosynthetica*, Vol. 6, pp. 387–93.

MOORE, R. T.; MILLER, P. C.; EHLERINGER, J.; LAWRENCE, W. 1973. Seasonal Trends in Gas Exchange Characteristics of Three Mangrove Species. *Photosynthetica*, Vol. 7, pp. 381–84.

MULLAN, D. P. 1931. Observations on the Water Storing Devices in the Leaves of some Indian Halophytes. *J. Ind. Bot. Soc.*, Vol. 10, pp. 187–9.

NAVALKAR, B. S. 1940. Studies in the Ecology of Mangroves. I: Determination of Osmotic Pressure of *Avicennia alba* Blum. *J. Univ. Bombay*, Vol. 8, pp. 58–74.

NEWMAN, E. I. 1974. Root and Soil Water Relations. In: E. W. Carson (ed.), *The Plant Root and its Environment*, pp. 363–440. Charlotteville, University of Virginia Press.

OERTLI, J. J. 1976. The States of Water in the Plant. In: O. L. Lange et al. (eds.), *Water and Plant Life. Ecological Studies*, Vol. 19. pp. 19–31. New York, Springer Verlag.
OERTLI, J. J.; ACEVES-NAVARRO, E.; STOLZY, L. H. 1975. Interpretation of Foliar Water Potential Measurements. *Soil Sci.*, Vol. 119, pp. 162–6.
O'LEARY, J. W. 1969. The Effect of Salinity on Permeability of Roots to Water. *Israel J. Bot.*, Vol. 18, pp. 1–9.
——. 1974. Salinity Induced Changes in Hydraulic Conductivity of Roots. In: J. Kolek (ed.), *Structure and Function of Primary Root Tissues. Proc. Symp. Tatranska Lomnica,* pp. 309–14. *Bratislava*, Slovak Academy of Science.
PANNIER, F. 1962. Estudio fisiológico sobre la viviparia de *Rhizophora mangle* L. *Acta Cientif. Venezolana*, Vol. 13, pp. 184–97.
PANNIER, R. F.; PANNIER, F. 1973. Relación entre la asimilación de carbono y potencial hídrico en plantas jóvenes de *Rhizophora mangle* L.: una contribución al problema de la tolerancia salina en halofitas. *Resum. III. Congreso Venezol. de Botánica, Cumana, Venezuela.*
PANNIER, F.; PANNIER, R. F. 1975. The Physiology of Vivipary in *Rhizophora*, In: G. Walsh, S. Snedaker and H. Teas (eds.) *Proc. Int. Symp. Biology and Management of Mangroves*, pp. 632–9. Gainesville, University of Florida.
——. 1976. Manglares: un enfoque fisiológico. *Biología* (México), Vol. 6 pp. 51–7.
——. 1977. Interpretación fisioecológica de la distribución de manglares en las costas del continente suramericano. *Interciencia*, Vol. 2, No. 3, pp. 153–62.
PARKINSON, K. J. 1968. Apparatus for the Simultaneous Measurement of Water Vapor and Carbon Dioxide Exchanges of Single Leaves. *J. Expl. Bot.*, Vol. 19 No. 61, pp. 840–56.
PHILIP, J. R. 1958. The Osmotic Cell, Solute Diffusibility and the Plant Water Economy. *Plant Physiol.*, Vol. 33, pp. 264–71.
RASCHKE, K. 1958. Über den Einfluss der Diffusionswiderstande auf die Transpiration und Temperatur eines Blattes. *Flora*, Vol. 146, pp. 546–78.
——. 1965. Das Seifenblasenporometer (zur Messung der Stomaweite an amphistomatischen Blattern). *Planta*, Vol. 66, pp. 113–20.
——. 1965. Eignung und Konstruktion registrierender Porometer für das Studium der Schliesszellenphysiologie. *Planta*, Vol. 67, pp. 225–41.
——. 1975. Stomatal Action. *Ann. Rev. Pl. Physiol.*, Vol. 26, pp. 309–40.
RICHTER. 1976. The Water Status in the Plant. In: O. L. Lange et al. (eds.), *Water and Plant Life. Ecological Studies*, Vol. 19, pp. 42–55. New York, Springer Verlag.
ROTH, R. 1961. Konstruktion und thermodynamische Eigenschaften des Piche-evaporimeters. *Arch. Metereol. Geophys. Bioklimatol. Ser. B.*, Vol. 11, pp. 108–25.
SADDLER, H. D. W.; PITMAN, M. G. 1970. An Apparatus for the Measurement of Sap Flow in Unexcised Leafy Shoots. *J. Expl. Bot.*, Vol. 21, pp. 1048–59.
SAMPSON, J. 1961. A Method of Replicating Dry or Moist Surfaces for Examination by Light Microscopy. *Nature*, Vol. 191, pp. 932–3.
SCHOLANDER, P. F. 1967. Osmotic Mechanism and Negative Pressure. *Science*, Vol. 134, pp. 1835–8.
——. 1968. How Mangroves Desalinate Water. *Physiol. Plant*, Vol. 21, pp. 251–61.
SCHOLANDER, P. F.; HAMMEL, H. T.; HEMMINGSEN, E.; GAREY, W. 1962. Salt Balance in Mangroves. *Plant Physiol.*, Vol. 37, pp. 722–9.
SCHOLANDER, P. F.; HEMMINGSEN, E. A.; BRADSTREET, E. D. 1964. Hydrostatic Pressure and Osmotic Potential in Leaves of Mangroves and Some Other Plants. *Proc. Nat. Acad. Sci.*, Vol. 52, pp. 199–25.
SCHOLANDER, P. E.; HAMMEL, H. T.; BRADSTREET, E. D.; HEMMINGSEN, E. A. 1965. Sap Pressure in Vascular Plants. *Science*, Vol. 148, pp. 339–46.
SCHOLANDER, P. F.; BRADSTREET, E. D.; HAMMEL, H. T.; HEMMINGSEN, E. A. 1966. Sap Concentration in Halophytes and Some Other Plants. *Plant Physiol.*, Vol. 41, pp. 529–32.
SHIMSHI, D. 1967. Some Aspects of Stomatal Behaviour, As Observed by Means of an Improved Pressure-drop Porometer. *Israel J. Bot.*, Vol. 16, pp. 19–28.
SHIMSHI, D.; LIVNE, A. 1967. The Estimation of the Osmotic Potential of Plant Sap by Refractometry and Conductometry: A Field Method. *Ann. Bot. N. S.*, Vol. 31, pp. 505–11.
SLATYER, R. O. 1960. Aspects of the Tissue Relationships of an Important Arid Zone Species (*Acacia aneura* F. Muell.) in Comparison with Two Mesophytes. *Bull. Res. Council Israel Sect. D.*, Vol. 8, p. 159.
SLATYER, R. O.; BIERHUIZEN, J. F. 1964. A Differential Psychrometer for Continuous Measurements of Transpiration. *Plant Physiol.*, Vol. 39, pp. 1051–6.

SLAVIK, B. 1974. Methods of Studying Plant Water Relations, *Ecological Studies*, Vol. 9, New York, Springer Verlag.
SPANNER, D. C. 1951. The Peltier Effect and its Use in the Measurement of Suction Pressure. *J. Expl. Bot.*, Vol. 2, pp. 145–68.
STEINKE, T. O. 1979. Apparent Transpirational Rhythms of *Avicennia marina* (Forsk.) Vierh. at Inhaca Island, Moçambique. *J. S. Afr. Bot.*, Vol. 45, No. 2, pp. 133–8.
STOCKER, O. 1928. Der Wasserhaushalt ägyptischer Wusten- und Salzpflanzen. *Botan. Abhandl.*, Vol. 13.
——. 1929. Das Wasserdefizit von Gefasspflanzen in verschiedenen Klimazonen. *Planta*, Vol. 7, p. 382.
——. 1929. Eine feldmethode zur bestimmung der momentanen transpiration-und evaporationsgrosse. *Ber. Deutsch. Bot. Ges.*, Vol. 47, pp. 126–36.
STEUDLE, E.; ZIMMERMANN, U.; LUTTGE U.; 1977. Effect of Turgor Pressure and Cell Size on the All Elasticity of Plant Cells. *Plant Physiol.*, Vol. 59, pp. 285–9.
TYREE, M. T. 1976. Negative Turgor Pressure in Plant Cells: Fact or Fallacy? *Can. J. Bot.*, Vol. 54, pp. 2738–46.
TYREE, M. T.; MACGREGOR, M. E.; PETROV, A.; UPENIEKS, M. I. 1978. A Comparison of Systematic Errors Between the Richard and Hammel Methods of Measuring Tissue Water Relations Parameters. *Can. J. Bot.*, Vol. 56, pp. 2153–61.
VAN ANDEL, O. M. 1952. Determination of the Osmotic Value of Exudation Sap by Means of the Thermoelectric Method of Baldes and Johnson. *Proc. Kon. Nederl. Akad. Wetensch.*, Vol. 55, pp. 40–8.
VIEWEG, G. H.; ZIEGLER, H. 1960. Thermoelektrische Registrierung der Geschwindigkeit des Transpirationsstromes. *Ber. Deutsch. Bot. Ges.*, Vol. 73, pp. 221–6.
VON FABER, F. C. 1913. Über Transpiration und osmotischen Druck bei den Mangroven. *Ber. Deutsch. Bot. Ges.*, Vol. 31, pp. 277–81.
WALTER, H. 1931. *Die Hydratur der Pflanze und ihre physiologische Bedeutung*. Jena, G. Fischer Verlag.
WALTER, H.; STEINER, M. 1936. Die Ökologie der ostafrikanischen Mangroven. *Zeitschr. Bot.*, Vol. 30, pp. 65–193.
WARING, R. H.; RUNNING, S. W. 1976. In: O. L. Lange et al. (eds.), *Water and Plant Life. Ecological Studies*, Vol. 19, pp. 189–202. New York, Springer Verlag.
WEATHERLEY, P. E. 1950. Studies in the Water Relations of the Cotton Plant. I: The Field Measurement of Water Deficit in Leaves. *New Phytologist*, Vol. 49, p. 81.
——. 1970. Some Aspects of Water Relations. *Adv. Bot. Res.*, Vol. 3, pp. 171–206.
——. 1974. The Hydraulic Resistance of the Root System. In: A. Kolek Jr. (ed.), *Structure and Function of Primary Root Tissues. Proc. Symp. Tatranska Lomnica*. Bratislava, Slovak Academy of Science.
WENZL. H. 1939. Die Bestimmung des Spaltoffnungzustandes nach dem Abdruckverfahren. *Jahrb. Wiss. Bot.*, Vol. 89, p. 122.
ZELITSCH, I. 1969. Stomatal Control. *Ann. Rev. Plant Physiol.*, Vol. 20, pp. 329–50.
ZIMMERMANN, U. 1977. Cell Turgor Pressure Regulation and Turgor Pressure-mediated Transport Processes. *Symp. Soc. Exptl. Biol.*, Vol. 31, pp. 117–254. Cambridge University Press.

14

Mangrove litter production and dynamics

Melvin S. Brown
Division of Biology and Living Resources
Rosenstiel School of Marine and Atmospheric Science
University of Miami
4600 Rickenbacker Causeway
Miami, Florida 33149, United States

Introduction

Mangrove-dominated intertidal habitats are the dominant ecosystem along sheltered subtropical and tropical coasts and are reported to comprise a biome of 43.7 million hectares (Rodin et al., 1975). They are one of the most productive ecosystems in the world in terms of gross primary productivity and litter production (Snedaker and Brown, 1981*a*). Odum (1969) describes mangrove ecosystems as interface or open systems because of their flow through pathways for transporting matter. These pathways are driven by physical (tides, terrestrial runoff and rainfall) and biological (litter production, decomposition, mineral uptake and faunal activities) factors that control the rate of matter import, export and storage (Pool et al., 1975). The result of these flows and transformations is the import of inorganic matter from terrestrial ecosystems and the export of both particulate and dissolved organic matter to adjacent marine ecosystems. Litter fall, consisting of both vegetative and reproductive structures, represents a fraction of net primary production that can be accumulated on the forest floor, remineralized through decomposition or exported. The importance of mangrove leaf litter in the maintenance of detrital-based food webs in the nearshore environment has been well documented in the scientific literature (e.g. detrital food web: W. E. Odum, 1971, Odum et al., 1973, Heald 1971; decomposition and nutritional value: Fell, 1973, Fell et al., 1975, Fell and Master, 1980; relationship to commercial fisheries: Macnae, 1974, Martosubroto and Naamin, 1977, Turner 1977).

The measurement of standing stock, turnover and production rates of litter represent a major step in understanding the structure and function of any forest ecosystem. The monitoring of mangrove litter fall has also given insight into the optimal harvest age for a managed forest (Gong Wooi-Khoon, personal communication), and the deteriorating health of a forest inundated with oil (Ariel Lugo, personal communication).

Definitions

Litter fall is the shedding of vegetation and reproductive plant structures and may be caused by senescence or stress, by mechanical factors, for example wind, by a combination of these, or by death and weathering of the whole plant (Kozlowski, 1973). Medwecka-Kornas (1970) differentiates between litter present at a given moment and that produced over a definite period as follows: standing crop (i.e. biomass) is litter present at a given moment in a definite area and should be expressed in weight or energy units per surface area (e.g. g/m^2, kcal/ha); litter production is the rate of litter shed within a definite time period (e.g. $g/m^2 \cdot day$, kg/ha·year). Turnover is defined by E. P. Odum (1971) as 'the ratio of throughput to content' (i.e. ratio of litter production to standing crop). Turnover may be expressed as the rate of the total amount of a substance released or entered into a compartment for a given period (e.g. g/h) or the time (reciprocal of the rate) required for complete replenishment of the compartment. The turnover concept is useful for comparing exchange rates between different compartments of an ecosystem.

Litter traps

CONSTRUCTION OF TRAPS

The most common method used for determining litter production in mangrove ecosystems is through the use of litter baskets or traps. The terms 'basket' and 'trap' are used synonymously throughout the literature and for the purpose of this chapter the more descriptive term 'trap' will be used. There are many designs for litter traps (square, circular, triangular) and some diversity of opinion as to which receptacles are best (Thompson and McGinnes, 1963). Saito and Shidei (1972) have shown, that for various softwood and hardwood stands, there is no difference in the amount of litter fall caught between square, circular or triangular traps of the same surface area. Thus, in regard to horizontal shape there seems to be no difference in collection efficiency, although there is an important relationship between the surface area of the litter trap and the maximum size of vegetative structures being collected (Saito and Shidei, 1972; Andrzejewski et al., 1968). The minimum size trap that should be considered is one whose length (diameter in the case of a circular trap) is twice the maximum length of the largest vegetative structure to be caught (Saito and Shidei, 1972). Additional important criteria to be considered are: litter should fall into the trap without any obstruction caused by trap design; once the litter is in the trap it should be retained; no contamination from ground sources should be experienced and there should be no significant weight loss due to decomposition of the sample (Newbould, 1967).

The litter traps which we and others have used (Snedaker and Brown, 1981*b*; Pool et al., 1975; Carter et al., 1973) successfully in mangrove and upland forests of southern Florida and the Caribbean are constructed of a square wooden frame with a plastic screen bottom (Fig. 14.1). The frame is constructed with four pressure-treated (preserved) wooden strips 50 cm long, 10 cm high and 2 cm wide. This yields a 2,500 cm^2 (0.25 m^2) interception area for litter. The bottom of the trap is covered with flexible, plastic coated, fibre-glass window screening with a mesh size of approximately 1 × 1 mm. We have experienced no detectable losses of mangrove litter with the use of this mesh size. One important point to consider in relation to mesh size is that it permits moisture loss so that the collected sample remains as dry as possible, thus reducing weight losses due to decomposition. The placing of the screen is very important. If the screen is placed tightly across the bottom of the basket a sizeable loss may be experienced owing to the heavier vegetative structures falling on to the screen and bouncing out of the trap (Randy Edwards, personal communication). It is best to shape the screen into a cone or bag-like receptacle. This will also minimize losses caused by high winds.

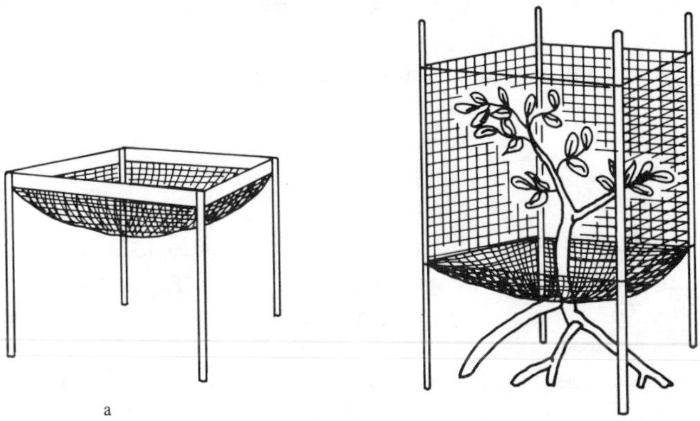

a

b

Figure 14.1
Litter traps. (a) typical substrate positioned litter trap; (b) enclosure-type litter trap used in southern Florida dwarf forest.

We have employed two methods for suspending the litter traps above the substrate. Wooden stakes, 2.5 × 2.5 cm, are attached to each corner of the trap and then secured in the substrate. The height of the trap should be above prop-root interference and the highest storm tide. The second method used is suspending the traps from the lower vegetative limbs of the trees with rope. Several precautions should be taken using this method. Care should be taken to ensure that the trap is: (a) positioned below the vegetative structures from which one wishes to collect; (b) unaltered in its horizontal position; (c) secured so it cannot be turned over by wind; and (d) unobstructed by the supporting ropes, which should not interfere in the collection of the litter.

CONSTRUCTION OF ENCLOSURES

A modification of the litter trap, in which the entire tree is enclosed, has been used on the dwarf mangroves in southern Florida (Fig. 14.1(b). These mangroves are approximately one metre in height and have a density and canopy configuration which makes the use of litter traps impractical and ineffective. The enclosures isolate and trap the total litter fall from one individual tree. The enclosure is constructed by placing four or more 2.5 × 2.5 cm stakes in the substrate, forming a polygon around the tree. The plastic screening (described above) is positioned around the trunk of the tree below the lowest branches and attached to the four stakes. Another piece of screening is wrapped round the stakes enclosing the tree. The screening is then secured to the stakes and to the screening positioned around the trunk. This forms an envelope around the tree and catches only the litter from the individual tree. The density of individuals (trees/m^2) must be known so that the litter fall caught in the enclosures can be expressed on a surface-area basis.

Litter traps and enclosures, which have been built according to the above-described methods out of wood treated with a preservative and fasteners made from non-corrosive material (monel, stainless steel, etc.), have lasted longer than four years in the field with minimal maintenance.

Location

The positioning of the litter traps within a forest stand is usually at the researcher's discretion (randomly or systematically). Our traps have usually been placed systematically along a straight line transect, five to ten metres apart. These transects have been located solely in homogeneous forest types or have transversed through several different types (see Lugo and Snedaker, 1974 for an explanation of forest types). The minimum number of traps needed seems to be between twenty (Newbound, 1967) and thirty-five (Medwecka-Kornas, 1970). We have used a minimum of twenty per sample site and have experienced excellent replication between two mangrove forests of the same type.

Collection

Litter should be collected at a frequency that minimizes decomposition. This depends on local climatic conditions. In the Caribbean area, a fourteen- to twenty-one-day collection interval has proved sufficient. The collected litter from each trap is placed into its own paper bag, labelled as to location, date, trap number and sealed to prevent loss. The litter is then dried in the bag at 70 °C to a constant weight (approximately seventy-two hours), sorted into compartments and weighed. The number of compartments the litter is separated into is determined by the questions sought, but I have found it immensely useful to sort the litter into every category imaginable. Figure 14.2 shows an example of the data recording sheet used by our laboratory. Leaves and seeds (reproductive organs) are sorted by species. Unidentifiable leaf fragments, wood and miscellaneous items (bracts, flowers, etc.) are also separated and their weights recorded. Space is provided for summing the total amount of litter in each trap as well as the totals for each compartment. This detailed separation supplies the researcher with the total amount of litter produced together with information about the phenological patterns. The traps also catch many dead insects and their frass. The quantification of this additional information proves useful in studies of arboreal herbivory.

Litter production
Dry weight - g/.25m^2

Date: Location:

Duration: Site:

		Subtotals			Mangrove subtotals by type								
					Red		Black		White		Buttonwood		
asket No.	Total Litter	Misc.	Wood	Leaf frag- ments	Leaves	Seed	Leaves	Seed	Leaves	Seed	Leaves	Seed	
/m^2 day													
.D. (N− 1)													
.E.													

Figure 14.2
Example of a laboratory data sheet used for recording dry weights of litter by compartments.

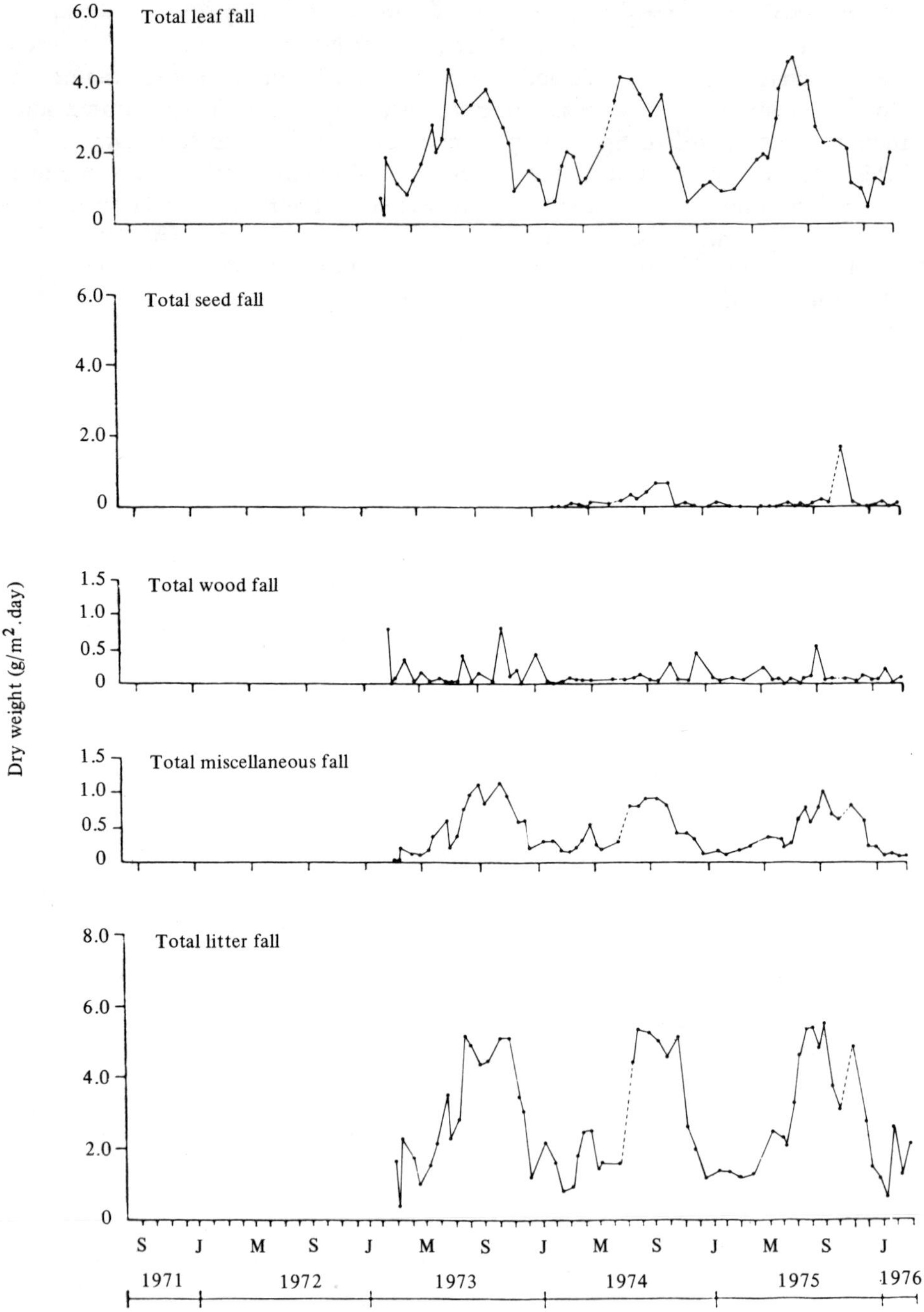

Figure 14.3
Graphical presentation of litter production by compartments for a mangrove overwash forest located in south-west Florida. (After Snedaker and Brown, 1981*b*.)

Presentation of data

Litter-production data are usually presented in the form of graphs. The x-axis (horizontal, independent variable) represents time, usually months, and the y-axis (vertical, dependent variable) is the rate of litter fall (e.g. $g/m^2 \cdot day$). The yearly totals (e.g. $g/m^2 \cdot year$) are presented in a table format. Figure 14.3 is an example of graph set showing mangrove litter production by compartments on a monthly basis.

Standing crop

The standing crop of litter can be determined simply by collecting all the litter from a known surface area of the forest floor. The surface area from which litter is collected, number of collections, location of collections and collection times should run parallel with the collection for litter production. All standing crop samples are collected, dried, sorted and weighed using the same method described above for litter fall. The period difference in the standing crop corrected for period incremental additions of litter fall is an indication of the amount of litter removed from the forest floor by export to adjacent systems, decomposition and consumption by fauna, or conversely, the accumulation of litter *in situ*.

Turnover

The turnover of mangrove litter has been determined by using the mathematical approach of Nye (1961) and Olson (1963). The turnover rate K is estimated by the ratio of input L (litter production) to the steady-state total X_{ss} (standing crop). The relationship is $K = L/X_{ss}$. The equation assumes steady-state (input = output), so that rate of change (dx/dt, where x is the standing crop and t is time) is zero.

References

ANDRZEJEWSKI, R.; BOROWSKI, S.; OLSZEWKSI, J. 1968. Determination of the Area of Organic Fall Catcher in Deciduous Forest. *Sylwan*, Vol. 112, No. 8, pp. 55–8. (English summary.)

CARTER, M. R.; BURNS, L. A.; CAVINDER, T. R.; DUGGER, K. R.; FORE, P. L.; HICKS, D. B.; REVELLS, H. L.; SCHMIDT, T. W. 1973. *Ecosystems Analysis of Big Cypress Swamp and Estuaries*. Atlanta, Ga., U.S. Environmental Protection Agency, Region IV. 478 pp. (EPA 904/9–74–002.)

FELL, J. W. 1973. Fungi Associated with the Degradation of Mangroves (*Rhizophora mangle*) Leaves in South Florida. In: H. Stevenson and R. Colwell (eds.), *Estuarine Microbial Ecology, Belle W. Baruch Symposium*, Vol. I, pp. 455–66. Columbia, University of South Carolina Press.

FELL, J. W.; MASTER, I. M. 1980. The Association and Potential Role of Fungi in Mangrove Detrital Systems. *Botanica Marina*, Vol. 23, pp. 257–63.

FELL, J. W.; CEFALU, R. C.; MASTER, I. M.; TALLMAN, A. S. 1975. Microbial Activities in the Mangrove (*Rhizophora mangle*) Leaf Detrital System. In: G. Walsh, S. Snedaker and H. Teas (eds.), *Proceedings of International Symposium on the Biology and Management of Mangroves, 8–11 October 1974. Honolulu*, Vol. 2, pp. 661–79. Gainesville, Institute of Food and Agricultural Sciences, University of Florida.

HEALD, E. J. 1971. The Production of Organic Detritus in a South Florida Estuary. *Sea Grant Tech. Bull.*, No. 6. University of Miami. 110 pp.

KOZLOWSKI, T. T. 1973. Extent and Significance of Shedding of Plant Parts, p. 1–44. In: T. T. Kozlowski (ed.), *Shedding of Plant Parts*. New York, Academic Press.

LUGO, A. E.; SNEDAKER, S. C. 1974. The Ecology of Mangroves. *Annual Review of Ecology and Systematics*, Vol. 5, pp. 39–64.

MACNAE, W. 1974. *Mangrove Forests and Fisheries*. FAO/UNDP Indian Ocean Fishery Programme. Indian Ocean Fishery Commission. 35 pp. (IOFC/Dev/74/34.)

MARTOSUBROTO, P.; NAAMIN, N. 1977. Relationship between Tidal Forests (Mangroves) and Commercial Shrimp Production in Indonesia. *Mar. Res. in Indonesia*, No. 18, pp. 81–6.

MEDWECKA-KORNAS, A. 1970. Litter Production, p. 139–43. In: J. Phillipson (ed.), *Proc. Symp. Methods of Study in Soil Ecology* (Paris, Unesco.)

NEWBOULD, P. J. 1967. *Methods for Estimating the Primary Productivity of Forests*. Oxford, Blackwell Scientific Publications. 62 pp. (IBP Handbook, No. 2.)

NYE, P. H. 1961. Organic Matter and Nutrient Cycles under Moist Tropical Forest. *Plant and Soil*, Vol. 13, No. 4, pp. 333–46.

ODUM, E. P. 1969. The Strategy of Ecosystem Development. *Science*, Vol. 164, pp. 262–70.

——. 1971. *Fundamentals of Ecology*. Philadelphia, W. B. Saunders Co. 574 pp.

ODUM, W. E. 1971. Pathways of Energy Flow in a South Florida Estuary. *Sea Grant Tech. Bull.*, No. 7. University of Miami. 162 pp.

ODUM, W. E.; ZIEMAN, J. C.; HEALD, E. J. 1973. Importance of Vascular Plant Detritus to Estuaries. In: R. Chabreck (ed.), *Proc. Symp. Coastal Marsh and Estuary Management*, pp. 91–114. Baton Rouge, Louisiana State University, Division of Cont. Education.

OLSON, J. S. 1963. Energy Storage and the Balance of Producers and Decomposers in Ecological Systems. *Ecology*, Vol. 4, No. 2, pp. 322–31.

POOL, D. J.; SNEDAKER, S. C.; LUGO, A. E. 1975. Litter Production in Mangrove Forests of South Florida and Puerto Rico. In: G. Walsh, S. Snedaker and H. Teas (eds.), *Proceedings of International Symposium on Biology and Management of Mangroves*, Vol. 1, pp. 213–37. Gainesville, Institute of Food and Agriculture Sciences, University of Florida.

RODIN, L. E.; BAZILEVICH, N. I.; ROZOV, N. N., 1975. Productivity of the World's Main Ecosystems, pp. 13–26. In: *Productivity of World Ecosystems*. Washington, D.C., National Academy of Science.

SAITO, H.; SHIDEI, T., 1972. Studies on Estimation of Leaf Fall under Model Canopy. *Bull. Kyoto Univ. Forests*, Vol. 43, pp. 162–85. (English summary.)

SNEDAKER, S. C.; BROWN, M. S., 1981*a*. Primary Productivity of Mangroves. In: C. C. Black and A. Mitsui (eds.), *CRC Handbook Series of Biosolar Resources*. Vol. I: *Basic Principles*, pp. 477–86. West Palm Beach, Florida, CRC Press.

——. 1981*b*. *Water Quality and Mangrove Ecosystem Dynamics*. Gulf Breeze, Florida. U.S. Environmental Protection Agency. Environmental Research Laboratory, 80 pp. (EPA–600/4–81–022. Prepared for the Office of Pesticides and Toxic Substances.)

THOMPSON, R. L.; MCGINNES, B. S. 1963. A Comparison of Eight Types of Mast Traps. *J. Forestry*, Vol. 61, pp. 679–80.

TURNER, R. E. 1977. Intertidal Vegetation and Commercial Yields of Penaeid Shrimp. *Trans. Am. Fish. Soc.*, Vol. 106, No. 5; pp. 411–16.

15

Litter decomposition and nutrient enrichment

Jack W. Fell and I. M. Master
Division of Biology and Living Resources
Rosenstiel School of Marine and Atmospheric Science
University of Miami
4600 Rickenbacker Causeway
Miami, Florida 33149, United States

R. G. Wiegert
Department of Zoology
University of Georgia
Athens, Georgia 30602,
United States

Introduction

Mangrove leaf litter in estuaries provides an important nutrient base for food webs leading to commercially important food fishes and invertebrates. Although poor in nutrients when they fall from the trees, mangrove leaves become nutritious due to the microbial enrichment process (Odum, 1971). This chapter will discuss methods for: (a) examining the processes of decomposition and the associated microbes; (b) measuring the nutrients that become available to the ecosystem; (c) designing laboratory techniques to ascertain the effects of natural and introduced perturbations; and (d) developing a predictive model to guide a research programme and to explain how the processes, inputs and outputs interact and change in response to environmental alterations. Many of these methods were developed for studies of *Rhizophora mangle* L. litter and should be applicable to other mangrove species.

A convenient indicator of the nutritional value of a food is the ratio of the carbon to nitrogen (C:N) content. The dietary requirement of protein for most animals is 16.5 per cent of the dry weight of the diet, which corresponds to a C:N intake of about 17:1. In cases where the C:N intake is above 17:1 there is a protein deficiency (Russell-Hunter, 1970). Examples of C:N ratios of agricultural crops are sugar beets 48:1, potatoes 30:1 and rice 31:1, whereas beef is 4.3:1 and soybeans 6:1. Rapidly dividing microbes such as diatoms have a C : N of 6:1, while bacteria and fungi may be even lower. An excellent review of C : N requirements is presented by Russell-Hunter (1970).

Nutritive enrichment of *R. mangle* leaf litter results from loss in carbon content and an increase in final nitrogen; the C:N changes from 120 in senescent leaves to 43 in partially decomposed leaves (Fell and Master, 1980). C:N decreases have also been reported during the decomposition of a variety of terrestrial aquatic plant litter (Barlöcher and Kendrick, 1974; Gosz et al., 1973; Howard and Howard, 1974). In coastal systems, Cruz and Gabriel (1974) reported a C:N reduction of 74 to 42 during the decomposition of black needle rush *Juncus roemerianus*, while fronds of the decomposing brown alga, *Fucus vesiculosus*, changed from 28 to 12 (Hunter, 1976). Not all plants demonstrate reduction in C:N; in general, plants with a C:N above 25:1 exert a nitrogen demand and cause nitrogen immobilization, whereas if the initial C:N is below the critical level there is no change (Parnas, 1975; Hodkinson, 1975). Thus, the high N genera *Zostera* (Thayer *et al.*, 1977) and *Thalassia* (Newell and Fell, unpublished) retain a C:N in the vicinity of the initial value (below 20) during decomposition.

Carbon represents 45 per cent of the dry weight of senescent *R. mangle* leaves; about half of this carbon is leached and half becomes particulate detritus (Fell and Master, 1980). Nitrogen is initially 0.2–0.4 per cent of the dry weight, increasing during decomposition (via microbial N immobilization) to 0.5 per cent. The average annual litter fall in a fringe forest of red mangrove is 896 g dry wt $m^{-2} \cdot yr^{-1}$ (Lugo and Snedaker, 1974) for a contribution to the ecosystem of 224 g C $m^{-2} \cdot yr^{-1}$ leachate and an equal amount of particulate detritus. Nitrogen contributed would be 4.5 g $m^{-2} \cdot yr^{-1}$.

The soluble carbon enters the food chain either as uptake by heterotrophs (Cooksey and Cooksey, 1978) which are then consumed by filter feeders, or by direct ingestion after flocculation. The proportion of flocculates:solubles has not been reported. Lush and Hynes (1973) reported abiotic precipitation of organic material in fresh-water; the rate varied with leaf species and water chemistry. They described the precipitates as either semi-transparent flakes or globular aggregations with a mean diameter of 25–50 μm. Wiebe and Pomeroy (1972) described similar organic particles in sea-water. Lush and Hynes (1973) found that abiotic flocs were invaded by bacteria and fungi; they thought that these precipitates might be important food sources to the benthos. During the decomposition of red mangrove leaves we observed formation of both types of flocs as described by Lush and Hynes (1973), Wiebe and Pomeroy (1972) and Johnson and Cooke (1980), as well as subsequent infestation of these aggregations by bacteria and fungi.

Paerl (1978) found a rapid bacterial conversion of dissolved organic carbon (DOC) to particulate organic carbon (POC) in fresh-water lakes, with a bacterial doubling time of 0.35 days. Paerl postulated that 50 per cent of the POC in the lake was derived from bacterial conversion of DOC and that this material constituted an important zooplankton food source. Abiotic conversion of DOC to POC by bubbles of air may or may not be important in sea-water (Johnson, 1976; Wallace and Duce, 1978; Johnson and Cooke, 1980). However, Wallace and Duce reported that the C:N of the particles is 4.2 to 8.5, suggesting a high food value. Flocculation increases with salinity from 0 to 20 per mil, but there is little additional removal at higher salinities (Sholkovitz, 1976).

Fungi may cause much of the change in C:N in some kinds of decomposing litter. They are particularly able to decompose leaves because they produce external enzymes which allow the hyphae to penetrate the litter. In constrast, bacteria tend to be restricted to surfaces and are particularly important where the material has been fragmented and the surface/volume ratio is high (Harley, 1971). Fungal decomposers of plant litter have been studied in soils (Jackson, 1975; Parkinson, 1975), forests (Kirk, 1973; Swift, 1977; Stark, 1972; Witkamp, 1974; Visser and Parkinson, 1975; DeBoois, 1976) and fresh-water (Barlöcher and Kendrick, 1974; Kaushik, 1975; Suberkropp and Klug, 1976; Barlöcher et al., 1978). In these terrestrial and fresh-water systems the fungi are essential conservers of nitrogen and other nutrients (Stark, 1972; Jackson, 1975; O'Neill, 1976). In contrast, the decomposition of the marine eelgrass *Zostera marina* may be the result of bacterial rather than fungal action (Newell, 1981).

The structure of the microbial communities will vary not only with substrate but also with variables such as season, water temperature, salinity, oxygen and length of time that the leaf has been decomposing. A general scheme of fungal population changes has been depicted for red mangrove leaves. Senescent leaves harbour species of *Nigrospora*, *Phyllosticta* and *Pestalotia*. During the early stage of decay *Phytophthora* is the most prevalent genus, in addition to *Drechslera* and *Gloeosporium*. In later stages of decay fungi such as *Gliocladium* and *Lulworthia* are prevalent (Fell and Master, 1973, 1975).

The interactions between nutrient flows and the associated microbes must be understood in order to ascertain how these processes function and how they are affected by external factors. Activities and interactions of the components in a complex ecosystem can be examined using laboratory models, keeping in mind the danger that such models may be far too simplified to mimic field conditions. Commonly, experiments consist in placing leaf litter into containers with sea-water with the various potential components of the mangrove system to be tested. A variety of problems should be considered. Nutrients such as nitrogen and phosphorous must be maintained at adequate levels. The pre-test treatment often eliminates or changes the normal microbial flora and additions of amibent sea-water may not ensure their re-establishment. Even purely physical alterations of the litter may seriously affect the results. Heat applied for sterilizing or drying can alter the carbon compounds in addition to changing the microflora; grinding structurally alters the leaves and changes the surface/volume ratio. Despite these difficulties, laboratory models are useful tools for preliminary studies of decomposition. A particularly helpful reference for the design and implementation of laboratory models is Giesy (1980).

Our studies (Fell and Hunter, 1979; Fell and Newell, 1980; Fell et al., 1980; Fell and Master, 1980) have emphasized the effect of fungi on nitrogen immobilization during decomposition of the red mangrove (*Rhizophora mangle*). We used partially sterilized leaves and sterilized controls in a semi-batch culture, i.e. a flask system in which the water was changed routinely (every twenty-four hours) to maintain nutrient levels and to reduce build-up of toxic compounds. The effect of varied concentrations of ambient nitrogen was tested in the presence of marine *Phytophthora* spp. Additions of nitrogen ranged from 0.025 to 17.8 mg-at N/1 as $(NH_4)_2SO_4$, with phosphorus constant at 161 μg-at P/1 (KH_2PO_4). The results demonstrated that the rapidly colonizing fungi are potential agents of nitrogen immobilization, i.e. nitrogen immobilization similar to that observed in the field occurred in the presence of fungi and inorganic nitrogen but not in sterile controls, or in the presence of fungi without added nitrogen. Both weight loss and change in carbon content were greater in the presence of fungi.

Methods

FIELD STUDIES

How and in what form(s) are mangrove leaves made available to the detritivores in the community? To answer this we must measure particulate and soluble fractions, carbon and nitrogen content and often look for specific organic compounds. A simple, effective way to follow decompositional changes in leaf litter is to use litter bags. The bags should be made of synthetic material because hemp and cotton fibres develop a fungal flora. Size of bags should vary with the number and size of leaves to be studied. One convenient size for red mangrove leaves is 100 × 30 cm with 50 leaves/bag. If there are too many leaves in a bag they will pack and develop anaerobic zones. Dividing the bag into compartments with one leaf/compartment helps prevent this problem. Numbering the compartments permits individual leaf identities to be maintained. Mesh size should allow small invertebrates free access to the leaves. The large invertebrates will be excluded, which is a major drawback of mesh bags. We have not observed large invertebrates feeding on the leaves of *Rhizophora mangle*, however this may be important in other mangrove species; Malley (1978) found that sesarmid crabs feed on decomposing black mangrove leaves in Thailand. In such cases, leaves can be tethered by the stem with a piece of light fishing line tied to a root or a stake. Decomposition is initially rapid then slow. Thus, retrieval of leaves should be frequent in the initial stage of the study and less frequent in later stages of decomposition. We recommend collections, at 0, 2, 4 and 7 days, then weekly for six weeks and after that biweekly until decomposition is complete, starting with senescent leaves picked from the trees.

DETERMINATIONS OF CARBON, NITROGEN AND DRY-WEIGHT CHANGES IN LEAVES

The surface of senescent and decomposing leaves has a community of bacteria, protozoans, fungi, diatoms, nematodes, polychaetes and other meiofauna plus accumulations of organic and inorganic material. A separate study of the interactions within this film is often appropriate. However, the film must be removed in order to measure the carbon and nitrogen content of leaves; this can be done by gently rubbing the leaves in fresh-water, which also removes the salt that interferes with weight-loss determination. Dry weight is obtained by heating at 104°C. Leaves should be ground, preferably in a Wiley mill with a No. 6 mesh sieve.

Individual leaves vary greatly in water, carbon and nitrogen content. For example, in one collection of red mangrove leaves we found that the C:N ranged from 55 to 180. To avoid using large sample numbers, we split the leaves longitudinally through the mid-rib. Reference halves of each leaf were analysed as a basis for predicting the original dry weight, as well as the percentage of C and N, in the experimental half. As a test of this procedure, we compared both halves of twenty leaves for C:N and dry weight (DW):fresh weight (FW). Linear regressions of predicted against measured value gave a high correlation between the two halves for each computed value. C:N sample correlation coefficient (r) = 0.96, slope (b) 1.01; C% $r = 0.94$, $b = 0.71$; N% $r = 0.95$; DW:FW $r = 0.95$, $b = 0.79$.

In addition to fresh and dry weight, carbon, and nitrogen content, measurements of surface area, length, width, ash content and thickness may be needed. Examples of using the split leaf technique in calculations are given below:

1. Predicted dry weight (PDW) of the experimental half (E) is the DW:FW of the reference half (R) multiplied by the fresh weight (FW) of the experimental half:

$$\text{PDWE} = \frac{\text{DWR}}{\text{FWR}} \times \text{FWE}$$

2. Weight loss (% WT LS) of the experimental half is the difference between the predicted and final dry weights expressed as a percentage:

$$\%\,\text{WT LS} = \frac{\text{PDWE} - \text{DWE}}{\text{PDWE}} \times 100$$

3. Final corrected nitrogen (N_{fc}) is the nitrogen content of the experimental half (N_f%E) expressed as a percentage of the original dry weight (PDWE):

$$N_{fc}\% = \frac{N_f\%\text{E} \times \text{DWE}}{\text{PDWE}}$$

4. Final corrected carbon (C_{fc}%): derived in the same manner as N_{fc}%.
5. Change in nitrogen (ΔN%) from the original expressed as a percentage: N_i is the initial nitrogen as measured in the reference half.

$$\Delta N\% = \frac{N_{fc} - N_i}{N_i} \times 100, \text{ or}$$

$$= \frac{(N_f\%E \times DWE) - (N_i\% \times PDWE)}{N_i\% \times PDWE} \times 100$$

6. Change in carbon ($\Delta C\%$): derived in the same manner as $\Delta N\%$.
7. Change in C:N ($\Delta C:N$): derived in the same manner as $\Delta N\%$ and $\Delta C\%$, also expressed as percentage change.[1] Data analysis requires customary statistical procedures, such as multiple regression and canonical correlation analyses (Nie et al., 1975).

IDENTIFICATION OF ASSOCIATED FUNGI

Many fungi on the surface layers of leaves may be present as either active or inactive propagules. Although these surface fungi have some effect on leaf decomposition, most decomposition occurs within the leaf. Surface fungi must be removed completely before the species within can be identified. Rinsing with sea-water and gently rubbing the leaves by hand will remove most of the surface fungi. The remainder can be eliminated by sterilization. Newell et al. (1981) and Newell and Fell (1980, 1982) reported the following procedure which will probably require some variation with leaves from different species of trees. Leaf discs (4.5 mm diameter) were placed for two minutes in 0.53 NaC10 and 1 per cent v/v detergent (such as Liquinox); two minutes in sterile tap-water; two minutes in sterile sea-water with continuous agitation of discs in sterilant and rinses. Sterilants and rinses were changed between sets of discs. Discs were then drained on sterilized blotting paper. The treated discs should be placed on a nutrient medium such as corn-meal agar (15 per mil sea-water and 0.014 per cent chloramphenicol), and incubated at room temperature.

The following references are useful for fungal identification: Ainsworth et al. (1973); Sparrow (1960); Ellis (1971, 1976); Kohlmeyer and Kohlmeyer (1971, 1979); Fell and Master (1975).

LABORATORY EXPERIMENTS

Experiments in the laboratory can use either semi-batch cultures (changing the test media at regular intervals) or a continuous flow system. In either case, build-up of toxic materials must be prevented and proper nutrient levels maintained. We use one-litre shaker flasks with filter sterilized (0.45 μm) 15 per mil sea-water (500 ml/flask) that is changed every 24 hours (Fell, et al., 1980; Fell and Master, 1980). Five leaves/flask is adequate. Preliminary experiments will define conditions required for reproducing the field processes. We used phosphate (161 μg-at P/L), nitrogen (400 μg-at N/L), and a fungal inoculum such as *Phytophthora vesicula*. Carbon and nitrogen in the leaves were used to indicate the stage of decomposition. Simultaneous measurement of the soluble carbon in the test medium permitted computation of a carbon budget. Correc-

1. All the above percentage data can also be calculated in quantitative terms, such as mg weight loss.

tion for carbon loss via respiration must be made. Combined with leaf-fall data from the field, such laboratory experiments permit an initial estimate of the carbon and nitrogen input from a mangrove forest. The laboratory system can be made more complex by addition of bacteria, meiofauna, etc., and by perturbation of temperature, nutrients, etc. Such data can lead to studies of the ultilization of the added materials by detritivores (Tenore and Rice, 1980).

Alternately, weaknesses of the laboratory measurements of decomposition of mangrove leaves can be pinpointed and the experiments revised to more accurately reflect the operation of the natural mangrove community. To effectively pursue either of these options, the preliminary information should be incorporated into a simulation model. Rapid, inexpensive experiments with such a model can then suggest profitable designs for further field and laboratory observations.

A CONCEPTUAL MODEL OF DECOMPOSITION

The major requirement of a conceptual model is that it be realistic, i.e. that the dynamic changes are simulated as a result of programmed interrelationships which are biologically definable and reasonable. This requirement excludes those predictive models that are wholly or largely constructed by statistical correlative techniques and in which parameters are chosen solely on the basis of improving the fit of the predictions to the data from the field. Within this set of constraints are a wide range of ecological models. We wish to provide an example of how one such model could be constructed for the mangrove detrital system. For an extensive discussion of ecological models and an entry into the recent literature, the reader is referred to Wiegert (1975) and Wiegert et al. (1981).

The steps taken to construct a simple model are: (a) choice of the appropriate energy/material flux and of the necessary compartments; (b) construction of a flow diagram; (c) enumeration of the vital controlling factors, both physical and biological; (d) verbal statements of the quantitative relationship of the compartments and controlling factors in so far as they interact to determine the fluxes between compartments and thus affect storage within compartments; and (e) transformation of the verbal statements into mathematical equations suitable for programming in a computer language. Subsequent simultaneous solution of the set of equations will predict dynamic behaviour by the compartments and thus by the system. Such behaviour is said to be the consequence of the interaction of system structure with system function. The system structure comprises the abstract compartment 'boxes' and the characteristics of the flux-rate coefficients and control parameters (Wiegart et al., 1981).

In Figure 15.1 the compartments and flow pathways of both carbon and nitrogen are shown. Although these are two separate models, the compartments of one model will regulate flows in the other model. Our major components, two in each submodel, are: X3-nitrogen in decomposing leaves, X4-soluble nitrogen in tidal water, X8-detrital carbon and X9-soluble carbon in tidal water. Within the leaves or particles of detritus represented by compartments X3 and X8 we recognize subdivisions. In X3 are contained the abiotic particulate nitrogen pool (X1) and the nitrogen in the standing crop of fungi (X2). Within the particulate detrital carbon compartment are: X5-soluble carbon compounds not yet leached (substantially all of the soluble N compounds are leached before leaf abscission), abiotic carbon X6 and biotic carbon X7.

Sixteen fluxes of carbon or nitrogen describe the major pathways of flow in the mangrove detrital system (Table 15.1). Two, F0003 and F0008, are forcing input fluxes determined by the rate of leaf production and the C:N content at the time of abcission of the leaf. This C:N content, plus the proportion of carbon that is leachable, also determines the apportionment of F0008 to X5 versus X6. Another two forcing inputs are F0004 and F0009. The former comprises the production of soluble N-compounds by N-fixers as well as inputs to the mangrove community from pollution and tidal exchange. Four exports are wholly (F0003, F0400, F0900) or largely (F0800) under control of tidal export. Flux F0800 includes also the CO_2 respired by the microbes in and on the detritus.

The remaining eight fluxes of nitrogen and carbon are controlled largely by factors from within the boundaries shown in Figure 15.1 (temperature is

TABLE 15.1. Fluxes and control functions of the conceptual model of a mangrove detritus community (for discussion of parameters, see text)

1.	F0003 = Constant specified over a given time
2.	F0004 = Constant specified over a given time (may be much more complex due to uncertain dynamics of N_2 fixers)
3.	F0102 = $T_{0102} \times P_{0102} \times X_2 \times f(X_1, X_2, X_4;$ temp.$)$
4.	F0204 = $AE_{02} \times X_2 + E_{0102} \times F_{0102} + E_{0402} \times F_{0402}$ [1]
5.	F0300 = $T_{0300} \times X_3$
6.	F0400 = Constant or variable (f (tidal exchange) over time)
7.	F0402 = $T_{0402} \times P_{0402} \times X_2 \times f(X_4, X_2;$ temp.$)$
8.	F0008 = Constant specified over time
9.	F0009 = Constant specified over time
10.	F0507 = $T_{0507} \times P_{0507} \times X_7 \times f(X_4, X_5, X_7;$ temp.$)$
11.	F0509 = $T_{0509} \times X_5$
12.	F0607 = $T_{0607} \times P_{0607} \times X_7 \times f(X_4, X_6, X_7;$ temp.$)$
13.	F0709 = $AE_{07} \times X_7 + E_{0507} + E_{0607} \times F_{0607} + E_{0907} \times F_{0907}$
14.	F0800 = $X_8 \times (T_{0800} + R_{0800})^1$
15.	F0900 = Constant or variable (f (tidal exchange) over time)
16.	F0907 = $T_{0907} \times P_{0907} \times X_7 \times f(X_4, X_7, X_9;$ temp.$)$

1. Leaf material (compartments $X_3 + X_8$) become detritus only after considerable time (about six weeks) and decomposition. The state of decomposition can be related to C:N (X_8/X_3): F0300 + F0800 occur only after C:N has dropped below 45:1. At this point, probably X_1 has gone to X_2 and $X_5 + X_6$ to X_7. Thus, the transfer rate (T) and respiratory rate (R) are variables.

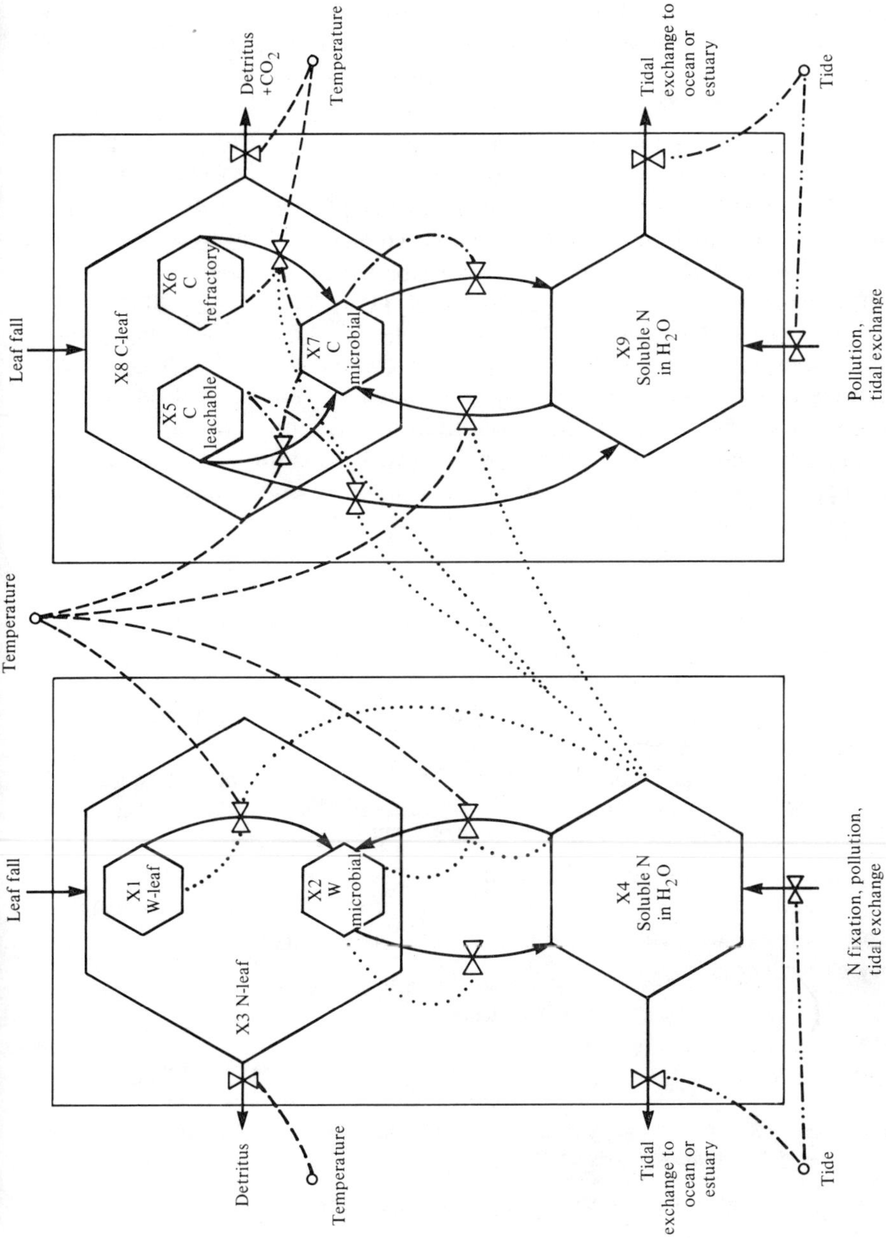

Figure 15.1
A conceptual model of carbon and nitrogen storages and fluxes associated with mangrove-leaf litter production, decomposition and export. Components of the model are outlined in Table 15.1 and discussed in the text.

the only exception). Fluxes F0102, F0402, F0507 and F0607 and F0907 are all flows to a biotic compartment. As such they are subject to a maximum rate of ingestion (T) per unit of recipient or consumer. This constraint follows from the simple observation that, all else being optimum (i.e. food, space, temperature, etc.), any living population, whether composed of one or many species, has a maximum exponential rate of growth that is genetically fixed, at least in the short run of ecological time. Therefore, ingestion by such an exponentially-growing (or 'processing' if all surplus is removed as fast as produced) population must be a constant per unit of the growing population. If all other conditions are not optimal the realized ingestion rate is reduced below the maximum. This is accomplished for the fluxes F0102, F0402, F0507, F0607 and F0907 by constraining each of the controlling parameters (*f*X1, *f*X7, *f*temp., etc.) between zero and one; 0 = no ingestion possible, i.e. starvation and 1 = maximum ingestion or optimal conditions. In all but one of these five fluxes an additional variable, feeding preference (P) is employed to apportion feeding between two or more resources, each of which could at some time be optimally abundant (for discussion of the details of calculating the value of this variable, see Wiegert, 1973).

The remaining three flows, F0204, F0509 and F0709, are determined by excretion rates (AE) and the proportion of ingestion that is egested (E). Because these flows are entirely a function of the donor compartment they need no other controls, though of course the excretion rates and egestion proportions may themselves be variable under the control of the kind of food, temperature and a number of other factors.

To sum up, a model such as that outlined here is a minimal statement of the biological relationships and parameters necessary to simulate decomposition and transfer of carbon and nitrogen in a mangrove-detrital system. By varying values of parameters and/or the equational structure, then simulating the dynamic behaviour with a computer, predictions leading to testable hypotheses are obtained. These in turn can suggest fruitful field and laboratory experiments.

References

AINSWORTH, G. C.; SPARROW, F. K.; SUSSMAN, A. S. 1973. *The Fungi, An Advanced Treatise 4A, A Taxonomic Review with Keys: Ascomycetes and Fungi Imperfecti*. New York, Academic Press, 621 pp.

BARLÖCHER, F.; KENDRICK, B. 1974. Dynamics of the Fungal Population on Leaves in a Stream. *J. Ecol.*, Vol. 62, No. 3; pp. 761–91.

BARLÖCHER, F.; KENDRICK, B.; MICHAELIDES, J. 1978. Colonization and Conditioning of *Pinus resinosa* Needles by Aquatic Hyphomycetes. *Arch. Hydrobiol.*, Vol. 81, pp. 462–74.

COOKSEY, K. E.; COOKSEY, B. 1978. Growth-influencing Substances in Sediment Extracts from a Subtropical Wetland: Investigation using a Diatom Bioassay. *J. Phycol.*, Vol. 14, pp. 347–52.

CRUZ, A. A. DE LA; GABRIEL, B. C. 1974. Caloric, Elemental and Nutritive Changes in Decomposing *Juncus roemerianus* Leaves. *Ecology*, Vol. 55, No. 4; pp. 882–6.

DEBOOIS, H. M. 1976. Fungal Development on Leaf Litter and Decomposition Potentialities of Some Fungal Species. *Rev. Ecol. Biol. Geol.* Vol.13, No. 3, pp. 437–48.

ELLIS, M. B. 1971. *Dematiaceous Hyphomycetes*. Kew, Commonwealth Mycological Institute, 608 pp.
——. 1976. *More Dematiaceous Hyphomycetes*. Kew, Commonwealth Mycological Institute. 507 pp.
FELL, J. W.; HUNTER, I. L. 1979. Fungi Associated with the Decomposition of the Black Rush, *Juncus roemerianus*, in South Florida. *Mycologia*, Vol. 71, No. 2, pp. 322–42.
FELL, J. W.; MASTER, I. M. 1973. Fungi Associated with the Degradation of Mangrove (*Rhizophora mangle* L.) Leaves in South Florida. In: H. L. Stevenson and R. R. Colwell (eds.), *Estuarine Microbial Ecology. Belle W. Baruch Symposium 1*, pp. 455–66. Columbia, University of South Carolina Press.
——. 1975. Phycomycetes (*Phytophthora* spp. nov. and *Pythium* sp. nov.) Associated with Degrading Mangrove (*Rhizophora mangle*) Leaves. *Can. J. Bot.*, Vol. 53, No. 24, pp. 2908–22.
——. 1980. The Association and Potential Role of Fungi in Mangrove Detrital Systems. *Botanica Marina*, Vol. 23, pp. 257–63.
FELL, J. W.; NEWELL, S. Y. 1980. Role of Fungi in Carbon Flow and Nitrogen Immobilization in Coastal Marine Plant Litter Systems. In: D. T. Wicklow and G. C. Carroll (eds.), *The Fungal Community: Its Organization and Role in the Ecosystem*, pp. 665–78. New York, Marcel Dekker.
FELL, J. W.; MASTER, I. M.; NEWELL, S. Y. 1980. Laboratory Model of the Potential Role of Fungi (*Phytophthora* spp.) in the Decomposition of Red Mangrove (*Rhizophora mangle*) Leaf Litter, p. 359–72. In: K. R. Tenore and B. C. Coull (eds.), *Marine Benthic Dynamics*. Columbia, University of South Carolina Press.
FELL, J. W.; CEFALU, R. C.; MASTER, I. M.; TALLMAN, A. S. 1975. Microbial Activities in the Mangrove (*Rhizophora mangle*) Leaf Detrital System. In: G. Walsh, S. Snedaker and H. Teas (eds.) *Proceedings of International Symposium on the Biology and Management of Mangroves, 8–11 October 1974, Honolulu*, Vol. 2, pp. 661–79. Gainesville, Institute of Food and Agricultural Sciences, University of Florida.
GIESEY, J. P. 1980. *Microcosms in Ecological Research*. Springfield, Va., Technical Information Center, U.S. Department of Energy. 1,110 pp.
GOSZ, J. R.; LIKENS, G. E.; BORMANN, F. H. 1973. Nutrient Release from Decomposing Leaf and Branch Litter in the Hubbard Brook Forest, New Hampshire. *Ecol. Monogr.*, Vol. 43, pp. 173–91.
HARLEY, J. L. 1971. Fungi in Ecosystems. *J. Appl Ecol.*, Vol. 8, No. 3, pp. 627–42.
HODKINSON, I. D. 1975. Dry Weight Loss and Chemical Changes in Vascular Plant Litter of Terrestrial Origin, Occurring in a Beaver Pond Ecosystem. *J. Ecol.*, Vol. 63, No. 1, pp. 131–42.
HOWARD, P. J. A.; HOWARD, D. M. 1974. Microbial Decomposition of Tree and Shrub Leaf Litter. 1: Weight Loss and Chemical Composition of Decomposing Litter. *Oikos*, Vol. 25, No. 3, pp. 341–52.
HUNTER, R. D. 1976. Changes in Carbon and Nitrogen Content during Decomposition of Three Macrophytes in Freshwater and Marine Environments. *Hydrobiologia*, Vol. 51, No. 2, pp. 115–28.
JACKSON, R. M. 1975. Soil Fungi. In: N. Walker (ed.), *Soil Microbiology*. pp. 165–80. New York, Halsted Press.
JOHNSON, B. D. 1976. Non-living Organic Particle Formation from Bubble Dissolution. *Limnol. Oceanogr.*, Vol. 21, No. 3, pp. 444–7.
JOHNSON, B. D.; COOKE, B. C. 1980. Organic Particle and Aggregate Formation Resulting from the Dissolution of Bubbles in Seawater. *Limnol. Oceanogr.*, Vol. 25, No. 4, pp. 653–61.
KAUSHIK, N. K. 1975. Decomposition of Allochthonous Organic Matter and Secondary Production in Stream Ecosystems. *Productivity of World Ecosystems*, pp. 90–5. Washington, D.C., National Academy of Sciences.
KAUSHIK, N. K.; HYNES, H. B. N. 1971. The Fate of the Dead Leaves that Fall into Streams. *Arch. Hydrobiol.*, Vol. 68, No. 4, pp. 465–515.
KIRK, T. K. 1973. The Chemistry and Biochemistry of Decay. In: D. Nicholas (ed.), *Wood Deterioration and its Prevention by Preservation Treatments*. pp. 149–81. Syracuse, N.Y., Syracuse University Press.
KOHLMEYER, J.; KOHLMEYER, E. 1971. *Synoptic Plates of Higher Marine Fungi. An Identification Guide for The Marine Environment*, 3rd ed. Lehre, Cramer. 87 pp.

——. 1979. *Marine Mycology. The Higher Fungi.* New York, Academic Press. 690 pp.
LUGO, A. E.; SNEDAKER, S. C. 1974. The Ecology of Mangroves. *Ann. Rev. Ecol. System*, Vol. 5, pp. 39–64.
LUSH, D. C.; HYNES, H. B. N. 1973. The Formation of Particles in Freshwater Leachates of Dead Leaves. *Limnol. Oceanogr.*, Vol. 18, No. 6, 968–77.
MALLEY, D. F. 1978. Degradation of Mangrove Leaf Litter by Tropical Sesarmid Crab *Chiromanthes onychophorum. Mar. Biol.*, Vol. 49; pp. 377–86.
NEWELL, S. Y. 1981. Fungi and Bacteria in or on Leaves of Eelgrass (*Zostera marina* L.) from Chesapeake Bay. *Applied and Environmental Microbiology*, Vol. 41, No. 55, pp. 1219–24.
NEWELL, S. Y.; FELL, J. W. 1980. Mycoflora of Turtlegrass (*Thalassia testudinum* Konig) as Recorded after Seawater Incubation. *Botanica Marina*, Vol. 23, pp. 257–63.
——. 1982. Surface Sterilization and the Active Mycoflora of a Seagrass. *Botanica Marina*, Vol. 25, pp. 339–46.
NEWELL, S. Y.; COOKSEY, K. E.; FELL, J. W.; MASTER, I. M.; MILLER, C.; WALTER, M. A. 1981. Acute Impact of an Organophosphorous Insecticide on Microbes and Small Invertebrates of a Mangrove Estuary. *Arch. Environm. Contam. Toxicol.*, Vol. 10, pp. 427–35.
NIE, N. H.; HULL, C. H.; JENKINS, J. G.; STEINBRENNER, K.; BENT, D. H. 1975. *SPSS. Statistical Package for the Social Sciences*, 2nd ed. New York, McGraw-Hill. 675 pp.
ODUM, E. P. 1971. *Fundamentals of Ecology*. Philadelphia, W. B. Saunders Co. 574 pp.
O'NEILL, R. V. 1976. Ecosystem Persistence and Heterotrophic Regulation. *Ecology*, Vol. 57, No. 6, pp. 1244–53.
PAERL, H. W. 1978. Microbial Organic Carbon Recovery in Aquatic Ecosystems. *Limnol. Oceanogr.*Vol. 23, No. 5, pp. 927–35.
PARKINSON, D. 1975. Terrestrial Decomposition. *Productivity of World Ecosystems*, pp. 55–60. Washington D.C., National Academy of Sciences.
PARNAS, H. 1975. Model for Decomposition of Organic Material by Microorganisms. *Soil Biol. Biochem.*, Vol. 7, pp. 161–9.
RUSSELL-HUNTER, W. D. 1970. *Aquatic Productivity*. New York, Macmillan. 306 pp.
SHOLKOVITZ. 1976. Flocculation of Dissolved Organic and Inorganic Matter during Mixing of River Water and Seawater. *Geochim. Cosmochim. Acta*, Vol. 40, pp. 831–45.
SPARROW, F. K., Jr. 1960. *Aquatic Phycomycetes*, 2nd ed. Ann Arbor, University of Michigan Press. 1,187 pp.
STARK, N. 1972. Nutrient Cycling Pathways and Litter Fungi. *Bioscience*, Vol. 22, No. 6, pp. 355–60.
SUBERKROPP, K.; KLUG, M. J. 1976. Fungi and Bacteria Associated with Leaves during Processing in a Woodland Stream. *Ecology*, Vol. 57, No. 4, pp. 707–19.
SWIFT, M. J. 1977. The Ecology of Wood Decomposition. *Sci. Prog. Oxford*, Vol. 64, pp. 175–99.
TENORE, K. R.; RICE, D. C. 1980. A Review of Trophic Factors Affecting Secondary Production of Deposit Feeders. In: K. R. Tenore and B. C. Coull (eds.), *Marine Benthic Dynamics*, pp. 325–40. Columbia, University of South Carolina Press.
THAYER, G. W.; ENGEL, D. W.; LACROIX, M. W. 1977. Seasonal Distribution and Changes in the Nutritive Quality of Living, Dead and Detrital Fractions of *Zostera marina*. L. *J. Exp. Mar. Biol. Ecol.*, Vol. 30, No. 2, pp. 109–27.
VISSER, S.; PARKINSON, D. 1975. Fungal Succession on Aspen Poplar Leaf Litter. *Can. J. Bot.*, Vol. 53, pp. 1640–51.
WALLACE, G. R.; DUCE, R. A. 1978. Transport of Particulate Organic Matter by Bubbles in Marine Waters. *Limnol. Oceanogr.*, Vol. 23, No. 6, pp. 1155–67.
WIEBE, W. J.; POMEROY, L. R. 1972. Microorganisms and their Association with Aggregates and Detritus in the Sea: A Microscopic Study. In: U. Melchiorri-Santolini and J. W. Hopton (eds.), *Detritus and its Role in Aquatic Ecosystems. Proc. IBP-Unesco Symposium Memoire dell-Istituto Italiano de Idrobiologia Suppl.*. Vol. 29, pp. 325–52.
WIEGERT, R. G. 1973. A General Ecological Model and its Use in Simulating Algal-fly Energetics in a Thermal Spring Community. In: P. W. Geier, L. R. Cark, D. J. Anderson and H. A. Nix (eds.) *Insects: Studies in Population Management*, pp. 85–102. Canberra (Ecological Society of Australia Memoirs, 1)
——. 1975. Simulation Models of Ecosystems. *Ann. Rev. Ecology & Systematics*, Vol. 6, pp. 311–38.
——. 1979. Simulation Experiments with a Fourteen-compartment Model of a *Spartina* Salt

Marsh. In: R. Dame (ed.) *Marsh Estuarine System Simulation*, pp. 7–39. Columbia, University of South Carolina Press.

Christian, R. R.; Wetzel, R. L. 1981. A Model View of the Marsh. In: L. R. Pomeroy and R. G. Wiegert (eds.), *The Ecology of a Salt Marsh*. pp. 183–218. New York, Springer Verlag.

Witkamp, M. 1974. Direct and Indirect Counts of Fungi and Bacteria as Indexes of Microbial Mass and Production. *Soil Science*, Vol 118, No. 3, pp. 150–5.

SC.83/XVIII.8/A